ACS SYMPOSIUM SERIES **888**

Chromogenic Phenomena in Polymers

Tunable Optical Properties

Samson A. Jenekhe, Editor
University of Washington

Douglas J. Kiserow, Editor
U.S. Army Research Office

Sponsored by the
ACS Division of Polymer Chemistry, Inc

American Chemical Society, Washington, DC

Library of Congress Cataloging-in-Publication Data

Chromogenic phenomena in polymers : tunable optical properties / Sampson A. Jenekhe, editor, Douglas J. Kiserow, editor ; sponsored by the ACS Division of Polymer Chemistry, Inc.

    p. cm.—(ACS symposium series ; 888)

    Includes bibliographical references and index.

    ISBN 0–8412 3830–8 (alk. paper)

    1. Photochromic polymers—Congresses. 2. Polymers—Optical properties—Congresses.

    I. Jenekhe, Sampson A. II. Kiserow, Douglas J., 1956- III. American Chemical Society. Division of Polymer Chemistry, Inc. IV. American Chemical Society. Meeting (223rd : 2002 : Orlando, Fla.) V. Series.

QD382.P45C48   2004
620.1′9204295—dc22   2004046275

The paper used in this publication meets the minimum requirements of American National Standard for Information Sciences—Permanence of Paper for Printed Library Materials, ANSI Z39.48–1984.

Copyright © 2005 American Chemical Society

Distributed by Oxford University Press

All Rights Reserved. Reprographic copying beyond that permitted by Sections 107 or 108 of the U.S. Copyright Act is allowed for internal use only, provided that a per-chapter fee of $27.25 plus $0.75 per page is paid to the Copyright Clearance Center, Inc., 222 Rosewood Drive, Danvers, MA 01923, USA. Republication or reproduction for sale of pages in this book is permitted only under license from ACS. Direct these and other permission requests to ACS Copyright Office, Publications Division, 1155 16th St., N.W., Washington, DC 20036.

The citation of trade names and/or names of manufacturers in this publication is not to be construed as an endorsement or as approval by ACS of the commercial products or services referenced herein; nor should the mere reference herein to any drawing, specification, chemical process, or other data be regarded as a license or as a conveyance of any right or permission to the holder, reader, or any other person or corporation, to manufacture, reproduce, use, or sell any patented invention or copyrighted work that may in any way be related thereto. Registered names, trademarks, etc., used in this publication, even without specific indication thereof, are not to be considered unprotected by law.

PRINTED IN THE UNITED STATES OF AMERICA

Foreword

The ACS Symposium Series was first published in 1974 to provide a mechanism for publishing symposia quickly in book form. The purpose of the series is to publish timely, comprehensive books developed from ACS sponsored symposia based on current scientific research. Occasionally, books are developed from symposia sponsored by other organizations when the topic is of keen interest to the chemistry audience.

Before agreeing to publish a book, the proposed table of contents is reviewed for appropriate and comprehensive coverage and for interest to the audience. Some papers may be excluded to better focus the book; others may be added to provide comprehensiveness. When appropriate, overview or introductory chapters are added. Drafts of chapters are peer-reviewed prior to final acceptance or rejection, and manuscripts are prepared in camera-ready format.

As a rule, only original research papers and original review papers are included in the volumes. Verbatim reproductions of previously published papers are not accepted.

ACS Books Department

Contents

Preface..xi

Overview

1. Chromogenic Effects in Polymers: An Overview of the Diverse
   Ways of Tuning Optical Properties in Real Time 2
   Samson A. Jenekhe and Douglas J. Kiserow

Electrochromic Polymers and Devices

2. Electrochromic Polyaniline Films from Layer-by-Layer
   Assembly...18
   Dean M. DeLongchamp and Paula T. Hammond

3. Electrochromic Devices Based on Ladder Polymer
   and Phenothiazine–Quinoline Copolymer Films................................34
   Jai-Pil Choi, Fernando Fungo, Samson A. Jenekhe,
   and Allen J. Bard

4. Novel Near-IR Electrochromic Ruthenium Complex Polymers..........51
   Pierre Desjardins and Zhi Yuan Wang

5. Far-IR-through-Visible Electrochromics Based on Conducting
   Polymers for Spacecraft Thermal Control and Military Uses:
   Application in NASA's ST5 Microsatellite Mission and in
   Military Camouflage..66
   P. Chandrasekhar, B. J. Zay, D. Ross, T. McQueeney, G. C. Birur,
   T. Swanson, L. Kauder, and D. Douglas

Photochromic and Related Stimuli-Responsive Polymers

6. Chromic Transitions and Nanomechanical Properties of Poly(diacetylene) Molecular Films 82
   Robert W. Carpick, Alan R. Burns, Darryl Y. Sasaki, M. A. Eriksson, and Matthew S. Marcus

7. Functional Amphiphilic and Bolaamphiphilic Poly(diacetylene) Assemblies with Controlled Optical and Morphological Properties 96
   Jie Song, Raymond C. Stevens, and Quan Cheng

8. Chromogenic Polymer Gels for Reversible Transparency and Color Control 110
   Arno Seeboth, Jörg Kriwanek, André Patzak, and Detlef Lötzsch

9. Modulation of Optical Properties of New Photosensitive Polymers: 3-D Optical Data Storage Media 122
   Kevin D. Belfield, Katherine J. Schafer, and Stephen Andrasik

10. Quantum Amplified Isomerization: A New Chemically Amplified Imaging System in Solid Polymers 135
    Douglas R. Robello, Joseph P. Dinnocenzo, Samir Farid, Jason G. Gillmore, and Samuel W. Thomas III

11. Chromicity in Poly(aryleneethynylene)s 147
    Uwe H. F. Bunz, James N. Wilson, and Carlito Bangcuyo

12. Novel Two-Photon Absorbing Conjugated Oligomeric Chromophores: Property Modulation by π-Center 161
    O.-K. Kim, Z. Huang, E. Peterman, S. Kirkpatrick, and C. S. P. Sung

13. Synthesis, Properties, and Applications of Photochromic Amorphous Molecular Materials and Electrochromic Polymers 173
    Yasuhiko Shirota, Hideyuki Nakano, Ichiro Imae, Yutaka Ohsedo, Yoshiaki Yasuda, Hisayuki Utsumi, Toshiki Ujike, and Toru Takahashi

Tunable Emission and Electroluminescence

14. Voltage-Tunable Multicolor Electroluminescence from Single-Layer Polymer Blends and Bilayer Polymer Films..................188
Maksudul M. Alam, Christopher J. Tonzola, Yan Zhu, and Samson A. Jenekhe

15. Blue Light Emitting Polymers and Devices........................201
Qibing Pei, S. Pyo, Shun-Chi Chang, and Yang Yang

16. Polarized Electroluminescence from Double-Layer LEDs with Active Film Formed by Two Perpendicularly Oriented Polymers..................211
A. Bolognesi, C. Botta, D. Facchinetti, C. Mercogliano, M. Jandke, P. Strohriegl, K. Kreger, A. Relini, and R. Rolandi

17. Tuning Optical and Electroluminescent Properties of Poly(thiophene)s via Post-Functionalization..................220
Steven Holdcroft, Yuning Li, George Vamvounis, Hany Aziz, and Zoran D. Popovic

18. Site-Isolated Luminescent Lanthanide Complexes with Polymeric Ligands..................233
Jessica L. Bender and Cassandra L. Fraser

19. Synthesis, Photophysical Property, and Electroluminescent Applications of Silicon-Based Alternating Copolymers..................247
H. K. Kim, N. S. Baek, K. L. Paik, Y. Lee, and J. H. Lee

20. Tunable Photoluminescence of Poly(quinoline)s in Polymer Blend Films and Silica..................264
S. W. Ho, W. Y. Huang, T. K. Kwei, and Y. Okamoto

Tunable Reflection and Photonic Band-Gap Structures

21. Electrically Switchable Reflectors of Chiral Gels..................278
Rifat A. M. Hikmet

22. Glassy Liquid Crystals for Tunable Reflective Coloration...........290
    Shaw H. Chen, Philip H. M. Chen, Dimitris Katsis,
    and John C. Mastrangelo

23. Tunable Near-IR Optical Properties from Trialkoxysilane-
    Capped Poly(methyl methacrylate)–Silica Waveguide
    Materials..307
    Ming-Hsin Wei, Chia-Hua Lee, and Wen-Chang Chen

24. Photochemical Control of Reflection Colors of Glass-
    Forming Non-Polymeric Cholesterics with Azobenzene
    Chromophore..320
    Nobuyuki Tamaoki and Masaya Moriyama

25. Photonic Papers: Colloidal Crystals with Tunable Optical
    Properties..329
    Hiroshi Fudouzi, Yu Lu, and Younan Xia

Chromic Polymers for Chemical Sensors and Biosensors

26. Synthesis and Tunable Chiroptical Properties of Amphiphilic
    Helical Polyacetylenes...340
    Kevin K. L. Cheuk and Ben Zhong Tang

27. DNA-Sensors Using a Water-Soluble, Cationic
    Poly(thiophene) Derivative...359
    Hoang-Anh Ho, Maurice Boissinot, Michel G. Bergeron,
    Geneviève Corbeil, Kim Doré, Denis Boudreau, and Mario Leclerc

28. Synthesis of Tunable Electrochromic and Fluorescent
    Polymers...368
    John D. Tovar and Timothy M. Swager

29. Enzymatically Synthesized Electronic and Photoactive
    Materials..377
    Jayant Kumar, Wei Liu, Soo-Hyoung Lee, Suizhou Yang,
    Sukant Tripathy, and Lynne Samuelson

30. **Explosive Detection by Fluorescent Electrospun Polymer Membrane Sensor**......388
    Xianyan Wang, Christopher Drew, Soo-Hyoung Lee,
    Kris J. Senecal, Jayant Kumar, and Lynne A. Samuelson

Indexes

Author Index......443

Subject Index......445

Preface

Chromogenic phenomena, exemplified by electrochromism, photochromism, thermochromism, piezochromism, and magnetochromism, (i.e., color changes induced respectively by applied electrical, optical, thermal, pressure, and magnetic fields), provide many mechanisms that facilitate real-time tunability of the optical properties of materials. Of these, photochromism and electrochromism have been more widely investigated in polymeric and organic materials. They currently find some commercial and defense applications such as self-adjusting sunglasses, filters in optical sensors, color-tunable coatings and paints, thermographic recording media in medical applications, self-adjusting car rearview mirrors, and color-switching clothing. However, the great promise of electrochromic, photochromic, and other chromogenic materials in applications ranging from large-area information displays, ultrahigh density optical memories, holographic recording, smart windows, "intelligent" materials systems, and photoresponsive transducers to large-area color-tunable wallpaper are yet to be fully realized. The challenges and barriers to achieving these large-scale and high-impact technological applications include the stability of the materials under long-term cycling, the speed with which optical changes can be effected, the amount of energy required to achieve optical tunability, satisfactory color contrasts, and the processability of the chromogenic materials into suitable forms, such as conformal coatings, multilayer films, and ordered nanoporous solids. These challenges are being addressed through the targeted synthesis and detailed structural understanding of chromogenic effects in polymers, the fundamental understanding of the physical and chemical mechanisms of coloration and the coloration dynamics, and the design and fabrication of devices and systems incorporating the materials.

Novel classes of chromogenic polymers have also emerged in the past several years and are currently of great research and technological interests, including tunable light emitting polymers and devices that are promising for flexible displays; polymers with tunable selective reflection spanning the visible and near IR; and polymer composites having one-, two-, or three-dimensional photonic band gaps. Self-assembly approaches to the synthesis, processing, patterning, and device applications of chromogenic polymers and hybrid materials have also emerged as major areas of research in the past several years. Nanostructured polymer systems having tunable electronic, optoelectronic, and photonic properties represent excellent model systems for exploring a range of new concepts of intelligent–self-repairing materials and systems, intelligent sensors–detectors, nanoelectromechanical systems, and various multifunctional devices.

Our goal in organizing the symposium was to provide an international forum to discuss important scientific and technological advances and future prospects in the broad field of chromogenic phenomena in polymers. The chapters included in the book were nearly all based on the invited presentations at the symposium. The main topics include electrochromic polymers and devices for the visible and IR; electroluminescent polymers and tunable emission; photochromic and stimuli-responsive polymers; optically switchable materials and devices; photonic band-gap materials; tunable multifunctional optical materials; bioinspired and biomimetic chromogenic materials; supramolecular chromism and self-assembly approaches to tunable optical materials; chromic polymers for chemical sensors and biosensors; photonic polymer–inorganic nanocomposites; mechanochromic and nanophotonic polymers; and polymers for imaging and high-density data storage.

A broad multidisciplinary group of researchers that include chemists, physicists, engineers, and materials scientists from academia, government, and industry participated in the symposium and contributed chapters to this book. We expect chemists and materials scientists who are interested in the synthesis and characterization of polymers with tunable electronic, optoelectronic, and photonic properties will find this book useful, along with physicists, biochemists, and engineers who are interested in the properties and device applications of chromogenic materials.

We acknowledge the U.S. Army Research Office for the generous financial support that enabled broad participation at the symposium and in the preparation of this book. We thank the referees for their important help in the critical assessment of the manuscripts and the authors for their contributions. We also thank John D. Wind and Jeanie Comstock at the University of Washington for assistance in the editing process. Finally, the help and patience of Stacy VanDerWall and Robert

W. Hauserman in acquisitions and Margaret Brown in editing and production of the ACS Books Department were essential in the publication of the book.

Samson A. Jenekhe
Department of Chemical Engineering
Benson Hall, Box 351750
University of Washington
Seattle, WA 98195–1750
jenekhe@u.washington.edu (email)

Douglas J. Kiserow
Branch Chief, Polymer Chemistry
U.S. Army Research Office
P.O. Box 12211
Research Triangle Park, NC 27709–2211
douglas.kiserow@us.army.mil (email)

Overview

Chapter 1

Chromogenic Effects in Polymers: An Overview of the Diverse Ways of Tuning Optical Properties in Real Time

Samson A. Jenekhe[1] and Douglas J. Kiserow[2]

[1]Department of Chemical Engineering, University of Washington, Seattle, WA 89195-1750
[2]Polymer Chemistry Branch, U.S. Army Research Office, P.O. Box 12211, Research Triangle Park, NC 27709-2211

In this overview chapter we introduce the subject of chromogenic phenomena in polymers and briefly review some of the recent advances in the field. Many external stimuli-responsive phenomena in polymers, including electrochromism, photochromism, photoelectrochromism, and piezochromism are discussed. Relatively new approaches to dynamic tunability of optical properties, such as mechanochromism, tunable electroluminescence and photonic band gap structures in polymers, are also discussed. There are good prospects that future advances in chromogenic polymers, particularly the integration of multiple chromogenic effects in the same material system, will usher in an era of "intelligent" materials for many technological applications.

The reversible change of the optical properties of a material, such as color, absorption spectrum, emission spectrum, or refractive index, by means of an external stimulus represents an example of a chromogenic phenomenon. Many chromogenic phenomena have been extensively documented and widely investigated in organic materials and polymers in the past three decades (*1-18*). The chromogenic effects associated with reversible color changes induced by applied electric, optical, thermal, pressure and magnetic fields are respectively known as electrochromism, photochromism, thermochromism, piezochromism and magnetochromism (*1-8*). The materials involved are said to be electrochromic, photochromic, thermochromic, piezochromic and magnetochromic, respectively. In addition to these, other chromogenic

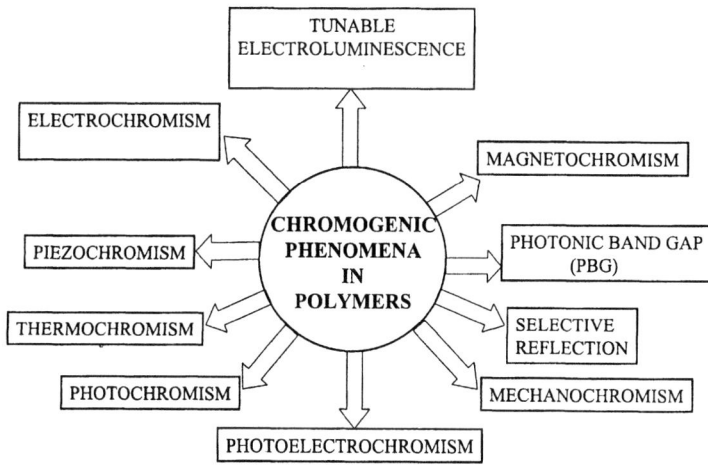

Figure 1. Main examples of chromogenic phenomena observed in organic polymers.

phenomena of growing interest in polymers are included in Figure 1. Different aspects of chromogenic phenomena in materials, including organic materials, have been covered in prior reviews and books (*1-12*). Here we focus mainly on chromogenic effects in polymeric materials.

It is useful to distinguish between *dynamic* or *real time* tunability of optical properties and *static* tunability. Nearly all of the different materials associated with the chromogenic phenomena cited in Figure 1 are such that the applied external stimulus in principle provides a ready means to reversibly tune or control their optical properties in real time. The optical properties of a polymeric material can also be changed in a static or trivial fashion, for example through change of the molecular structure or composition by synthesis or the solvent medium. The phenomenon involved in the change of the color of a polymer in solution when the solvent is changed is called solvatochromism and the material is said to be solvatochromic (*12*). This latter effect is an example of static tunability which will not be discussed further.

Only several of the chromogenic phenomena cited in Figure 1 will be briefly discussed below because of space limitation. Recent advances in nearly all of these areas can be found in the following chapters of this book. Thermochromism, for example, occurs widely in many π- and σ-conjugated polymers and thermochromic polymers are discussed in some prior reviews (*1-3, 8*). We conclude from our review of the literature that the development of chameleon-like "intelligent" materials is feasible and can be facilitated by further advances in multifunctional chromogenic polymers.

Photochromic Polymers

A molecule that undergoes a reversible photoinduced change between two states with different optical properties such as color or absorption spectrum is considered to be photochromic. The chromogenic effect, photochromism, can be illustrated by the photochemical transformation of molecule **A** to **B** upon absorption of light:

$$A \xrightleftharpoons[h\nu_2 \text{ (or } \Delta)]{h\nu_1} B$$

The back reaction can occur photochemically with absorption of longer wavelength or thermally. Normally, species **A** has an absorption spectrum that is blue shifted from that of species **B**. Besides the reversible changes in the absorption spectrum, and hence color, the emission spectrum, refractive index, dielectric constant and other physicochemical properties of a photochromic material may also be tuned or controlled by light (*1-5*).

Numerous photochromic organic molecules are known and have been extensively investigated in solution, as solids and most importantly in polymers. Important examples of photochromic molecules include aromatic azo compounds, spirobenzopyran and derivatives, spirooxazines, fulgides, fulgimides and diarylethenes (*2-5*). The synthesis and photochromic properties of these and other classes of photochromic compounds are described in many reviews and books (*1-5*). It is the incorporation of these diverse photochromic molecules into polymers, either through physical blending or covalently, that has substantially advanced their technological applications in many areas. One of the well known commercial applications of photochromic polymers is in plastic lenses of variable optical density; the self-adjusting lenses darken under sunlight and become relatively clear inside (*3-5*). Among the many other applications that exploit the photochromic properties of polymers include optical filters for cameras, ultrahigh density (~10^{12} bits/cm^3) 3D optical memories (*2b,13*), real time holography, fluid flow visualization, displays and security printing inks (*3-5,9-11,13*).

Achieving high photochemical cycling rates (>10^6 cycles) without fatigue remains one of the major challenges in using photochromic polymers for high performance applications such as optical memories and displays. Detailed studies of the dynamics of the photocoloration and the reverse photo- or thermal-bleaching processes of the most promising photochromic polymers are of interest. Development of new photochromic polymer systems with substantially improved fatigue resistance is essential to success in currently known applications of the materials. Another area of future direction of research is the coupling of photochromism with other chromogenic effects or important physical property in the same material. Photoelectrochromism which facilitates the optical control of electrochromism is an early example of this (*6,14a*). Photomagnetism wherein the magnetic properties of a material are controlled by light absorption is another (*14b,14c*).

Electrochromic Polymers and Devices

An electrochromic (EC) material undergoes a reversible change in color or other optical property under an applied electrochemical potential (*6*). An EC material is also referred to as an *electrochrome*. The underlying phenomenon, electrochromism, arises from the different electronic structures and thus optical properties of an EC material in the different redox states accessible by switching the applied voltage. An electrochrome that changes color upon oxidation is an anodic EC material. Similarly, a cathodic EC material is one that changes color upon reduction. Although other classes of EC materials are known, including inorganic metal oxides (WO_x, TiO_2) and organic small molecules (*1,2,6*), we focus here on conducting polymer-based EC materials and electrochromic devices (ECDs) made from them. Applications of such electrochromic devices include large area displays, variable transmittance ("smart") windows and variable reflectance mirrors.

π-Conjugated polymers are well suited to be EC materials because their oxidation or reduction is accompanied by substantial changes in optical absorption spectra in the visible and near infrared regions. Indeed, there is a vast literature on conducting polymer-based EC materials and devices (*6,7,15-18*). In their neutral forms conjugated polymers have optical band gaps (E_g), arising from π-π* optical transitions, in the visible to near infrared range (~1-3 eV). New optical transitions due to the formation of polaron and bipolaron charged states are observed on oxidation or reduction of the materials. Since the optical spectra of a conjugated polymer vary continuously with the degree of oxidation or reduction, multiple colors are possible and are commonly observed with conducting polymer ECDs.

Typical conjugated polymers that have been explored as EC materials are shown in Figure 2. Polythiophenes (**1**), polypyrroles (**2**), polyaniline (**3**), poly(3,4-ethylenedioxythiophene) (**4**) and related derivative (**5**) are examples of anodically coloring electrochromic materials (*6,7,15-18*). Ladder poly(benzimidazolebenzophenanthroline) (**6**, BBL) and polyphenylquinoline (**7**) are examples of cathodically coloring EC materials (*18b, 19*).

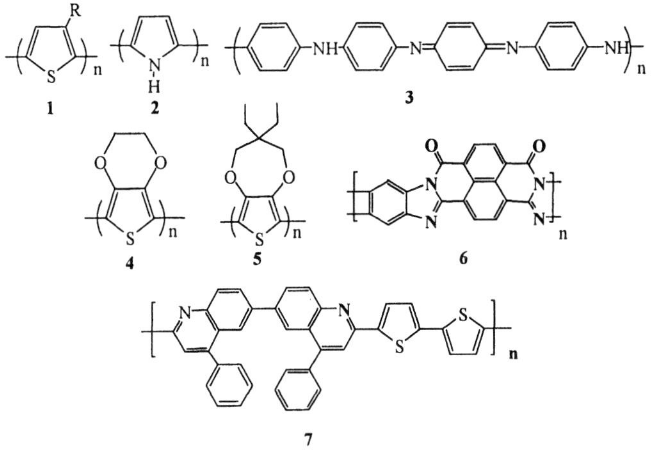

Figure 2. Examples of conjugated polymer EC materials.

Tunable Emission and Electroluminescence

Many conjugated polymers in solution or as thin films show strong photoluminescence (PL) under photoexcitation. Examples of some photoluminescent conjugated polymers are shown in Figure 3. Poly(p-phenylenevinylene) (PPV) thin films emit green light whereas its derivative, poly(2-methoxy-5-(2'-ethylhexyloxy)-1,4-phenylenevinylene) (MEH-PPV), exhibits orange-red photoluminescence as are poly(4-phenylquinoline) (PPQ) thin films. In contrast, poly(9,9'-dioctylfluorene) (PFO) thin films emit blue light. By sandwiching a PPV thin film between a transparent indium tin oxide (ITO) conductor and aluminum and applying a voltage, as shown in Figure 3, electroluminescence (EL) from conjugated polymers was discovered in 1990 (*20*). The resulting EL emission spectrum of PPV was identical to the PL emission spectrum. A few years earlier, similar thin film light emitting diodes (LEDs) were demonstrated with a small molecule organometallic compound (aluminum quinolate, Alq$_3$) (*21*). Hundreds of organic molecules and polymers have since been found to be similarly useful as emissive materials for LEDs (*20-24*). Substantial progress has been made in developing organic and polymer LEDs of various colors for flat panel displays (*20-24*).

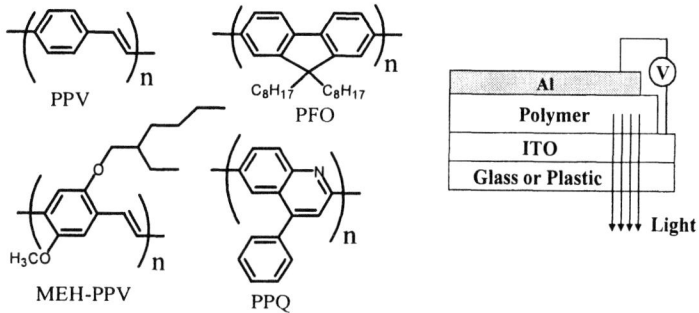

Figure 3. Examples of electroluminescent polymers and schematic of a simple polymer light emitting diode.

An organic or polymer LED (Figure 3) normally emits one color at all applied voltages. The color of light emitted from such a diode is determined by the electronic structure of the emissive polymer layer. However, if the LED is constructed from a multicomponent polymer system with two or more emissive components, *voltage-tunable multicolor emission* can result from the LED (23,24). Tunable multicolor EL has thus been achieved in LEDs fabricated from nanophase-separated blends (24), bilayer thin films (23), multilayer thin films (23), and nanophase-separated block copolymers. In such LEDs a different EL spectrum, and thus color, can be obtained at different applied voltages as illustrated in Figure 4. For example, a bilayer LED constructed from a green-emitting PPV layer and an orange/red-emitting PPQ layer, ITO/PPV/PPQ/Al, exhibited many different colors under applied bias voltages of 6-13V as shown in Plate 1 (23b).

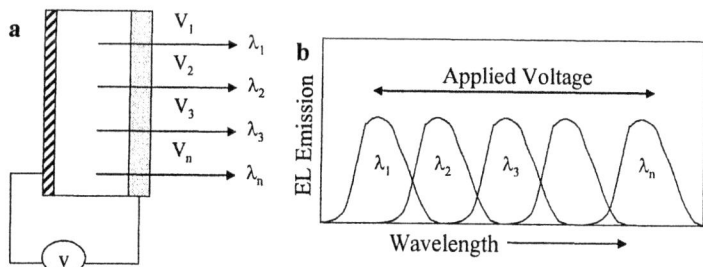

Figure 4. Schematic of a voltage-tunable multicolor polymer LED and the resulting EL emission spectra.

The voltage-tunable multicolor EL emission from bilayer polymer LEDs, such as in Plate 1, arises from variation in the EL emission intensity coming from the different layers and a physical mixing of the EL emission from both polymers. A similar mechanism explains the multicolor EL emission from nanophase-separated blends (24) and other multicomponent EL polymer systems

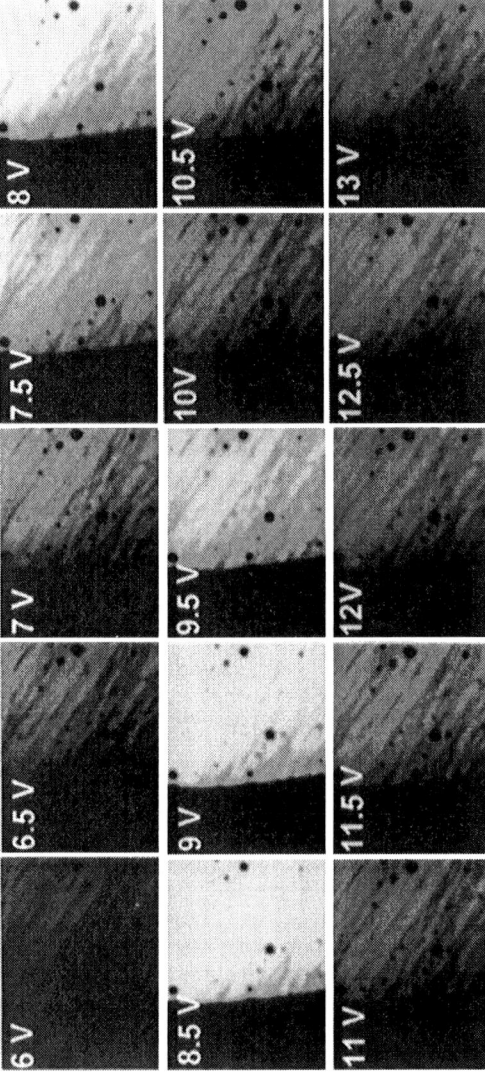

Plate 1. Multicolor emission from ITO/PPV(35 nm)/PPQ(35 nm)/Al diodes at various bias voltages (from ref. 23b). (See page 1 of color insert.)

(23). The relative film thicknesses of bilayers and multilayers, and domain sizes in the case of blends, play a critical role in the ability to achieve multicolor emission from LEDs based on multicomponent emissive materials. The reason for this is due to the electric field-dependence of the charge carrier mobilities in the materials (23). The interchromophore photophysics also places a bound on achievable multicolor emission in such multichromophoric systems (25). In particular, energy transfer, exciplex formation, and photoinduced electron transfer among the different components must be minimized in order to obtain multicolor emission from the different components (23-25). Blends of polyfluorene (PFO) and MEH-PPV, for example, show one-color (orange) EL because of the very efficient energy transfer from the blue-emitting PFO to the smaller energy gap MEH-PPV.

Voltage-tunable multicolor EL emission from polymer systems, which is briefly described here, represents a novel chromogenic effect that has emerged from recent studies of emissive polymer semiconductors. There is no analogous effect in inorganic semiconductors. Besides flat-panel displays, other applications including flexible displays, traffic lights, and biological imaging can be envisioned for real time tunable multicolor polymer LEDs.

Mechanochromic Polymers

Color change induced by applied mechanical stress in a material is termed mechanochromism. Such mechanically-induced color changes have been observed in free-standing and substrate-supported polydiacetylene (PDA) films (26,27). Both tensile stress and shear stress can effect chromic transitions in PDA films. Examples of polymers or polymer segments that facilitate mechanochromism are shown in Figure 5. Elastomeric segmented poly(urethane-co-diacetylene) copolymer films were shown to undergo reversible color changes

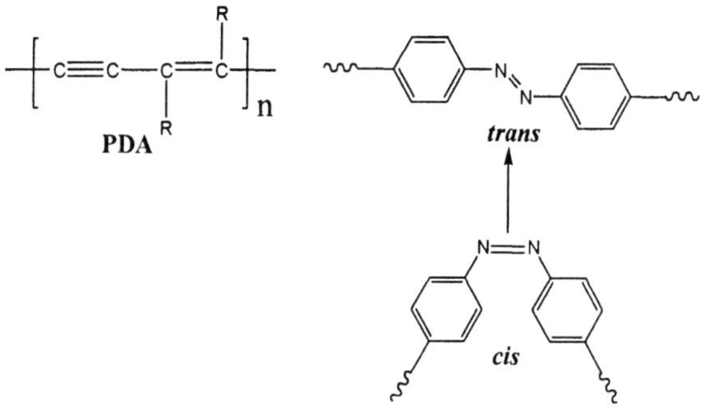

Figure 5. Examples of mechanochromic polymers.

when stretched to strains of 250-300%. The blue unstretched elastomer film changed to red or yellow depending on the degree of strain. The resulting red or yellow film returned to its original blue color when the tensile strain was removed. Visible absorption spectroscopy showed that the stress-induced color changes are due to order-disorder transitions of the conjugated PDA chains within domains of the hard segments of the polyurethane elastomer.

Recent work has extended mechanochromism in the polydiacetylenes to the nanoscale. Studies of Langmuir-Blodgett films of PDAs on mica or silicon oxide substrates have shown that the tip of an atomic force microscope (AFM) can be used to induce local changes in color, demonstrating nanometer scale mechanochromism. The blue-to-red change in color of PDA monolayers was caused by the shear stress on the film due to tip/film contact and friction. This chromic transition is related to the mechanochromism observed in the bulk films (26) as well as previous observations of solvatochromism and thermochromism in many different polydiacetylenes (28). The nanoscale color changes were irreversible. Potential uses of this nanoscale mechanochromism include optical sensor/detection of contact, friction, and adhesion if the mechano-optical effect can be made reversible through improved materials.

A thermoplastic polyurethane elastomer that undergoes irreversible change in optical absorption with tensile strain has been demonstrated as a strain-recording "smart" material (29). An azobenzene chromophore that undergoes cis-trans isomerization was covalently embedded in the polyamide segment of the elastomer. Deformation due to applied tensile stress converts the azobenzene chromophore to the trans form and consequently an increase in the long wavelength absorption. Irreversibility allows retention of information about the strain. Intended applications of such an irreversible mechanochromic polymer include components of smart fiber-reinforced polymer composites that are able to respond to a changing environment and send out warning signals prior to structural failure (29). Obvious potential systems applications include smart skins on bridges or aircraft wings and panels (29).

Piezochromic Polymers

The change of optical properties with pressure is known as piezochromism (1,30-38). The change of color by a solid under compression at high pressures exemplifies this chromogenic effect. Because significant changes in optical properties occur only at relatively high pressures, the possible application of piezochromic materials is unclear. Unlike many other classes of chromogenic materials, studies of piezochromism in conjugated molecules and polymers are thus largely motivated by interest in gaining fundamental information about how interchain interactions influence the electronic structure and optical properties of the materials.

Two underlying mechanisms of color change in a piezochromic material have been advanced (*1*). In a polymorphic crystalline solid that has one phase at ambient pressure and another at high pressure, the corresponding colors or optical properties of the two phases can be observed by pressure-induced phase transition. Alternatively, pressure-induced changes in molecular geometry and intermolecular interactions of the molecular solid can facilitate the continuous variation of optical properties with pressure. The latter mechanism appears to

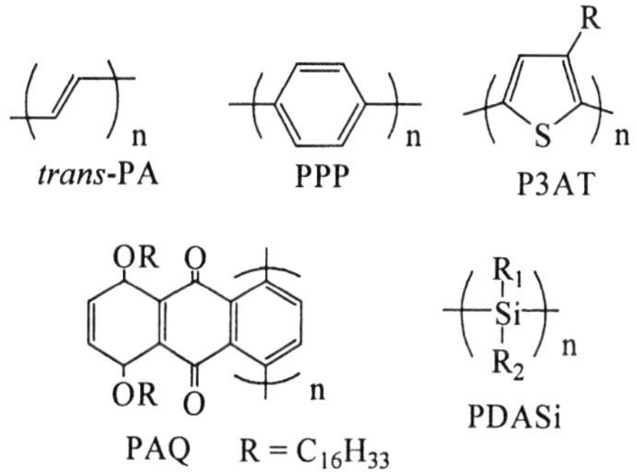

Figure 6. Examples of polymers exhibiting piezochromism.

account for the observed piezochromism in many σ– and π–conjugated polymers that have been reported (*30-38*).

Examples of polymers known to exhibit piezochromism are shown in Figure 6. Studies of the optical absorption spectra of undoped *trans*-polyacetylene (*trans*-PA) under pressure from ambient to about 5 GPa at 300K showed that it exhibited reversible piezochromism (*30*). The optical absorption edge band gap (E_g) decreased from 1.4 eV at ambient pressure to 0.85 eV at 5 GPa. However, pressure-induced reactions occurred in the 5-8 GPa range, leading to an apparently cross-linked material that was transparent to visible light above this range of pressures. In contrast, crystalline samples (30 μm initial thickness) of poly(*p*-phenylene) (PPP) were found to be robust with no reaction at pressures as high as 20 GPa (*31*). The measured optical absorption spectra of PPP films under pressure from ambient to 20 GPa revealed reversible piezochromism. The optical band gap E_g decreased monotonically from 2.74 eV at ambient to 2.19 eV at 20 GPa. This represents a 0.55 eV reduction in E_g value at 20 GPa compared to ambient. This pressure-induced red shift in the absorption spectrum of PPP was explained by the increase in interchain interactions and decrease in the torsion angle between rings (*31*).

Reversible piezochromism has been observed in regioregular and regiorandom poly(3-alkylthiophene)s (*32-35*). Optical absorption spectra of solutions of poly(3-hexylthiophene) (P3HT) in toluene were observed to exhibit large bathochromic shifts with increasing hydrostatic pressure, compared to ambient pressure. The optical band gap of P3HT in toluene solution was 2.4 eV under ambient pressure but was reduced to 1.7 eV at 8 kbar (0.8 GPa) (*32*).

Similar results were obtained for other poly(3-alkylthiophene)s with alkyl chain lengths exceeding butyl. These piezochromic properties of poly(3-alkylthiophene)s were explained by pressure-induced change of the polymer conformation and particularly the reduced torsion angle between rings at high pressure (*32*). In accord with the observations in solution, the optical absorption and photoluminescence (PL) spectra of P3HT films showed reversible piezochromism up to 0.8 GPa (*33*). Essentially similar results and conclusions were reached for regioregular and regiorandom poly(3-octylthiophene) (*35*). Interestingly, the observed thermochromism of P3HT under ambient was completely suppressed at 1.4 GPa (*33*). Studies of the piezochromic properties of P3HT to much higher pressures revealed new effects, including hysteresis in the absorption spectra and dc conductivity data under compression and decompression (*34*). The absorption maximum energy and optical band gap decreased to a minimum at about 3 GPa and were followed by an increase with further increase in pressure. The dc conductivity of the undoped P3HT films under pressures of up to 5 GPa had a maximum at about 2 GPa, followed by a large decrease. The observed increase in optical band gap or blue shift in the absorption spectrum after the 3 GPa minimum was explained by P3HT backbone bending at the very high pressures (*34*).

Conjugated poly(5,8-dihexadecyloxyanthraquinone-1,4-diyl) (PAQ) films were observed to exhibit reversible piezochromism up to 11 GPa. The orange-yellow films at ambient pressure progressively change to dark red at 11 GPa (*36*). The associated optical band gap decreased from 2.42 eV (ambient) to 1.5 eV at 11 GPa. The observed piezochromism was interpreted as due to pressure-induced shortening of the π-π interchain distance. Interestingly, piezochromism was not observed in the homologous polyanthraquinones with shorter dialkoxy side chains. Polydiacetylenes (PDAs) with various side groups have reported piezochromism in solution, as gels and as films (*37*). For example, a yellow solution of poly(nBCMU) at ambient pressure turns red at high pressures (1.6-3.8 GPa) (*37*). Piezochromism has also been observed in various σ-conjugated poly(di-n-alkylsilane)s (PDASi) (*38*). Both cases of pressure-induced red shift and blue shift have been observed for different polysilanes (*38*). These piezochromic properties have also been understood in terms of pressure-induced changes in the conformation of the polysilanes. In general, it is very likely that most conjugated polymers, particularly those bearing side chains, will show piezochromic properties at sufficiently high pressures. An important question for workers in this area is: can applications be found for this interesting chromogenic phenomenon?

Polymer-Based Photonic Band Gap Composites

The development of materials or structures that possess a photonic band gap (PBG) and the development of devices exploiting this effect are currently of intense and wide interest (*39-44*). A PBG material or photonic crystal is an artificial dielectric composite that has a periodic modulation of the refractive index on a distance scale of half an optical wavelength (*39*). Consequently, a photonic crystal has frequency bands for which light cannot propagate but can be confined. A quarter-wave-stack optical interference filter is a simple one-dimensional (1-D) example of a photonic crystal. Interest in two-dimensional (2-D) and three-dimensional (3-D) PBG materials stems from the dramatic modification of the propagation properties of light within such a material. Many novel optical properties, including complete confinement of light of certain frequencies, substantial modulation of the refractive index, modulation of spontaneous emission and tunable reflectivity, are predicted for 3-D PBG materials. Applications ranging from zero-threshold lasers, sensors, optical communication and optical switching devices to ultrahigh density 3-D optical memories are envisioned (*39-44*). 3-D photonic crystals, with photonic band gaps in the visible to near infrared spectral range, have been made from a variety of polymer composites (*40-44*). These include polymer-air composites created from opals, inverse opals and block copolymers, polymerized crystalline colloidal arrays, homopolymer/block copolymer blends and two-photon photopolymerized structures (*40-44*).

Acknowledgments

We acknowledge the support by the U. S. Army Research Office TOPS MURI Program (Grant no. DAAD19-01-1-0676), the NSF STC (Grant no. DMR-0120967) and the Boeing-Martin Professorship Endowment.

References

(1) Greenberg, C. B.; Crano, J. C.; Drickamer, H. G. In *Encyclopedia of Chemical Technology*, 4th ed., Vol. 6; Wiley: New York, 1993; pp. 312-343.
(2) (a) Granquist, C. G. *Critical Rev. Solid State Mater. Sci.* **1990**, *16*, 291. (b) Irie, M. *Chem. Rev.* **2000**, *100*, 1685.
(3) Crano, J. C.; Guglielmetti, R. J., Eds. *Organic Photochromic and Thermochromic Compounds*; Vol. 1, Plenum: New York, 1999.
(4) Dürr, H.; Bouas-Laurent, H., Eds. *Photochromism: Molecules and Systems*; Elsevier: Amsterdam, 1990.
(5) McArdle, C. B., Ed. *Applied Photochromic Polymer Systems*; Blackie: New York, 1992.
(6) Monk, P. M. S.; Mortimer, R. J.; Rosseinsky, D. R. *Electrochromism: Fundamentals and Applications*; VCH: Weinheim, Germany, 1995.
(7) Mortimer, R. J. *Electrochim. Acta* **1999**, *44*, 2971.
(8) Leclerc, M. *Adv. Mater.* **1999**, *11*, 1491.

(9) Berkovic, G.; Krongauz, V.; Weiss, V. *Chem. Rev.* **2000**, *100*, 1741.
(10) Kawata, S.; Kawata, Y. *Chem. Rev.* **2000**, *100*, 1777.
(11) Xie, S.; Natansohn, A.; Rochon, P. *Chem. Mater.* **1993**, *5*, 403.
(12) Suppan, P.; Ghoneim, N. *Solvatochromism*; Royal Society of Chemistry: Cambridge, UK, 1997.
(13) Parthenopoulos, D. A.; Rentzepis, P. M. *Science* **1989**, *245*, 843.
(14) (a) Bechinger, C.; Ferrer, S.; Zaban, A.; Sprague, J.; Gregg, B. A. *Nature* **1996**, *383*, 608. (b) Yamamoto, T.; Umemura, Y.; Sato, O.; Einaga, Y. *Chem. Mater.* **2004**, *16*, 1195. (c) Sato, O. *Acc. Chem. Res.* **2003**, *36*, 692.
(15) Schwendeman, I.; Hickman, R.; Sommez, G.; Soloducho, J.; Musgrave, R.; Reynolds, J. R. *Chem. Mater.* **1997**, *9*, 1578.
(16) Chandrasekhar, P.; Zay, B. J.; Birur, G. C.; Rawall, S.; Pierson, E. A.; Kauder, L.; Swanson, T. *Adv. Funct. Mater.* **2002**, *12*, 95.
(17) (a) DeLongchamp, D. M.; Hammond, P. T. *Adv. Funct. Mater.* **2004**, *14*, 224. (b) DeLongchamp, D. M.; Kastantin, M.; Hammond, P. T. *Chem. Mater.* **2003**, *15*, 1575.
(18) (a) Sapp, S. A.; Sotzing, G. A.; Reynolds, J. R. *Chem. Mater.* **1998**, *10*, 2101. (b) Fungo, F.; Jenekhe, S. A.; Bard, A. J. *Chem. Mater.* **2003**, *15*, 1264.
(19) (a) Agrawal, A. K.; Jenekhe, S. A. *Macromolecules* **1993**, *26*, 895. (b) Agrawal, A. K.; Jenekhe, S. A. *Chem. Mater.* **1996**, *8*, 579.
(20) Burroughes, J. H.; Bradley, D. D. C.; Brown, A. R.; Marks, R. N.; Mackay, K.; Friend, R. H.; Burns, P. L.; Holmes, A. B. *Nature* **1990**, *347*, 539.
(21) Tang, C. W.; Van Slyke, S. A. *Appl. Phys. Lett.* **1987**, *51*, 913.
(22) (a) Kraft, A.; Grimsdale, A. C.; Holmes, A. B. *Angew. Chem. Int. Ed.* **1998**, *37*, 402. (b) Zhang, X.; Shetty, A. S.; Jenekhe, S. A. *Macromolecules* **1999**, *32*, 7422.
(23) (a) Jenekhe, S. A.; Zhang, X.; Chen, X. L.; Choong, V.-E.; Gao, Y.; Hsieh, B. R. *Chem. Mater.* **1997**, *9*, 409. (b) Zhang, X.; Jenekhe, S. A. *Macromolecules* **2000**, *33*, 2069.
(24) (a) Zhang, X.; Kale, D. M.; Jenekhe, S. A. *Macromolecules* **2002**, *35*, 382. (b) Alam, M. M.; Tonzola, C. J.; Jenekhe, S. A. *Macromolecules* **2003**, *36*, 6577. (c) Berggren, M.; Inganas, O.; Gustafsson, G.; Rasmusson, J.; Andersson, M. R.; Hjertberg, T.; Wennerstrom, O. *Nature* **1994**, *372*, 444.
(25) (a) Jenekhe, S. A.; Osaheni, J. A. *Science* **1994**, *265*, 765. (b) Osaheni, J. A.; Jenekhe, S. A. *Macromolecules* **1994**, *27*, 739.
(26) Nallicheri, R. A.; Rubner, M. F. *Macromolecules* **1991**, *24*, 517.
(27) (a) Carpick, R. W.; Sasaki, D. Y.; Burns, A. R. *Langmuir* **2000**, *16*, 1270. (b) Burns, A. R.; Carpick, R. W.; Sasaki, D. Y.; Shelnutt, J. A.; Haddad, R. *Trib. Lett.* **2001**, *10*, 89.
(28) Chance, R. R.; Patel, G. N.; Witt, J. D. *J. Chem. Phys.* **1979**, *71*, 206.
(29) Kim, S. J.; Reneker, D. H. *Polym. Bull.* **1993**, *31*, 367.

(30) Brilliante, A.; Hanfland, M.; Syassen, K.; Hocker, J. *Physica* **1986**, *139 and 140B*, 53.
(31) Hanfland, M.; Brilliante, A.; Syassen, K.; Stamm, M.; Fink, J. *J. Chem. Phys.* **1989**, *90*, 1930.
(32) Yoshino, K.; Nakao.; Onoda, M. *Jpn. J. Appl. Phys.* **1989**, *28*, L323.
(33) Hess, B. C.; Kaner, G. S.; Vardeny, Z. *Phys. Rev. B* **1993**, *47*, 1407.
(34) Iwasaki, K. –I.; Fujimoto, H.; Matsuzaki, S. *Synth. Met.* **1994**, *63*, 101.
(35) Kaniowski, T.; Niziol, S.; Sanetra, J.; Trznadel, M.; Pron, A. *Synth. Met.* **1998**, *94*, 111.
(36) Yamamoto, T.; Muramatsu, Y.; Lee, B. L.; Kokubo, H.; Sasaki, S.; Hasegawa, M.; Yagi, T.; Rutota, K. *Chem. Mater.* **2003**, *15*, 4384.
(37) (a) Variano, B. F; Sandroff, C. J.; Baker, G. L. *Macromolecules* **1991**, *24*, 4376. (b) Lacey, R. J.; Batchelder, D. N.; Pitt, G. D. *J. Phys. C: Solid State Phys.* **1984**, *17*, 4529.
(38) Song, K.; Miller, R. D.; Wallraff, G. M.; Rabolt, J. F. *Macromolecules* **1991**, *24*, 4084.
(39) (a) Yablonovitch, E. *Phys. Rev. Lett.* **1987**, *58*, 2059. (b) John, S. *Phys. Rev. Lett.* **1987**, *58*, 2486. (c) Yablonovitch, E. *J. Opt. Soc. Am. B* **1993**, *10*, 283.
(40) Fink, Y.; Winn, J. N.; Fan, S. H.; Chen, C. P.; Michel, J.; Joannopoulos, J. D.; Thomas, E. L. *Science* **1998**, *282*, 1679.
(41) (a) Edrington, A. C.; Urbas, A. M.; DeRege, P.; Chen, C. X.; Swager, T. M.; Hadjichristidis, N.; Xenidou, M.; Fetters, L. J.; Joannopoulos, J. D.; Fink, Y.; Thomas, E. L. *Adv. Mater.* **2001**, *13*, 421. (b) Xia, Y. N.; Gates, B.; Li, Z. Y. *Adv. Mater.* **2001**, *13*, 409. (c) Jenekhe, S. A.; Chen, X. L. *Science* **1999**, *283*, 372.
(42) (a) Siwick, B. J.; Kalinina, O.; Kumacheva, E.; Miller, R. J. D. *J. Appl. Phys.* **2001**, *90*, 5328. (b) Gates, B.; Xia, Y. N. *Adv. Mater.* **2001**, *13*, 1605.
(43) (a) Ozaki, M.; Shimoda, Y.; Kasano, M.; Yoshino, K. *Adv. Mater.* **2002**, *14*, 514. (b) Debord, J. D.; Lyon, L. A. *J. Phys. Chem. B* **2000**, *104*, 6327. (c) Foulger, S. H.; Jiang, P.; Ying, Y.; Lattam, A. C.; Smith, D. W.; Ballato, J. *Adv. Mater.* **2001**, *13*, 1898.
(44) (a) Sun, H. B.; Mizeikis, V.; Xu, Y.; Juodkazis, S.; Ye, J. Y; Matsuo, S.; Misawa, H. *Appl. Phys. Lett.* **2001**, *79*, 1. (b) Liguda, C.; Böttger, G.; Kuligk, A.; Blum, R.; Eich, M.; Roth, H.; Kunert, J.; Morgenroth, W.; Elsner, H.; Meyer, H. G. *Appl. Phys. Lett.* **2001**, *78*, 2434.

Electrochromic Polymers and Devices

Chapter 2

Electrochromic Polyaniline Films from Layer-by-Layer Assembly

Dean M. DeLongchamp and Paula T. Hammond*

Department of Chemical Engineering, Massachusetts Institute of Technology, 77 Massachusetts Avenue, Cambridge, MA 02139

A survey of layer-by-layer (LBL) assembled poly(aniline) (PANI)-based electrochromic films is presented. PANI was LBL assembled with a wide variety of counterpolymers by both electrostatic and hydrogen bonding complexation forces. Post-assembly analysis included electrochemical and optical evaluation such as cyclic voltammetry, potential step measurements, and spectroelectrochemistry. Counterpolymer identity influenced PANI electrochemistry and spectra; more strongly acidic partner polymers shifted equilibrium between the differently colored PANI base and salt forms. However, PANI directs overall film performance, with variations in thickness and roughness accounting for most contrast and switching speed differences between films. These films feature high contrast due to their substantial thickness and low roughness. This work provides a basis for the development and evaluation of multiply colored, high contrast, and fast switching electrochromics based on the LBL assembly PANI and other readily available water-soluble polymers.

Introduction

There has been recent commercial and academic interest in next-generation display technologies. The ideal technology should provide low-cost, full-color displays in a variety of formats, including flat panels, large area boards, and even flexible "electronic paper" *(1-3)*. Electrochromic materials may be ideally suited to satisfy these applications *(4-6)*. Electrochromics change visible absorbance color in response to electrochemical oxidation or reduction. Electrochromic display elements consist of thin-film electrochemical cells with the same architecture as a power-storage cell and perform well in practically any flexible or rigid geometry. In addition, electrochromics are inexpensive and cell manufacturing tolerances are quite forgiving, making possible ultra-low cost or even disposable dynamic displays. A few challenges remain before the capabilities of these materials can be fully commercialized, in particular somewhat poor contrast and slow switching speed. Once these obstacles are overcome, full commercial exploitation of this promising technology will be possible, resulting in high-performance displays, solar windows, and even electro-optical switches.

This work has applied the processing technique of layer-by-layer (LBL) assembly to solve these remaining challenges. LBL assembly is a recently developed type of assisted self-assembly that creates macromolecular composites in ultra thin films with a fine degree of control. The classical LBL technique involves the exposure of a charged substrate alternately to dilute aqueous solutions or dispersions containing materials of opposite attractive affinities *(7)*. Upon each exposure, surface affinity reversal is achieved provided that the assembled species possess a large number of affinity sites and that complexation is sufficiently kinetically irreversible on the time scale of the exposure step. Composites of practically any material system including light emitting polymers *(8)*, inorganic nanoparticles *(9)*, and biological materials *(10)* can be assembled on any charged substrate. Using this technology, one can tailor the composition and morphology of electrochromic films on the nanoscale, combining existing electrochromic polymers into high-performance and easily-applied electrochromic composites *(11-14)*. This ability to create fine-grained composite films with excellent smoothness and unlimited thickness on practically any substrate makes LBL assembly more powerful and flexible than traditional electrochromic film fabrication techniques such as electropolymerization or spin-coating. In this study, we have explored some of the composites that can be obtained with the LBL assembly a common, readily available electrochromic polymer.

Poly(aniline) (PANI) is the most commonly utilized conjugated conducting polymer, and it has attracted much attention due to a simple and low cost preparation coupled with high electronic conductivity and good environmental stability *(15,16)*. Investigations of PANI and PANI derivatives have led to applications in corrosion protection *(17-19)*, hole transport in light emitting diodes *(20,21)*, and a wide variety of microelectronic device applications. Importantly, PANI exhibits electrochromic behavior upon oxidation and reduction as do most other conjugated polymers, though PANI electrochromism is exceptionally stable to repeated switching *(22)*. As an electrochromic polymer, PANI switches from a semi-transparent yellow, reduced state termed leucoemeraldine PANI to a darker, oxidized green emeraldine salt or blue emeraldine base (depending on pH) *(23,24)*. Even greater optical absorption is possible at higher oxidation states – for example the nigraniline base exhibits a blue-black color – but repeated cycling to the oxidative potentials required to attain these states inevitably causes degradation due to hydrolysis of the imine group and subsequent chain scission *(22-24)*.

LBL films containing PANI have been constructed based on electrostatic interactions and hydrogen bonding interactions, and the conductivity and doped absorption of these materials was described by Rubner and co-workers *(25,26)*. This advance provided a method for applying robust, conformal PANI thin films onto essentially any planar or nonplanar substrate. Other efforts have expanded the range of PANI-containing LBL composites *(27-29)*. In some cases, these studies have touched on the electrochromic nature of PANI *(28)*, but until recently there has been no in-depth study of the electrochromic properties in particular. Recently we presented a complementary coloring electrochromic device with anodic coloration from a PANI-containing LBL film *(12)*.

Here we present a survey of the electrochromic properties of PANI within various LBL film architectures. As a pH-sensitive polycation and a strong H-bonding donor/acceptor, PANI can be assembled into a wide variety of LBL composites with different chemical identities; molecular structures of the counterpolyions employed to achieve this variation are depicted in Figure 1. The polyanions poly(2-acrylamido-2-methyl-1-propanesulfonic acid) (PAMPS), the perfluoronated ionomer Nafion®, the polysaccharide carageenan, and strongly sulfonated, self-doping PANI (SPANI) can be LBL assembled with PANI based on electrostatic interactions. H-bonding donor/acceptors such as poly(acrylic acid) (PAA, here assembled at low pH so that it is fully protonated), poly(acrylamide) (PAAm), and poly(ethylene oxide) (PEO), can be assembled with PANI via H-bonding interactions. Finally, in order to explore the possibility of an all-polycation LBL film, we assembled PANI with a H-bonding polycation, ethoxylated poly(ethylene imine) (ePEI), assessing the capability of a strong H-bonding LBL system to overcome electrostatic repulsion between two charged polycations. This wide range of partner polymer character provides a wealth of variety in the properties of the finished PANI composites, providing the opportunity to assess the relative contributions of such properties as hydrophilicity and dielectric constant on the behavior of PANI coloration and electrochemical switching.

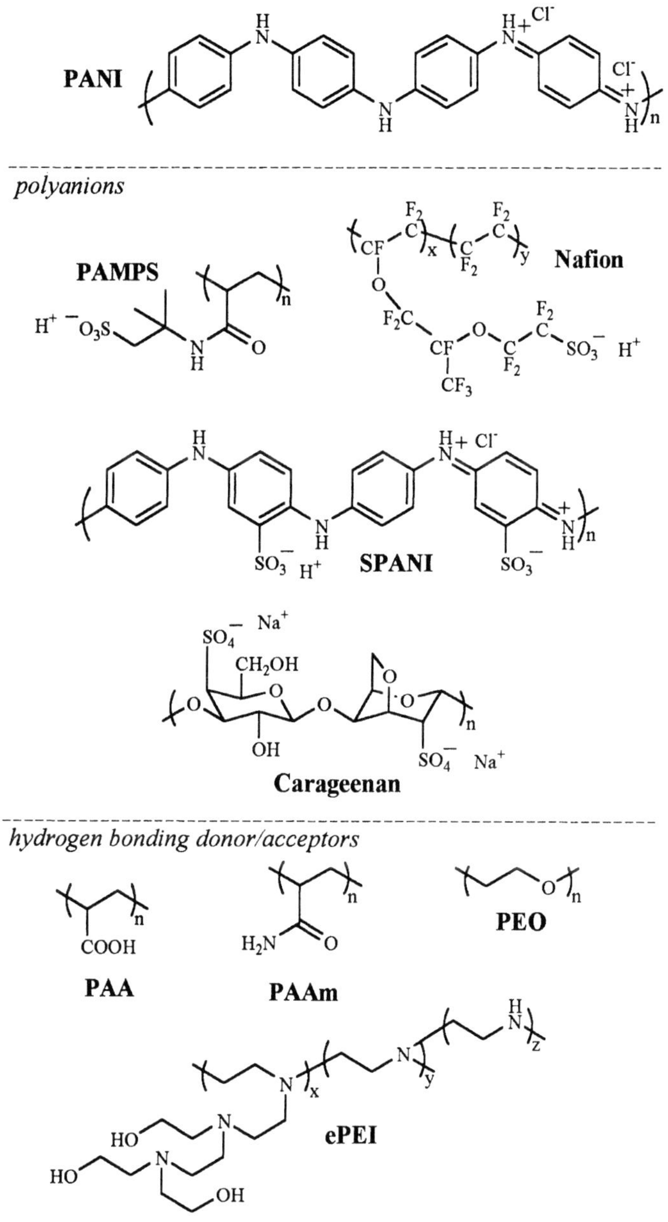

Figure 1. Chemical structures of the polyelectrolytes and nonionic polymers employed in this study.

Experimental

Materials. The electroactive polymer was PANI (Aldrich). PANI films were LBL assembled with PAMPS (Aldrich), Nafion 117 (Fluka), Carageenan (Fluka?), SPANI (Aldrich), PAA (Aldich MW), PAAm Aldrich), and ePEI (MW, Aldrich). Polymer solutions were made using Milli-Q (Millipore deionized, >18.2 Ωcm, 0.22 μm filtered) water, and pH adjusted using NaOH or HCl. The PANI solution was 10mM (all polymer solution concentrations are respect to the molecular weight of the repeat unit). PANI solutions were formulated using a 1:9 dimethylacetamide and water solvent pair as described by Rubner and co-workers (25,26). PAMPS and NAFION solutions were formulated at 2mM, while all other polymer solutions wre formulated at 20 mM. The pH of all deposition baths was adjusted to pH 2.5. ITO-glass substrates with dimensions 0.7 cm × 5 cm (Delta Technologies, 6 Ω/square) were cleaned by ultrasonication in a series of solvents: dichloromethane, methanol, acetone, and Milli-Q water for 15 minutes each, followed by a 5-minute oxygen plasma etch (Harrick PCD 32G) to provide a clean, hydroxyl-rich surface.

Assembly. Film assembly was automated with a Carl Zeiss HMS DS-50 slide stainer. The substrates were exposed to PANI solution for 15 minutes, followed by copious water rinsing for 4 minutes in three consecutive Milli-Q water baths, and then exposed to polyanion solution for 15 minutes and again rinsed. This cycle was repeated for 15 layer pairs for each PANI-containing LBL system.

Measurement. Film thickness and roughness measurements were performed using a Tencor P10 profilometer using a 2 μm stylus and 5 mg stylus force. Electrochemical analysis was performed using an EG&G 263A potentiostat/galvanostat. These measurements were performed in a flat cell of 30 mL volume and approximately 0.3 cm^2 working electrode area. The electrolyte used was aqueous 0.1 M sulfuric acid with a pH of approximately 1.1. The counterelectrode was 4 cm^2 platinum foil, and reference was a K-type saturated calomel electrode. Cyclic voltammetry was performed with potential limits of –0.2 V and 0.6 volts, at scan rates of 25, 50, 100, and 200 mV/s. Double potential step chronoamperomtery was performed by stepping between –0.2 V and 0.6 V vs. SCE, with 5 seconds per step and 10 seconds per cycle, with approximately 20 cycles performed sequentially before the measurement cycle. Spectral characterization was performed on a rail-mounted Oriel UV-Vis spectrophotometer with a 75 W Xe lamp, 300 L/mm, 300 nm blaze grating and InstaSpec IV CCD. For spectroelectrochemistry, potential control was provided by EG&G 263A, with the polymer-coated ITO-glass substrate positioned in a quartz cell and immersed in electrolyte, along with a platinum wire counter electrode, and SCE reference.

Results and Discussion

Assembly. The PANI-containing films assembled in this study featured a linear increase in thickness with the number of layer pairs deposited, as was described previously by Rubner for some systems of this type *(25,26)*. Film thickness and roughness are shown in Table 1. Thickness of electrostatic films varied from 2.5 to 5.1 nm per layer pair, well in agreement with previous work on the PANI/poly(styrene sulfonate) system *(25,26)*. The thickest film assembled by electrostatics was found to be PANI/Nafion, most likely owing to Nafion's high equivalent weight and its possible deposition as stable agglomerates rather than extended chains. The Nafion deposition solution is an alcohol/water mixture that is a poor solvent in which Nafion can be easily destabilized and precipitated by the addition of salt. Despite this possible agglomerate deposition, PANI/Nafion films are quite smooth and defect-free.

Table 1. Thickness results for different PANI LBL composites

system	final thickness (per layer pair) (nm)		R_q rms roughness (nm)
(PANI/PAMPS)$_{15}$	53.0	(3.5)	2.6
(PANI/Nafion®)$_{15}$	77.2	(5.1)	2.3
(PANI/carageenan)$_{15}$	60.8	(4.1)	3.6
(PANI/SPANI)$_{15}$	37.9	(2.5)	2.5
(PANI/PAA)$_{15}$	111.9	(7.5)	8.3
(PANI/PAAm)$_{15}$	452.5	(30.2)	16.9
(PANI/PEO)$_{15}$	221.9	(14.8)	8.9
(PANI/ePEI)$_{15}$	13.6	(0.9)	1.5

H-bonded films assembled in this study are far thicker than those studied earlier *(26)*. In particular, the PANI/PAAm and the PANI/PEO systems are two to four times thicker. This difference may be due to the use of an ITO substrate rather than silane- or PEI-treated silica. The hydroxyl surface of ITO provides strong Lewis acidity that has stronger attractive interactions with polyimines than the Brönsted acid silica surface, especially at lower pH conditions. Due to this advantage, PANI deposits in thicker layers onto ITO – an effect which propagates throughout the film in H-bonded systems because H-bond growth is less self-limiting than electrostatic growth. Greater thickness could also stem from differences in nonionic polymer dipping solution pH, which was ambient in the previous work *(26)*, yet adjusted to 2.5 in our work, which could have resulted in stronger interactions with a fully protonated PANI surface. The roughness of H-bonded systems is greater than that of electrostatics, possibly due to instabilities arising from the deposition of the extremely thick layers.

The PANI/ePEI system is thinner than any other system due to the inherent electrostatic repulsion between PANI and ePEI, which both have protonated imine groups at the assembly pH of 2.5. Despite the repulsion, H-bonding between PANI amine/imine groups and ePEI hydroxyls and amines is sufficient to facilitate the growth of a very thin film. This example is a surprising result of the complex interplay of attractive/repulsive forces in LBL films.

Cyclic voltammetry. The general electrochemical behavior of PANI-containing LBL composites were assessed using cyclic voltammetry (CV). The potential limits of –0.2 to 0.6 V vs. SCE were chosen to reduce oxidative degradation while still accessing the full color change between the pale yellow leucoemeraldine base to the saturated green/blue emeraldine salt/base (22). The CV results are shown in Figure 2. In general, the electroactive behavior of PANI is similar to that reported from other sources, with strong oxidation and reduction peaks centered at 0.11-0.12 V vs. SCE. This potential is consistent with the desired transition in this pH range (30). There is no strong shift in the redox potential when PANI is complexed with stronger polyacids (e.g. Fig 2a,b) or weaker polyacids (e.g. Fig 2e), indicating that the polyanions ionically crosslinked with PANI are primarily in salt form and thus do not provide a local low pH environment, which would shift the redox potential of this transition to higher values for greater acidity (30).

While the PANI is undergoing reduction and oxidation throughout the CV experiment, ions travel through the thickness of the film to balance the changing electrostatic charge of PANI. For neat PANI and H-bond LBL PANI films, the mobile species would be small polyanions – in our case sulfate anions. PANI within electrostatic LBL films is paired with polyanions, so the mobile species in those systems would be small cations - in our case protons - moving in and out of association with the polyanionic dopant (31). The speed of this ionic exchange can be qualitatively assessed from the oxidative and reductive CV peak currents at different scan rates, as shown in Figure 3. The peak height increases linearly with scan rate in all cases. This linear increase indicates that the redox reaction is confined to the thin film and not limited by diffusion. There is a very slight nonlinearity in the response of the very thick H-bond LBL films (notably PANI/PAAm), indicating a small resistance to counterion diffusion. Even very thick LBL films present an open and ion permeable morphology, a result that has also been found in other electroactive polymer LBL systems (13). These results contrast with studies that show severely restricted ion permeation in highly-charged LBL films due to intrinsic charge compensation and a lack of free ion exchange sites (32). Though electrolyte exposure increased the exchange site number, the ionic strength required for significant permeation was much higher than the ionic strength of our own electrolytes (32). Facile ion permeation in these PANI-containing systems may be attributed to the low cationic charge density of PANI, which may naturally assemble into a more loopy, open morphology.

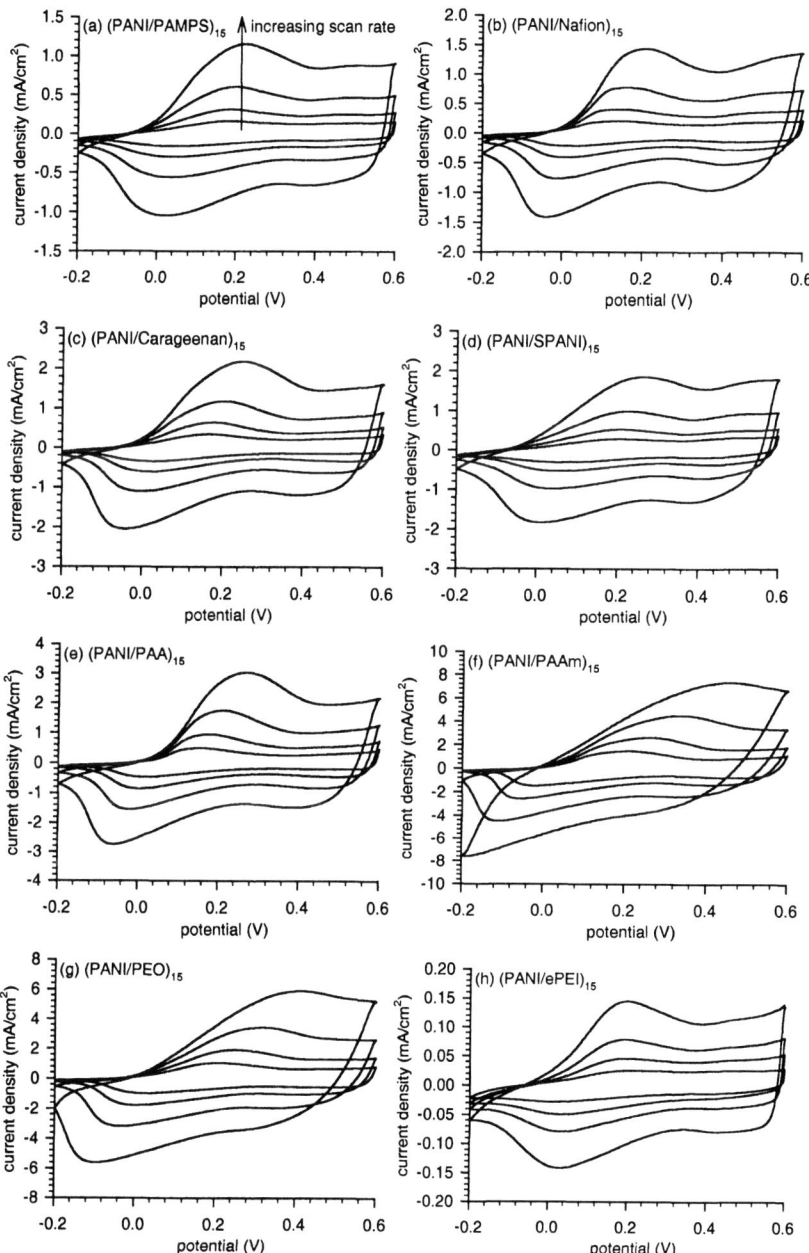

Figure 2. Cyclic voltammograms of 15 layer pair PANI LBL films. Scans were at 0.025, 0.05, 0.1, and 0.2 V/s; peak height increases with scan rate.

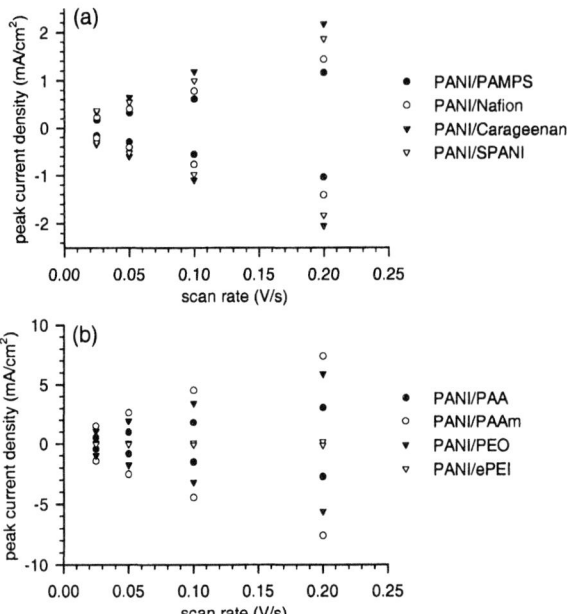

Figure 3. Oxidative (positive) and reductive (negative) peak currents for (a) electrostatic and (b) H-bond PANI LBL films linearly increase with scan rate.

Another transport phenomenon occurring within the films during CV switching is the movement of the redox front - the transfer of electrons at the ITO/PANI interface and within the film interior. A qualitative characterization of this phenomemon may be made by examining the hysteresis between CV oxidation and reduction peaks. For most systems the hysteresis between these peaks grows with increasing scan rate, indicating a non-Nernstian condition at the reactive front caused by charge transfer resistance within the LBL film structure. The hysteresis is greatest for the thickest LBL films, where the outermost PANI layers are isolated and less electrochemically available. The large individual layer thickness may limit redox propagation by insulating PANI between thick layers of resistive polymer. The thinnest film, PANI/ePEI, shows a small hysteresis that does not increase, indicating that the PANI within this film is immediately electrochemically available.

Double potential step chronoamperometry. The electrochemical technique of double potential step chronoamperomtery (DPSCA) was used to investigate the electrochemical accessibility of PANI and the composite switching. A square wave between –0.2 V and 0.6 V vs. SCE was applied while monitoring current. The dynamics of current change and the integrated charge injected/withdrawn are be plotted as a function of time in Figure 4.

Electrostatically assembled PANI films switch faster than hydrogen-bonded films, an effect due primarily to film thickness that may also be influenced by the different mobile species (sulfate vs. proton). The final charge capacity scales well with expectations based on thickness, indicating similar PANI "loading" in all LBL composites. PANI/SPANI has a higher charge capacity than other systems due to the inclusion of a reactive SPANI polyanion. The PANI/PAAm system has less charge density than may be expected, indicating that outer PANI layers in this extremely thick composite may not be fully accessible.

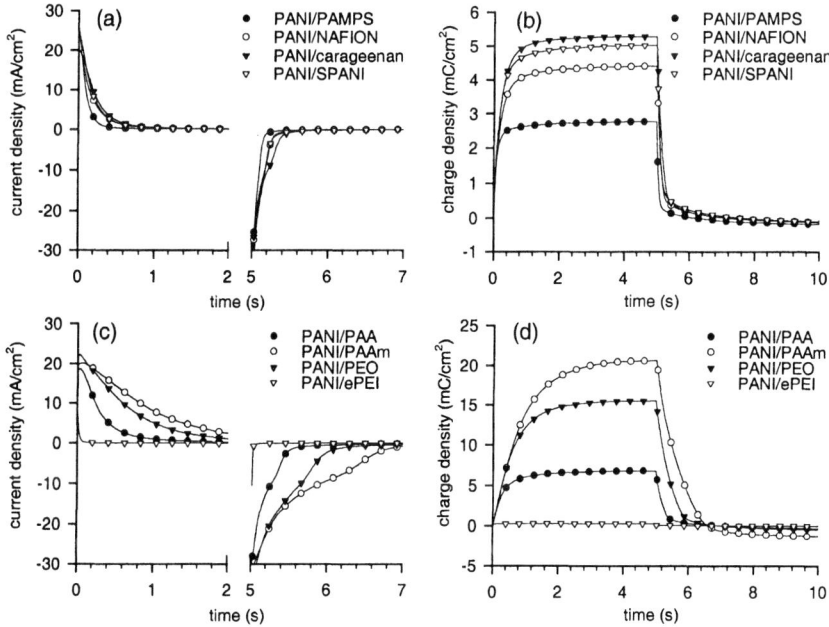

Figure 4. Current and charge switching profiles for (a,b) electrostatic and (c,d) H-bond PANI LBL films. Step duration was 5 s, limits 0.6 V and –0.2 V.

Spectroelectrochemistry. The full electrochromic properties of the PANI LBL assembled films were examined using spectroelectrochemistry, taking a UV-Vis "snapshot" of each film at equilibrated potentials between –0.6 and 0.2 V, with results shown in Figure 5. At very cathodic potentials, leucoemeraldine PANI exhibits a single absorbance maximum at 340 nm, with essentially no additional absorbance in the visible region. At more anodic potentials, emeraldine PANI evinces a 700 nm peak, with broad visible absorbance.

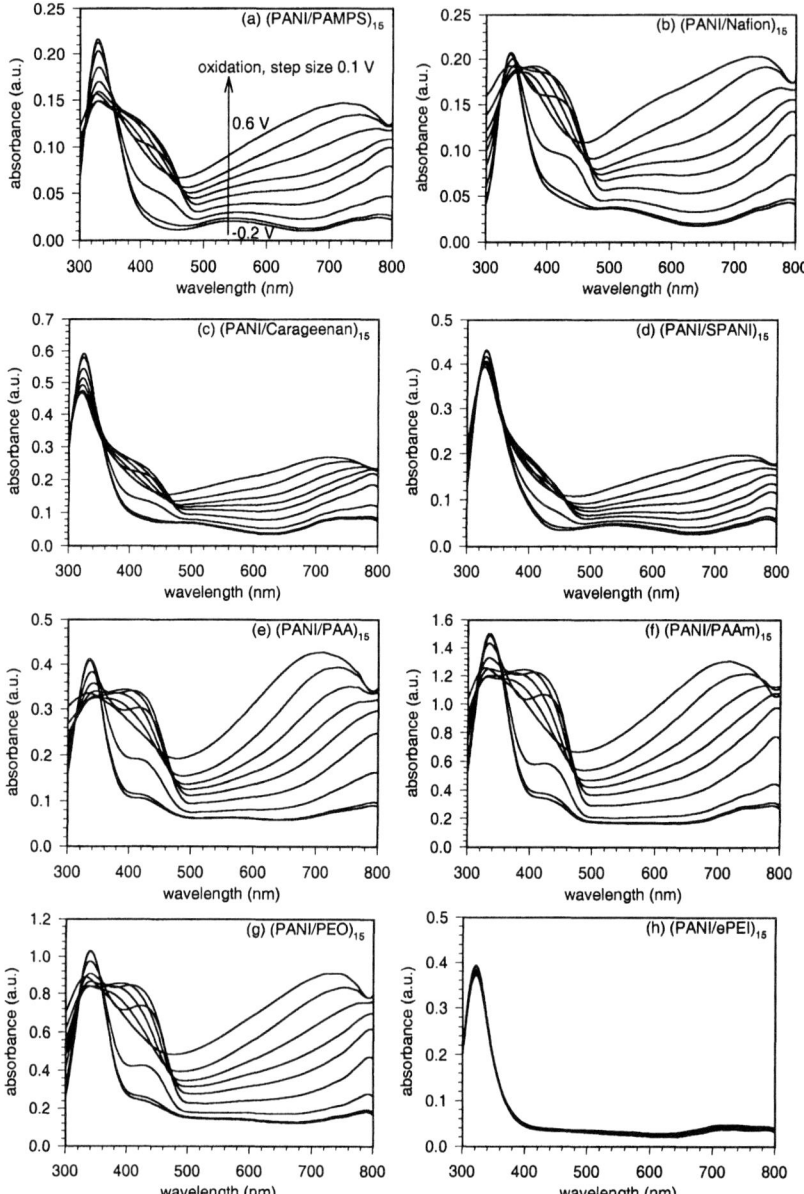

Figure 5. Spectroelectrochemistry of 15 layer pair PANI LBL films. Step size was 0.1 V; potential limits were 0.6 V to -0.2 V.

The primary spectral differences between the various LBL assembled composites described in Figure 5 are the intensities of leucoemeraldine and emeraldine peaks. In general, the emeraldine peak absorbance scales mostly linearly with the film thickness and Faradaic charge capacity, but there is some small variation in extinction coefficient. The calculated extinction coefficients are shown in Table 2. It should be noted that the concentrations shown in Table 2 are not concentrations of PANI monomer, but rather the concentrations of redox centers (as determined from square wave switching in Figure 4) that may be distributed over several monomer units and might be considered a macro-chromophore. As can be seen, the concentration of these centers for PANI/SPANI is approximately double that seen for the other composites, as would be expected because both polycation and polyanion are redox-active.

Table 2. PANI film redox center concentration and extinction.

system	PANI centers (mmol/cm^3)	ε (M^{-1}cm^{-1}) at 700 nm
(PANI/PAMPS)$_{15}$	5.5	4618
(PANI/Nafion)$_{15}$	6.0	4264
(PANI/carageenan)$_{15}$	9.1	4814
(PANI/SPANI)$_{15}$	13.7	3627
(PANI/PAA)$_{15}$	6.5	5575
(PANI/PAAm)$_{15}$	5.0	5306
(PANI/PEO)$_{15}$	7.4	5141
(PANI/ePEI)$_{15}$	7.6 *	3473 *

* questionable data from ultrathin film

What is especially notable from the trends in Table 2 is that the electrostatically assembled PANI films consistently display extinction lower than the hydrogen bond assembled PANI films. In fact, the extent of acidity can be correlated to some degree with the caliber of extinction; Nafion the superacid has the lowest extinction of the redox-inert sulfonic acids. Another factor introduced by the strong polyacids is the appearance of a small peak at 500-550 nm that can be seen clearly in the –0.2 V absorbance spectrum of the electrostatically assembled PANI films in Figure 5. Together, these two phenomena suggest that the acidity of the counterpolymer has a direct influence on the properties of PANI. This acidity may influence the equilibrium between emeraldine salt and emeraldine base in the oxidized composite; at the pH of the electrolyte, both should be present but emeraldine salt should dominate. The emeraldine base form is more prevalent at higher pH conditions and features greater absorbance in the red region of the spectrum, therefore displaying a more blue color, while the emeraldine salt form is more prevalent at lower pH conditions and features greater absorbance in the 400 nm range, resulting in a greener color. In PANI films assembled with strong polyacids, the additional acidity provided by the acid appears to shift the equilibrium to the emeraldine salt so that less absorbance is observed in the 700 nm range. In PANI films assembled with hydrogen-bonding polymers, there is no additional acidity and the oxidized PANI takes on an equilibrium composition influenced entirely by the electrolyte that features a greater amount of the emeraldine base.

The emeraldine peak that is preserved in the reduced electrostatic PANI films must be due to continued doping by the polyacid even when the film is polarized to a potential that should result in complete conversion to leucoemeraldine PANI. This phenomenon has been noted for PANI|PAMPS interfaces and is one of the reasons these materials are employed in tandem in electrochemical cells *(33-35)*. This effect may be enhanced by polyacid enthalpic resistance to protonation under these pH conditions and the potential entropic loss due to re-association of a free proton with the polyacid.

A final phenomenon that should be noted is that the extinction of PANI/SPANI is poorer than any of the other composites. SPANI coloration has been found to be inferior to that of PANI, which in general explains this lowered absorbance*(29,36)*. PANI/ePEI is not sufficiently absorbing for full detection.

Potential Step Absorptometry. The dynamics of switching were investigated using in situ fast spectral scans during the DPSCA waveform. Absorbance switching of an isolated wavelength is shown in Figure 6. In general, the absorbance switching mirrors the charge switching described in Figure 4.

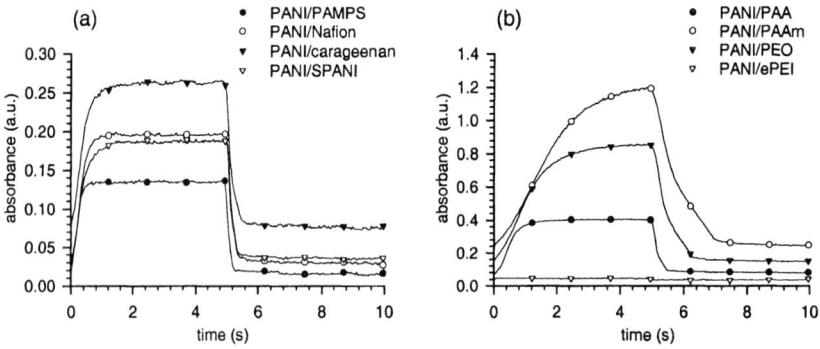

Figure 6. Absorbance switching profiles at 700 nm for (a) electrostatic and (b) H-bond PANI LBL films. Step duration was 5 s, limits 0.6 V and –0.2 V.

Electrochromic performance. Primary performance metrics for electrochromic polymer films are response time and contrast. Contrast can be evaluated by many measures; maximum transmittance change is often used. Contrast and response times for the PANI LBL composites surveyed herein are presented in Table 3. Response time is based on the time required for 90% of full transmittance change. In general, films of similar thickness performed similarly, indicating that their commonality - PANI - must control the morphology and internal structure of the final films, especially as the counterpolymers have such a diverse range of dielectric constant, hydrophilicity, and acidity. The maximum contrast was exhibited by PANI/PAAm and PANI/PEO. With similar extinction coefficient and PANI loading, this maximum contrast is achieved in films near a simple optimum thickness, which may lie between 250 and 400 nm. The best combination of fast switching and high contrast was found in PANI/PAA, which is thinner than both PANI/PAAm and PANI/PEO. Contrasts achieved from these LBL composites are high when compared to other electrochromic polymer films, due to freedom from defects even in very thick films. For example, PANI films created using electropolymerization *(37)* or spin casting *(38)* feature 25-50% contrast even at 300-500 nm thickness due to higher extinction in the nominally bleached state.

Table 3. Electrochromic performance of 15 layer pair PANI LBL films.

system	bleach / color time (s)	$\Delta\%T_{max}$ (bleach – colored, loc.)
(PANI/PAMPS)$_{15}$	0.25 / 0.41	26% (98% - 72%, 688 nm)
(PANI/Nafion®)$_{15}$	0.41 / 0.62	31% (95% - 64%, 682 nm)
(PANI/carageenan)$_{15}$	0.50 / 0.74	34% (92% - 58%, 649 nm)
(PANI/SPANI)$_{15}$	0.41 / 0.83	29% (94% - 62%, 687 nm)
(PANI/PAA)$_{15}$	0.50 / 0.91	49% (87% - 38%, 687 nm)
(PANI/PAAm)$_{15}$	2.0 / 2.2	61% (68% - 7%, 649 nm)
(PANI/PEO)$_{15}$	1.3 / 1.7	62% (78% - 14%, 686 nm)
(PANI/ePEI)$_{15}$	<0.15 / <0.05	2% (93% - 91%, 688 nm)

Conclusions

A survey of the electrochemical and optical properties of a wide variety of PANI LBL assembled electrochromic films has been presented. Different counterpolymers do influence the electrochemical and spectral properties. For the first time it has been shown that the acidity of the LBL counterpolymer can be used to directly manipulate the coloration of PANI films by influencing local pH conditions and shifting the equilibrium between PANI emeraldine salt and emeraldine base forms. However, PANI appears to control the overall morphology - in particular the connectivity of the electroactive species and the ion mobility environment - because films of very different composition yet similar thickness possess similar electrochemical kinetics and switching behavior. These PANI composites switch extremely fast and display high contrast owing to the high thickness that can be achieved with low roughness in LBL assembled films. PANI as a LBL-capable electrochromic polycation is an excellent candidate for incorporation into "dual electrochrome" electrode concepts wherein electrochromic polycations and polyanions are combined to achieve strongly enhanced contrast and multiple colors *(14)*.

Acknowledgement

The authors gratefully thank the DOD NDSEG program and the ARO TOPS MURI for support. This work was also supported by the MIT MRSEC Program of the National Science Foundation under award DMR 94-00334.

References

(1) Dagani, R. *Chem. Eng. News* **2001**, *79*, 40-43.
(2) Seeboth, A.; Schneider, J.; Patzak, A. *Sol. Energy Mat. Sol. Cells* **2000**, *60*, 263-277.
(3) Green, M. *Chem. Ind.* **1996**, *17*, 641-644.
(4) Monk, P. M. S.; Mortimer, R. J.; Rosseinsky, D. R. *Electrochromism: Fundamentals and Applications*; Weinheim: New York, 1995.
(5) Mortimer, R. J. *Electrochim. Acta* **1999**, *44*, 2971-2981.
(6) Bach, U.; Corr, D.; Lupo, D.; Pichot, F.; Ryan, M. *Adv. Mater.* **2002**, *14*, 845-848.
(7) Decher, G. *Science* **1997**, *277*, 1232.
(8) Fou, A. C.; Onitsuka, O.; Ferreira, M.; Rubner, M. F.; Hsieh, B. R. *J. Appl. Phys.* **1996**, *79*, 7501-7509.
(9) Lvov, Y.; Ariga, K.; Onda, M.; Ichinose, I.; Kunitake, T. *Langmuir* **1997**, *13*, 6195.
(10) Lvov, Y.; Ariga, K.; Ichinose, I.; Kunitake, T. *J. Chem. Soc.-Chem. Commun.* **1995**, 2313-2314.
(11) Laurent, D.; Schlenoff, J. B. *Langmuir* **1997**, *13*, 1552-1557.
(12) DeLongchamp, D.; Hammond, P. T. *Adv. Mater.* **2001**, *13*, 1455-1459.
(13) Cutler, C. A.; Bouguettaya, M.; Reynolds, J. R. *Adv. Mater.* **2002**, *14*, 684-688.
(14) DeLongchamp, D. M.; Kastantin, M.; Hammond, P. T. *Chem. Mater.* **2003**, *15*, 1575-1586.
(15) Diaz, A. F.; Logan, J. A. *J. Electroanal. Chem.* **1980**, *111*, 111-114.
(16) MacDiarmid, A. G. *Synth. Met.* **1997**, *84*, 27-34.
(17) Deberry, D. W. *J. Electrochem. Soc.* **1984**, *131*, C302-C302.
(18) Sekine, I.; Kohara, K.; Sugiyama, T.; Yuasa, M. *J. Electrochem. Soc.* **1992**, *139*, 3090-3097.
(19) Trochnagels, G.; Winand, R.; Weymeersch, A.; Renard, L. *J. Appl. Electrochem.* **1992**, *22*, 756-764.
(20) Gustafsson, G.; Cao, Y.; Treacy, G. M.; Klavetter, F.; Colaneri, N.; Heeger, A. J. *Nature* **1992**, *357*, 477-479.
(21) Yang, Y.; Heeger, A. J. *Appl. Phys. Lett.* **1994**, *64*, 1245-1247.
(22) Kobayashi, T.; Yoneyama, H.; Tamura, H. *J. Electroanal. Chem.* **1984**, *161*, 419-423.

(23) Huang, W. S.; Humphrey, B. D.; MacDiarmid, A. G. *J. Chem. Soc.-Faraday Trans.* **1986**, *82*, 2385-2400.
(24) Huang, W. S.; Macdiarmid, A. G. *Polymer* **1993**, *34*, 1833-1845.
(25) Cheung, J. H.; Stockton, W. B.; Rubner, M. F. *Macromolecules* **1997**, *30*, 2712-2716.
(26) Stockton, W. B.; Rubner, M. F. *Macromolecules* **1997**, *30*, 2717-2725.
(27) Onoda, M.; Yoshino, K. *Jpn. J. Appl. Phys. Part 2 - Lett.* **1995**, *34*, L260-L263.
(28) Li, D.; Jiang, Y.; Li, C.; Wu, Z.; Chen, X.; Li, Y. *Polymer* **1999**, *40*, 7065-7070.
(29) Sarkar, N.; Ram, M. K.; Sarkar, A.; Narizzano, R.; Paddeu, S.; Nicolini, C. *Nanotechnology* **2000**, *11*, 30-36.
(30) MacDiarmid, A. G.; Humphrey, B. D.; Huang, W.-S. *J. Chem. Soc., Faraday Trans. 1* **1986**, *82*, 2385-2400.
(31) Schlenoff, J. B.; Ly, H.; Li, M. *J. Am. Chem. Soc.* **1998**, *120*, 7626-7634.
(32) Farhat, T. R.; Schlenoff, J. B. *Langmuir* **2001**, *17*, 1184-1192.
(33) Jelle, B. P.; Hagen, G.; Sunde, S.; Odegard, R. *Synth. Met.* **1993**, *54*, 315-320.
(34) Jelle, B. P.; Hagen, G.; Nodland, S. *Electrochim. Acta* **1993**, *38*, 1497-1500.
(35) Goff, A. H.-L.; Bernard, M.-C.; Zang, W. *Electrochim. Acta* **1998**, *44*, 781-796.
(36) Baba, A.; Park, M. K.; Advincula, R. C.; Knoll, W. *Langmuir* **2002**, *18*, 4648-4652.
(37) Yang, S. C. *Proceedings of SPIE-The International Society for Optical Engineering* **1988**, *IS 4*, 335-365.
(38) Genies, E. M.; Lapkowski, M.; Santier, C.; Vieil, E. *Synth. Met.* **1987**, *18*, 631-636.

Chapter 3

Electrochromic Devices Based on Ladder Polymer and Phenothiazine–Quinoline Copolymer Films

Jai-Pil Choi[1], Fernando Fungo[1], Samson A. Jenekhe[2], and Allen J. Bard[1]

[1]Department of Chemistry and Biochemistry, The University of Texas at Austin, Austin, TX 78712
[2]Department of Chemical Engineering, University of Washington, Seattle, WA 98195-1750

> We describe the morphological, electrochemical, spectroelectrochemical characterization, and electrochromic device properties of several polymer films: ladder poly(benzobisimidazobenzophenanthroline) (BBL), semiladder poly(benzobisimidazobenzophenanthroline) (BBB), and poly(2,2'-[10-methyl-3,7-phenothiazylene]-6,6'-bis[4-phenylquinoline]) (PPTZPQ). Cyclic voltammograms of polymer films showed a number of oxidation and reduction waves with color changes that were a function of the potential. The electrochromic effect was spectroelectrochemically characterized and the behavior of PPTZPQ was interpreted as the combination of the properties of the constituent donor and acceptor moieties. A number of different polymer electrochromic devices were constructed using plastic-indium tin oxide and different solvents and electrolytes. We demonstrate the feasibility of constructing flexible all-plastic electrochromic devices with interesting color changes upon potential cycling. Electrochromic devices based on BBL or BBB films had switching lifetimes of 3×10^4 to 10^5 circles. Plastic electrochromic cells combining PPTZPQ and BBL films as complementary electrochromic materials showed reversible switching between black and red.

Introduction

A chemically reversible electrochemical process producing a color change is known as electrochromism. The color change results from the generation of different electronic absorption transitions within the material in the visible region (*1, 2*). An electrochromic device (ECD) is formed by a cell, consisting of two optically transparent electrodes (OTE), separated by a solid (often polymeric) gel or liquid electrolyte. The electrochromic material (EM) can be present in the electrolyte or in the form of a film on the OTEs (Figure 1). A quantity of charge is reversibly exchanged between a working electrode (primary electrode) and a counter-electrode (secondary electrode) by switching of potential, resulting in a cyclic color change (*3*).

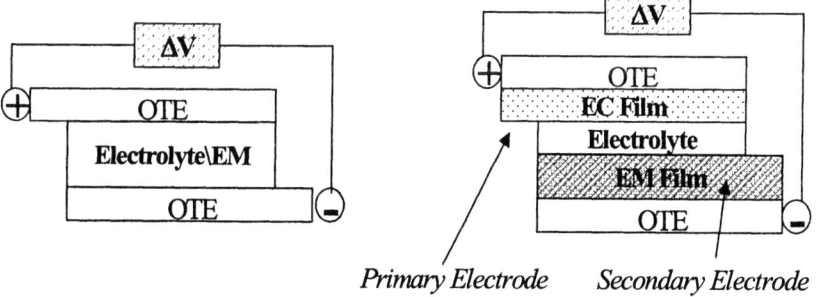

*Figure 1. Diagrams of two type of ECDs, **(left)** the electrochromic material, EM, is present in electrolyte support; **(right)** the EM is deposited over the electrode as a film.*

The potential uses of this type of device are very broad, for example electrochromic windows, which control the radiant energy transfer in buildings and cars, improving energy efficiency (*1, 3*), electrochromic rear-view mirror systems in cars, flat panel displays, and smart paint. The prospects for these kinds of applications have generated a high demand for new materials with improved electrochromic response, which is reflected in the increasing number of publications on the topic in recent years (*4*).

Many different types of materials have been described and used in the construction of electrochromic devices, starting with the inorganic systems based on transition metal oxides (e.g. WO_3 (*5*), V_2O_5 (*6, 7*) and TiO_2 (*8*)) and organic systems based on viologen, anthroquinone, and phenazine derivatives (*3, 9, 10*). The construction of electrochromic devices based on inorganic materials has attained significant results. However, current research interests are focused toward the use of polymeric materials where there is more flexibility in the molecular design and improved molar absorptivity, lower production costs, and more appropriate mechanical properties are possible. These characteristics make polymer materials excellent candidates for designing tunable optical materials for electrochromic device applications.

To develop electrochromic polymers, it is necessary to obtain polymers that are reasonably conducting with low band gap energies so that the change of energy produced in passing from one redox state to another occurs with the band gaps in the range of energies in the visible region of the electromagnetic spectrum. Low band gap polymers can be obtained by maximizing the π-extended-conjugation within the conjugated polymer backbone (*11*); for example, ladder-type structures with coplanar conformations between the polymer's consecutive repeat units (*12*).

Another strategy employed in producing a lower band gap in conjugated polymers involves constructing [AD]-type alternating copolymers where the A and D units are strong electron-accepting and electron-donating moieties, respectively (*11b, 13, 14*). An appropriate choice of potentials for the reduction and oxidation of the A and D units allows control of the energies of the effective HOMO and LUMO levels, which determine the polymer's oxidative and reductive properties. The alternating donor-acceptor (A-D) arrangements permit a high degree of intramolecular charge transfer (ICT) within the conjugated framework of the polymers (*13, 15*). This intramolecular charge transfer produces a successive zwitterion-like interaction with high double bond character between the repeat units. The stabilizing effects of the quinoidal-like forms within the polymer backbone produce a decrease in the band gap energy. Common examples of polymers previously used in electrochromic devices include polyaniline (*16-19*), polythiophenes (*20, 21, 22*), and polypyrroles (*22, 23*).

We report here morphological, electrochemical and spectroelectrochemical studies of semiladder poly(benzobisimidazobenzophenanthroline) (BBB), ladder poly(benzobisimidazobenzophenanthroline) (BBL) and poly(2,2'-[10-methyl-3,7-phenothiazylene]-6,6'-bis[4-phenylquinoline]) (PPTZPQ); the structures are shown in Figure 2. We also describe the construction and characterization of the electro-optical properties of all-polymer electrochromic devices using PPTZPQ as one electrode and BBL, BBB or V_2O_5 as the counter electrode.

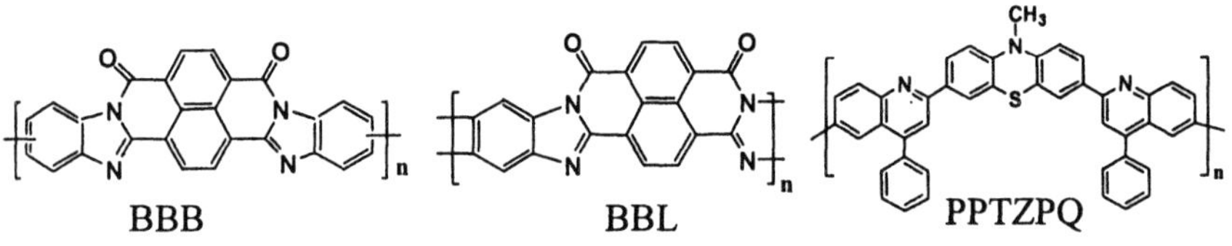

Figure 2. Structures of electrochromic polymers.

Results and Discussion

Film Characterization

A better understanding of intrinsic properties of electrochromic polymer films (such as the electrical, optical and structural properties) and of the behavior of these films at different interfaces (such as electrode surface and electrolyte solution) will permit improvement in the design (electrode materials, composition of support electrolyte, contacts, etc.) and performance of ECDs. Useful information that is needed includes:

i) The topography of the surface (degree of uniformity), presence of channels or pinholes (permeability), and film thickness.

ii) The behavior of the monomer unit and its properties and their effect on the polymer behavior (redox potentials, electron transfer rate constants, optical behavior).

iii) Electrochemical stability and degradation mechanism.

iv) Carrier mobility and ion transport through the film upon oxidation and reduction.

v) The electronic structure and optical changes that occur during redox switching.

This information can be obtained using various techniques. In situ UV-vis spectroscopy, electrochemistry and surface studies have been used to characterize electrochromic polymer films upon oxidation and reduction. We present here the investigation of BBB, BBL and PPTZPQ.

Electrochemical Characterization

Cyclic voltammograms of ITO electrodes spin-coated with BBB and PPTZPQ films are shown in Figure 3. Quinto et al. (*24*) found reproducible and stable cyclic voltammograms for the reduction of BBL and BBB polymers under anaerobic conditions and significant color changes during the potential scans. The BBL film is initially violet and becomes blue during the first reduction, whereas at the second reduction wave, the color of the film changes to red. The BBB film is pink in its neutral state and it becomes orange and yellow in two successive reduction steps. BBL and BBB do not show a useful anodic response and hence these materials can be considered as cathodic electrochromic materials. The PPTZPQ film shows a good anodic response with a clear and homogeneous change of color from yellow to red; while its cathodic behavior is unstable (25).

Film conductivity and charge transport were studied in the redox process by performing CV scan rate analysis, scanning electrochemical microscopy (SECM) and direct resistance measurements. The cyclic voltammetry (CV) was performed with an array of four Pt microbands. BBL and BBB films showed appreciable changes in the film conductivity after reduction, which is consistent with the ladder nature of the polymer backbone of these polymer (24). On the other hand, PPTZPQ showed a different behavior; it showed low conductivity when the film was oxidized over an array of four Pt microbands, but we observed a positive feedback approach curve in SECM experiments which is evidence of conduction of electrons through the PPTZPQ film by a redox exchange (hopping) mechanism, i.e. redox conductivity where the mobile electrons are localized in fixed sites and are transported under force of the concentration gradients by thermally activated hopping or self-exchanges between occupied and unoccupied sites (25).

Morphological and Structural Characterization

AFM was used to image the film surface and determine its thickness (Plate 1). The image reveals an amorphous structure for PPTZPQ with a rough surface and large depressions (BBL showed similar surface morphology, AFM not shown). This surface morphology suggests the presence of channels or pinholes in the film. The topography of the BBB and PPTZPQ films showed a rough surface with clusters 50-100 nm in diameter. The BBB film (Figure 4) appeared less rough than the PPTZPQ and BBL films, and it is possible to recognize a structure of well-organized parallel rows, possibly because of interpolymer chain interactions.

To investigate the film permeability in its neutral form, we performed CV with a mediator in solution over bare ITO and PPTZPQ/ITO. The redox potential of the mediator was chosen within the potential window where the polymer does not show an electrochemical response, and therefore, behaves solely as a blocking layer on the ITO electrode. If the thickness of the film is of the order of the diameter of the hole, the behavior of the electrode can be treated as that of ultramicroelectrode arrays and the response is modulated by the polymer film thickness, the size and distribution of the pores, and the time scale of the experiment. The reduction behavior of the methyl viologen (MV^{2+}) on

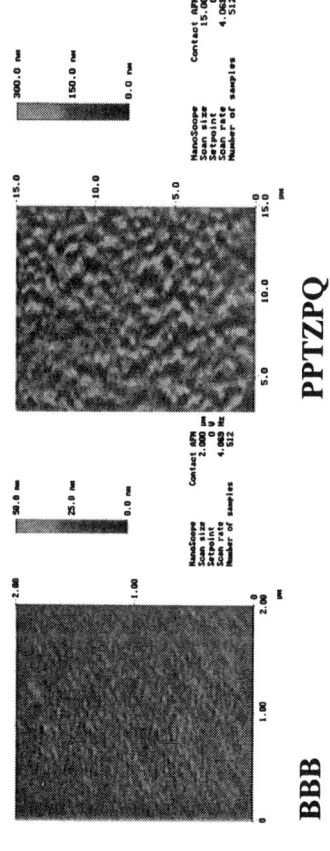

Plate 1. AFM images of BBB and PPTZPQ films spin-coated over an ITO electrode.

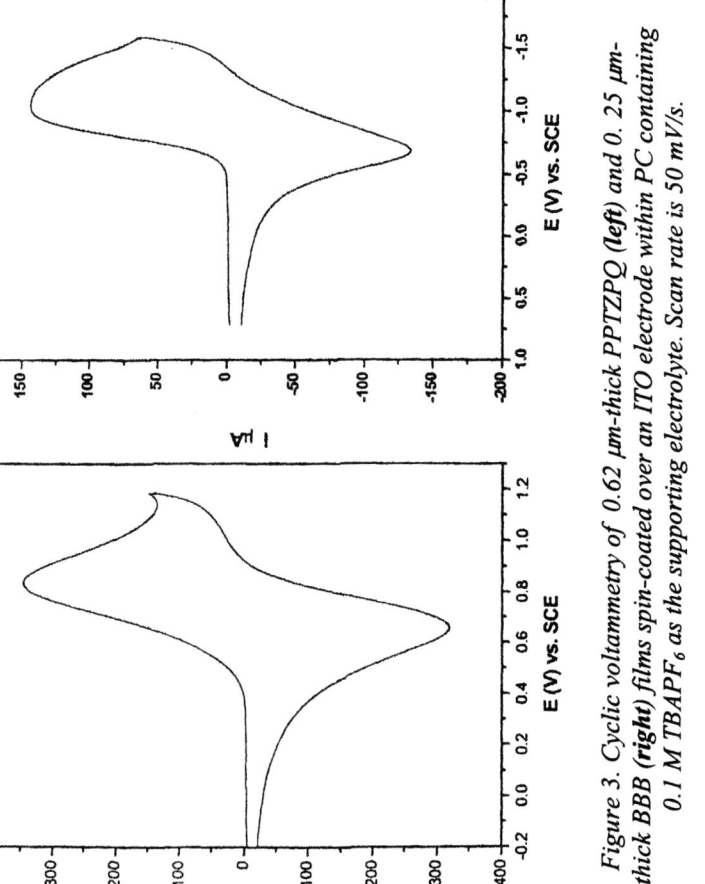

Figure 3. Cyclic voltammetry of 0.62 μm-thick PPTZPQ *(left)* and 0.25 μm-thick BBB *(right)* films spin-coated over an ITO electrode within PC containing 0.1 M TBAPF$_6$ as the supporting electrolyte. Scan rate is 50 mV/s.

bare ITO and ITO/PPTZPQ is shown in Figure 4. The smaller current peak for the coated electrode in comparison with the bare electrode is evidence of a high coverage of the electrode surface with pores spaced far apart in relation to the radius of the pores. In other words, the distance between the pores is such that, in the timescale of the experiment, the growth of a diffusion layer for an individual pore does not overlap with that of a neighboring pore producing a decrease in the current proportional to the uncovered area. This result does not indicate a high concentration of pores suggested by the AFM results. The smallest thickness attainable with PPTZPQ was of the order of 0.3 μm. This is thick enough so that many of the pores do not extend completely through the film from ITO to the solution. This combination of an imaging technique like AFM with the electrochemical analysis represents a useful approach in the study of film surface permeability.

Spectroelectrochemical Characteristics

Spectroelectrochemical analyses of the BBB, BBL, and PPTZPQ polymer films were performed to study the electronic structure and to examine the spectral changes that occur during redox switching; both are important for electrochromic applications. BBL spectra showed an increase in absorbance at 524 nm, a decrease at 583 nm, and two not well-defined isosbestic points at 375 and 550 nm. BBB spectra did not show clear isosbestic points. For the reduced form of BBB, the maximum peak was at 503 nm, and as the applied potential passed over the first voltammetric feature, the film exhibited an absorption peak at 576 nm. These electrochromic effects were stable and reversible under anaerobic and anhydrous conditions; the BBL and BBB films continued to exhibit strong color changes after more than 500 successive cycles in MeCN (*24*).

Figure 5 shows the absorption spectra of PPTZPQ as a function of potential. Neutral PPTZPQ exhibits two absorption peaks, one near 290-300 nm (not shown) and the other at ~425 nm. The former band can be assigned to a π-π* transition whereas the lowest-energy band, which is less intense, is largely of charge transfer character (*26, 27*). The latter absorption band is responsible for yellow color of the the neutral form of PPTZPQ film. As the potential was increased to positive values, the charge transfer peak at ~425 nm underwent a hypochromic shift concomitant with the growth of new absorption bands. The spectra show two isosbestic points, one at 400 nm and the other at 480 nm (Figure 5). A larger increase in absorption is observed at higher energies (~350 nm) compared to the charge transfer transition, and a peak at lower energies (540 nm) produces a red color in the oxidized polymer film.

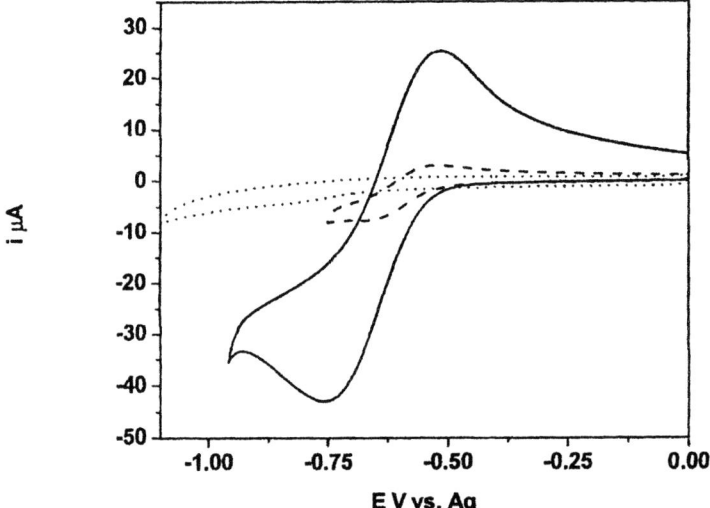

Figure 4. Cyclic voltammograms of 1 mM MV^{2+} in PC, 0.1 M $TBAPF_6$, on a bare ITO electrode (solid line), PPTZPQ-coated ITO electrode without MV^{2+} (dotted line), and PPTZPQ coated ITO electrode with MV^{2+} (dashed line).

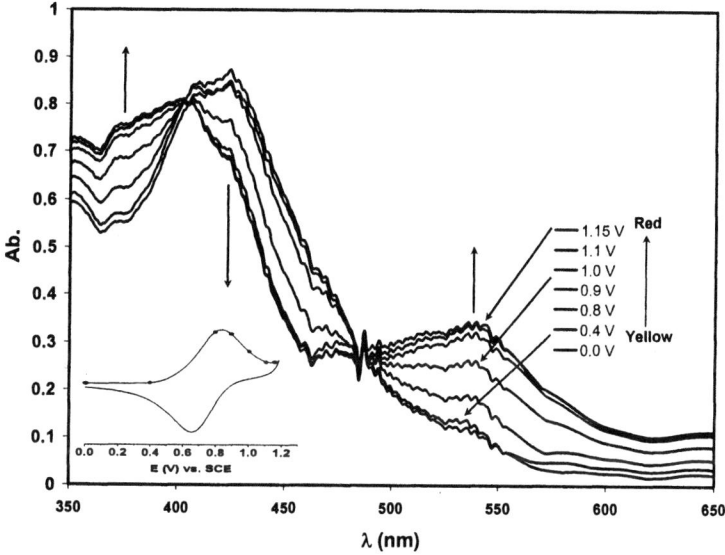

Figure 5. Electronic absorption spectra of PPTZPQ film spin-coated on an ITO electrode at various applied E. Inset shows the potentials at which spectra were obtained.

We can interpret the overall process as follows; the phenothiazine-quinoline (PTZ-Q) repeat unit in the neutral polymer shows a charge transfer-type absorption (PZT-Q →PZT$^{•+}$-Q$^{•-}$) at 425 nm. Following oxidation, the polymer produces phenothiazine radical cation (PZT$^{•+}$-Q). Charge transfer is then not possible and this results in a decrease of absorption at 425 nm. At the same time, the absorption of PZT$^{•+}$ and Q moieties results in the new peak at 540 nm and the increase in absorption at lower wavelengths (Figure 5) *(25)*.

Device Construction and Characterization

An electrochromic device is basically a two-electrode cell as mentioned in the introduction (Figure 1). In the construction of a two-electrode cell with only a single electrochromic material, there may be a problem with degradation reactions of the electrolyte that occur because of the absence of an effective counter-electrode reaction. This limits operating lifetimes through destruction of the electrolyte and a build-up of degradation products in the cell. The use of a reversible electrochromic material as a counter electrode or secondary electrode helps to avoid this problem. The secondary electrode, in electrochromic terms, can either be optically neutral or switch in a complementary mode to the working electrode (e.g. the counter-electrode is anodically colored while the working electrode is cathodically colored). In this way, the electrochemical stability and efficiency of the cell is greatly improved.

Electrochromic devices currently manufactured as rear view mirrors involve at least one, and in most cases more than one, solution-based electrochromic material. These species are free to diffuse to the surface of the electrodes and undergo oxidation or reduction with a color change when the device is activated. The solution phase electrochromic devices show self-erasing processes, so that they require continuous passage of current to maintain the colored state of the device (28). On the other hand, it is difficult to construct large-area ECDs with liquid electrolytes because the hydrostatic pressure of the liquid can cause leaks with attendant environmental and health hazards. Solid electrochromic devices have been developed to address these problems. In general, the solid ECDs are built by depositing an EM on OTE as thin films and using gel or polymeric electrolytes (Figure 1). The low conductivities that characterize many polymer electrolytes can be improved by employing a plasticizer, to yield a solid or gel electrolyte with high conductivity coupled with good mechanical properties (28). The conductivities of polymer gel electrolytes can be enhanced by either changing the ratio of conducting ion and gel polymer. Usually, the conductivity decreases with increasing amounts of the gel polymer because ionic migration becomes slower. As shown in Figure 6, the conductivity of the gel electrolyte based on polyethylene oxide (PEO) increased linearly with wt % of $LiClO_4$ up to 11.8 wt %. In addition, a gel electrolyte based on PEO gives a better conductivity (3.7 mS/cm) with the same concentration of $LiClO_4$ than one based on polymethylmethacrylate (PMMA) (2.0 mS/cm).

Desired characteristics in an ECD are: color tunable capacity, good color contrast, durability, lifetime, low operating voltage, fast response, flexibility, and low cost manufacturability. Thin films of PPTZPQ, BBB and BBL had stable anodic electrochromic properties and showed good color changes when oxidized and reduced, suggesting their use as an anodically and cathodically coloring electrochromic material. We built and characterized four solid ECDs using a combination of these polymers and vanadium pentoxide (V_2O_5) as a counter electrode. We investigated devices based on glass ITO substrates with two kinds of polymer electrolytes, polymethyl methacrylate (PMMA) as gel electrolyte and poly(ethylene oxide)-propylene carbonate (PEO-PC) as solid electrolyte, as well as cells with a plastic substrate coated with ITO.

PPTZPQ, BBL and BBB Electrochromic Devices using V_2O_5 as the Counter Electrode

Three electrochromic cells were built using a V_2O_5 film as the counter electrode with either PPTZPQ, BBB or BBL polymer films as the

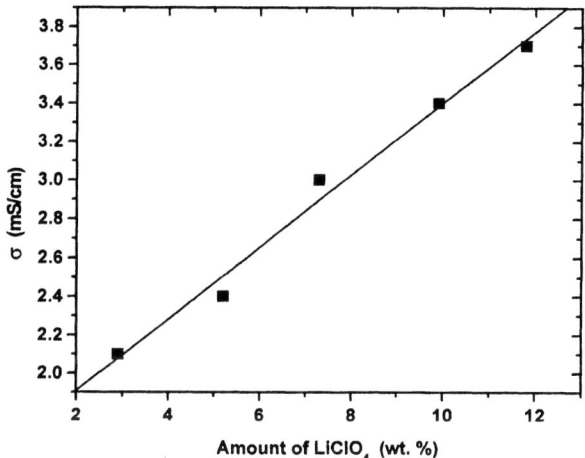

Figure 6. Dependence of conductivity of PEO gel electrolytes on the amount of LiClO$_4$. All gel electrolytes contain the same amount of PEO.

electrochromic material. The V$_2$O$_5$ film is a well known electrochromic material (*3, 7*). It is yellow and, with the reductive injection of lithium ions into the V$_2$O$_5$ film, forms Li$_x$V$_2$O$_5$, with a pale blue color. The color changes are rather weak and V$_2$O$_5$ films are usually used simply as transparent counter electrodes.

Voltammetric studies showed that the PPTZPQ- PEO/PC/LiClO$_4$-V$_2$O$_5$ cell stably operated with an applied voltage of 0.0 to 1.5 V under aerobic conditions. Cell testing was carried out by monitoring the absorbance changes at 560 nm as a function of time during repeated potential steps between 0.0 V and 1.5 V (Figure 7). The current transients in Figure 7, show that the oxidation process proceeds more slowly than the reduction process. The oxidation current does not reach zero during the 20 s amplitude potential pulses, while the reduction is almost complete within the same time period. However, the amount of charge, Q, in each current curve in both processes was very similar, in agreement with the stable oxidation of PPTZPQ and reduction of its oxidized form. The faster reduction can be attributed to an easier charge transport through the oxidized film. The oxidation of the neutral film starts at the insulating polymer-electrode interface and requires the influx of compensating anions. The reduction process on the other hand starts with the charged state. In addition, the difference in the anion mobility in the oxidation-reduction process in combination with the charge transport can govern this conduct. The kinetics of the overall process is reflected in the coloration response time (Figure 7), which is approximately 20 s.

The coloration efficiency was determined by using the equation $\eta=\Delta A/Q\ Q_d$ (*3*). This is obtained by determining the injected/ejected charge as a function of the electrode area (Q_d) and the change in absorbance (ΔA) during a redox step of the device. The η value may be regarded as the electrode area colored to unit absorbance by unit charge. The absorbance change at 560 nm and under the

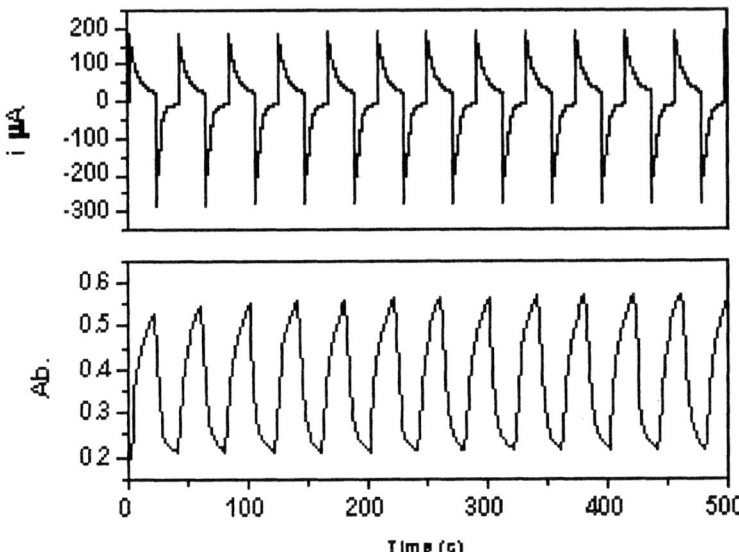

Figure 7. Electrochromic switching, current and optical responses to potential step of between 0.0 V and 1.5 V recorded at 560 nm for PPTZPQ/PEO-PC/V_2O_5 device.

same potential switching conditions as Figure 7 yielded a red coloration efficiency, η, of ~ -316 cm^2C^{-1}.
This value is higher than those of electrochromic inorganic materials, such as WO_3. It is also higher than that obtained with electrochromic cells employing polyaniline (*18, 29*) and polypyrrole (*30*) as the anodic coloring material and WO_3 as the cathodic coloring material. This value is, however, lower than that obtained in dual electrochromic cells with complementary electrochromic polymers, with total device η-values ranging from 250 to 1413 cm^2 C^{-1} (*21, 30, 31, 32*). Both of the electrochromic cells showed stable operation for hours, but their performance started to decrease irreversibly after a few days of cycling, probably because of crystallization of PEO.

BBL and BBB ECDs exhibited well-defined, reversible color changes as shown by the electronic absorption spectra of ECDs, taken at various potentials (Figure 8). The BBL cell showed a change from purple to dark greenish yellow, while the BBB cell showed pink to dark greenish yellow change. However, it was difficult to distinguish the intermediate colors, such as reddish green for BBL and green, which were observed in spectroelectrochemistry in a 0.1 M $LiClO_4$/PC solution. The background color of V_2O_5 and the added thickness of glass substrates caused by two electrodes and the gel electrolyte may account for this difference. The BBL ECD showed maximum absorbance at 560 nm before reduction. The absorption maximum at 560 nm shifted to 460 nm as the film was reduced, and a new broad absorbance band appeared at around 700nm as shown in Figure 8.

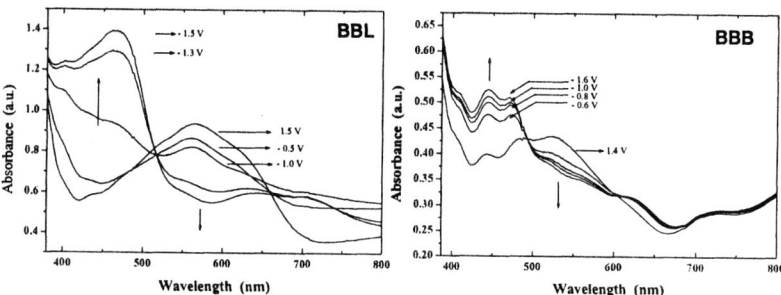

Figure 8. Electronic absorption spectra of BBL and BBB/ PMMA/V_2O_5 devices at various applied potentials.

The coloration efficiency (η) was calculated using absorbance changes (ΔA) at 560 nm for BBL and at 522 nm for BBB. The η values for BBB ECDs decreased as the BBB film thickness increased, whereas an opposite trend was observed for BBL ECDs. Therefore, the electrochemical redox reaction probably does not occur efficiently with a thick BBB film. Lifetime measurements were performed by switching potentials from neutral to reduced state. ECDs with PEO and PMMA-based polymer electrolytes produced lifetimes with more than 30,000 switching cycles. Despite their better conductivity, PEO-based polymer electrolyte yielded ECD with shorter lifetimes than those with the PMMA-based electrolyte, since an ECD with PMMA lasted for more than 100,000 switching cycles. Passivation of V_2O_5, such as the formation of PEO-V_2O_5 compound (7), may be possible during cycling and cause this shorter lifetime.

PPTZPQ-BBL Plastic Electrochromic Device

By combining two electrochromic materials whose neutral states are transparent in the visible, with one electrode anodically colored while the other is cathodically colored, allows fabrication of a device that can switch between highly transmissive and absorptive states. These kinds of ECDs are used as smart windows. In the case of electrochromic materials where the neutral state shows absorption in the visible, it is possible to choose as the secondary electrode a material whose neutral state absorption is complementary to that of the primary electrode, and its reduced form shows the same color as that of the oxidized form of the primary electrode. In other words, the combination of the spectra of two complementary colors leads to absorption over the whole visible spectrum turning the device black. Upon potential switching between the two redox states, the electrochromic cell will turn from black to the characteristic color of the original forms.

A material that meets these conditions and can be used as a complementary electrode to PPTZPQ is BBL, which is violet (the complementary color of yellow) in the neutral state, while both are red in their fully oxidized and reduced forms, respectively. The film spectra of both polymers are shown in Figure 9.

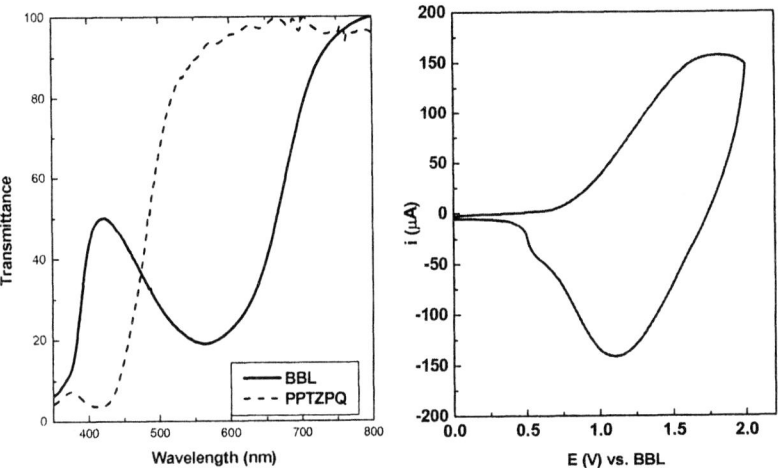

Figure 9. (left) Visible spectra collected in transmittance mode of PPTZPQ and BBL over ITO film in their neutral colored states. (right) Two-electrode voltammograms for PPTZPQ/PEO-PC- TBAPF$_6$/BBL cell. Scan rate 50 mV/s.

The combination of the two polymer films produces high absorbance over almost the whole visible spectrum and a black color appears when the two neutral films are superimposed. The construction of an all-plastic, two electrode cell was then successfully accomplished by sandwiching PEO/PC/TBAPF$_6$ as a polymer electrolyte between films of PPTZPQ and BBL deposited over plastic ITO. The charge capacity of the cathode and anode was equalized ($Q^{PPTZPQ}/Q^{BBL} \sim 1$) by adjusting the film thickness (*33*). The cell was operated within a voltage range where reversible switching was found. The driving voltage limits were obtained from cyclic voltammetry measurements of these films in solution cells. In Figure 10 (right) a PPTZPQ-BBL cell current-potential curve is shown, where PPTZPQ is taken as the working electrode. The operating voltage was 0.0 V to +1.8 V (vs. BBL). Within this voltage range the device showed good stability and a black to red color change under anaerobic conditions. When the applied potential was greater than this range, the electrochromic response was gradually lost. Degradation was also observed when the device was operated in the presence of oxygen.

Conclusions

The construction of all-plastic PPTZPQ-based electrochromic devices with interesting color changes showed the potential use of PPTZPQ as an electrochromic material in the building of practical devices. Electrochrmoic devices with PMMA-based electrolytes and BBL or BBB coloring material had lifetimes of over 10^5 switching circles whereas those based on PEO electrolyte had switching lifes of $\sim 3 \times 10^4$. We have also demonstrated an all-plastic electrochromic cell constructed from PPTZPQ and BBL films as complementary electrochromic materials. Further work is still necessary to improve the spectroelectrochemical characteristics of these cells by optimizing their charge capacity ratio and the composition of the polymer electrolytes.

Acknowledgments

We are grateful to the U.S. Army Research Office TOPS MURI (DAAD19-01-1-0676) and Consejo Nacional de Investigaciones Científicas y Técnicas (CONICET-Argentina) for their support. We acknowledge the contributions of Soley Ozer in preliminary experiments with PPTZPQ.

References

(1) Platt, J. R. J. *J. Chem. Phys.* **1961**, *34*, 862.
(2) Mortimer, R. *J. Chem. Soc. Rev.* **1997**, *26*, 47.
(3) Monk, P. M. S.; Mortimer, R. J.; Rosseinsky, D. R. *Electrochromism: Fundamentals and Applications*, VCH, Weinheim (1995).
(4) For a recent reviews see, *Electrochim. Acta* **2001**, *46*, 1919-2289.
(5) Faughnan, B. W.; Crandall, R. S., in Pankove, J. I. (ed.) "Display Devices", Springer-Verlag, Berlin, 1980. Chapter 5.
(6) Park, H. K.; Smyrl, W. H.; Ward, M. D. *J. Electrochem. Soc.* **1995**, *142*, 1068.
(7) Livage, J. *Chem. Mater.* **1991**, *3*, 578.
(8) Ohzuku, T.; Hirai, T. *Electrochim. Acta* **1982**, *27*, 1263.
(9) Reichman, B.; Fan, F. R.; Bard, A. J. *J. Electrochem. Soc.* **1980**, *127*, 333.
(10) Mortimer, R. J. *J. Electrochem. Soc.* **1991**, *138*, 633.
(11) (a) Brédas, J. L. *J. Chem. Phys.* **1985**, *82*, 3808. (b) Roncali, J. *Chem Rev.* **1997**, *97*, 173.

(12) (a) Osaheni, J. A.; Jenekhe, S. A. *Chem. Mater.* **1996**, *7*, 672. (b) Alam, M. M.; Jenekhe, S. A. *J. Phys. Chem. B* **2002**, *106*, 11177. (c) Jenekhe, S. A.; Johnson, P. O. *Macromolecules* **1990**, *23*, 4419.
(13) Zhang, Q. T., Tour, J. M. *J. Am. Chem. Soc.* **1998**, *120*, 5355.
(14) Agrawal, A. K.; Jenekhe, S. A. *Macromolecules* **1993**, *26*, 895.
(15) Yamamoto, T.; Zhou, Z.; Kanbara, T.; Shimura, M.; Kizu, K.; Maruyama, T.; Nakamura, Y.; Fukuda, T.; Lee, B. L.; Ooba, N.; Tomaru, S.; Kurihara, T.; Kaino, T.; Kubota, K.; Sasaki, S. *J. Am. Chem. Soc.* **1996**, *118*, 10389.
(16) Marcel, C.; Tarascon, J. M. *Solid State Ionic* **2001**, *143*, 89.
(17) Rodrigues, M. A.; De Paoli, M. A.; Mastragostino, M. *Electrochim. Acta,* **1991**, *36*, 2143.
(18) Panero, S.; Scrosati, B.; Baret, M.; Cecchini, B.; Masetti, E. *Solar Energy Mater. Solar Cells* **1995**, *39*, 239.
(19) Tassi, E. L.; De Paoli, M. A. *Electrochim. Acta* **1994**, *39*, 2481.
(20) Dietrich, M.; Heinze, J.; Heywang, G.; Jonas, F. *J. Electroanal. Chem.* **1994**, *369*, 87.
(21) Sapp, S. A.; Sotzing, G. A.; Reynolds, J. R. *Chem. Mater.* **1998**, *10*, 2101.
(22) De Paoli, M. A.; Casalbore-Miseli, G.; Girotto, E. M.; Gazoti, W. A. *Electrochemica Acta* **1999**, *44*, 2983.
(23) De Paoli, M. A.; Zanelli, A.; Mastragostino, M.; Rocco, A. *M Journal of Electroanal. Chem.* **1997**, *435*, 217.
(24) Quinto, M.; Jenekhe, S. A.; Bard, A. *J. Chem. Mater.* **2001**, *13*, 2824.
(25) Fungo, F.; Jenekhe, S. A.; Bard, A. J. *Chem. Mater.* **2003**, *15*, 1264.
(26) (a) Lai, R. Y.; Fabrizio, E. F.; Lu, L.; Jenekhe, S. A.; Bard, A. J.; *J. Am. Chem. Soc.* **2001**, *123*, 9112. (b) Lai, R. Y.; Kong, X.; Jenekhe, S. A.; Bard, A.J *J. Am. Chem. Soc.* **2003**, *125*, 12631.
(27) Jenekhe, S. A.; Lu, L.; Alam, M. M. *Macromolecules,* **2001**, *34*, 7315.
(28) Byker, H. J. *Electrochim. Acta* **2001**, *46*, 2015.
(29) (a) Jelle, B. P.; Hagen, G.; Odegard, R. *Electrochim. Acta* **1992**, *37*, 1377.
(30) Arbizzani, M. M.; Zanelli, A. *Solar Energy and Solar Cells* **1995**, *39*, 213.
(31) Rauh; R. D.; Wang, F.; Reynolds, J. R.; Meeker, D. L. *Electrochim. Acta.* **2001**, *46*, 2023.
(32) Michalak, F.; Aldebert, P. *Solid State Ionics* **1996**, *85*, 265.
(33) Chen, L. C.; Ho, K. C. *Electrochim. Acta* **2001**, *46*, 2159.

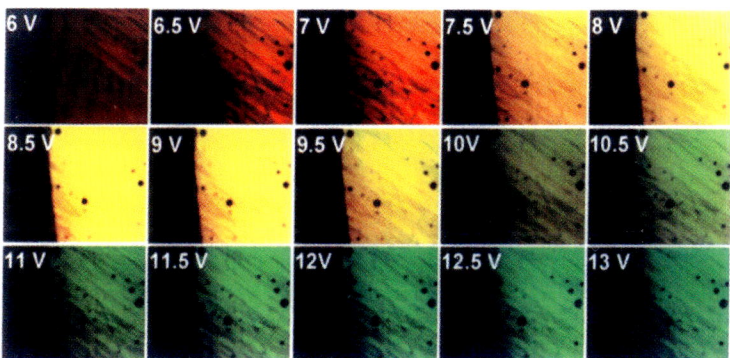

Plate 1.1. Multicolor emission from ITO/PPV(35 nm)/PPQ(35 nm)/Al diodes at various bias voltages (from ref. 23b).

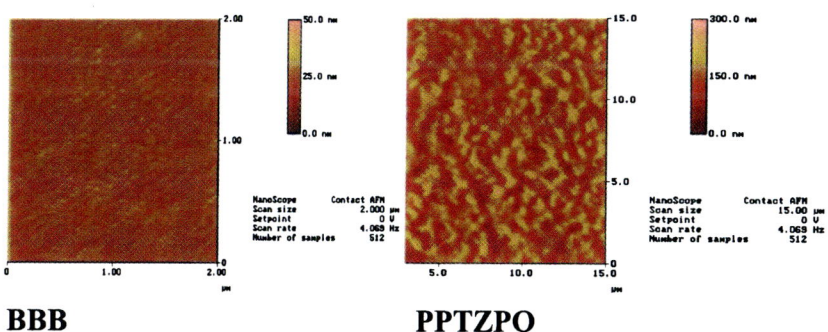

Plate 3.1. AFM images of BBB and PPTZPQ films spin-coated over an ITO electrode.

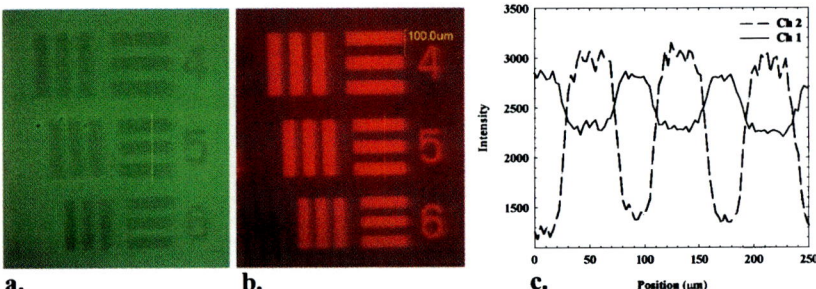

Plate 9.1. Two-photon fluorescence images of photosensitive films developed (via 350 nm broadband exposure, 4.4 mW/cm^2) through Air Force targets. Image recorded by channel 1 (a), image recorded by channel 2 (b), and fluorescence intensity by scanning an x,y line across one set of three-member elements (c).

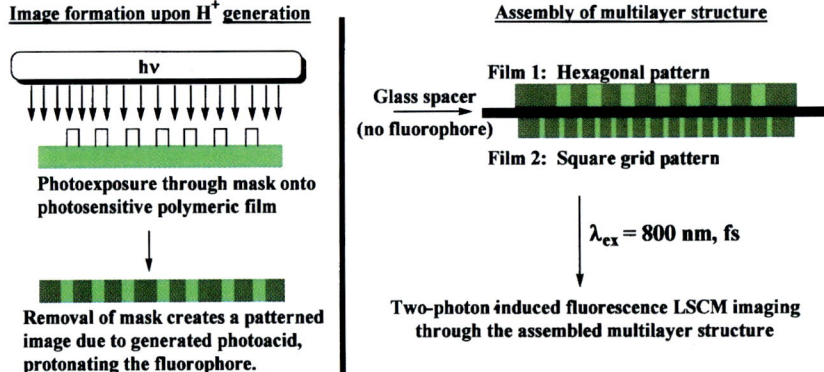

Plate 9.2. Image formation (upon photoacid generation) within photosensitive polymer films for assembly of multi-layered structures. Two-photon fluorescence LSCM imaging using fs pulsed near-IR pump allows for 3-D volumetric imaging of the layered structure.

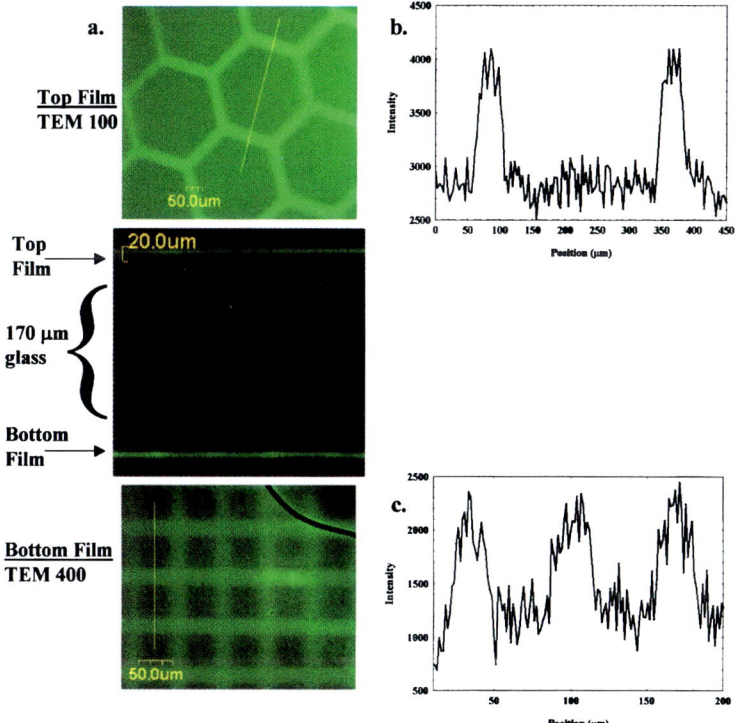

Plate 9.3. Two-photon fluorescent images of multi-layered films developed via 350 nm, broadband irradiation (6.0 mW/cm^2) by exposure through TEM hexagonal and square grid masks (a). Fluorescence intensity plots for a line scan across a region (as defined by the yellow line across the image area) provides image readout in one layer (b), and changing the depth (z position) for image (signal) readout in the lower layer within a multi-layered system (c).

*Plate 11.1. Thermochromicity of dialkyl-PPE **4a**.
Left: Film at room temperature. Right: Film at 140° C.*

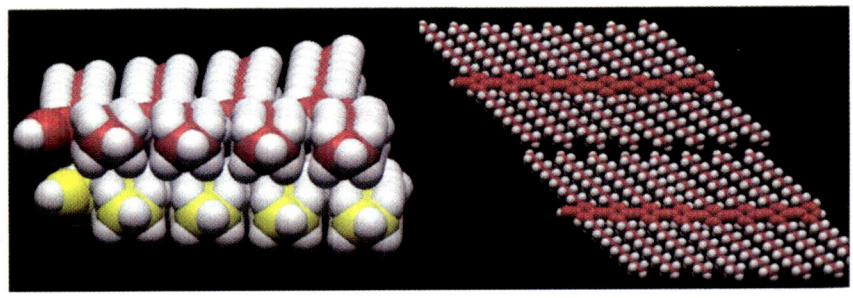

*Plate 11.2. Solid state structure of didodecyl-PPE **4a** at ambient temperature.*

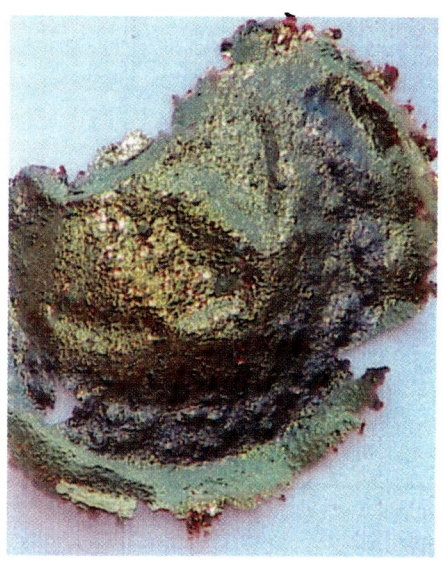

*Plate 11.3. Thick film of a sample of **8a**. Visible is the merocyanine luster.*

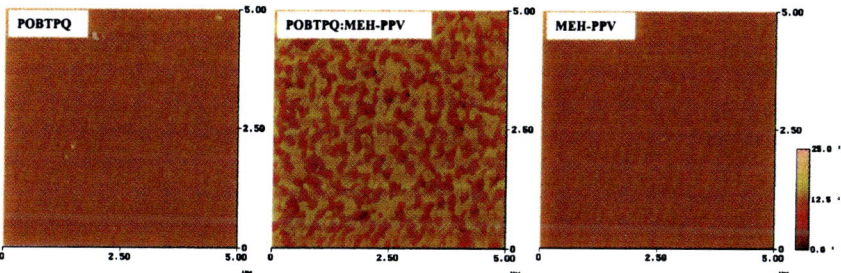

Plate 14.1. Tapping-mode AFM phase images of POBTPQ, POBTPQ:MEH-PPV blend (60:40 wt%) and MEH-PPV on polished silicon substrate.

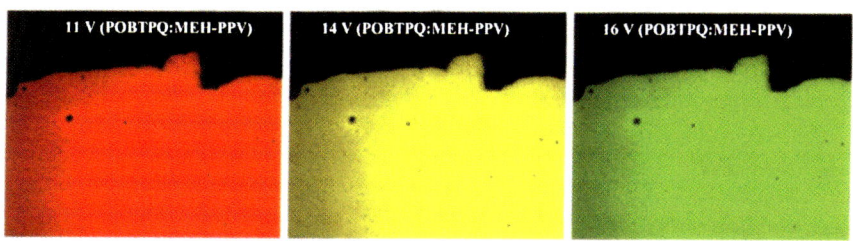

Plate 14.2. EL micrographs (x10) of a POBTPQ:MEH-PPV blend (60:40 wt%) diode at 11, 14, and 16 V.

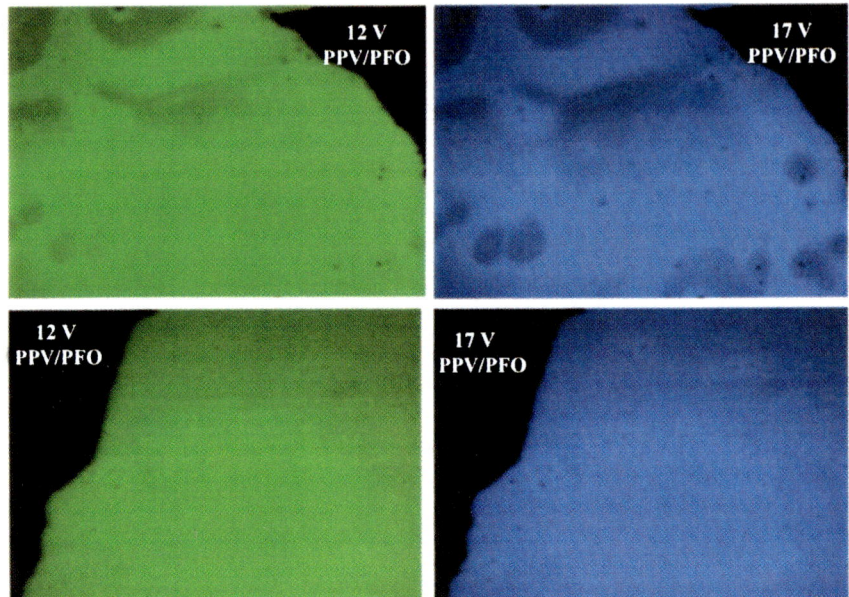

Plate 3. EL micrographs (× 10) of a bilayer ITO/PPV(30 nm)/PFO(50 nm)/Al diode.

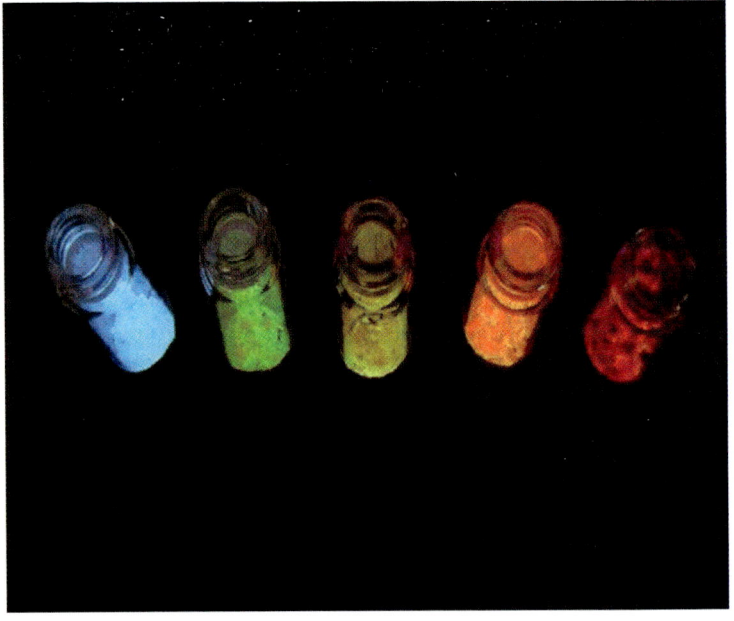

Plate 20.1. Emitting colors under UV (365 nm) radiation of PPQ-Silica gels. (a) 0.0001 wt% PPQ/TEOS; (b) 0.001 wt%; (c) 0.1 wt%; (d) 1 wt%; (e) 5 wt%
A small amount of solvent was still remained in the gel.

Plate 21.1. Photographs of a patterned cholesteric gel at various applied voltages.

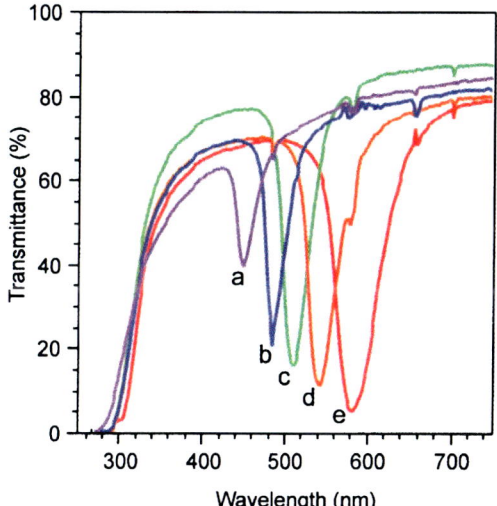

Plate 25.1. UV-Vis transmission spectra taken from a photonic paper (assembled from 175-nm PS beads) before (curve, a) and after (curves, b-e) it had been swollen with silicone liquids of different molecular weights (and viscosities): b) T12 (M_w=2000, 20 cSt), c) T11 (M_w=1250, 10 cSt), d) T05 (M_w=770, 5 cSt), and e) T00 (M_w=162, 0.65 cSt).

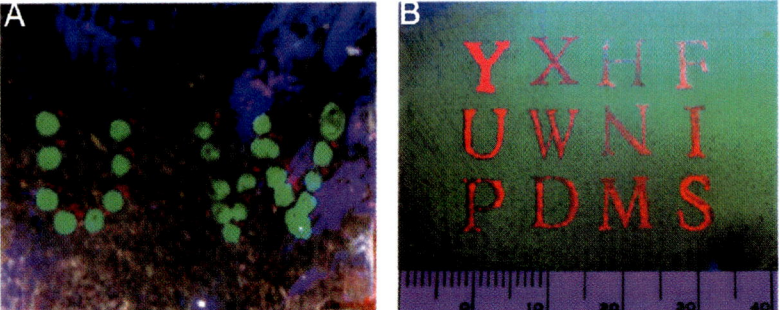

Plate 25.2. (A) A photograph of two dotted letters written on a photonic paper by delivering octane droplets to its surface using a Pilot pen. (B) A photograph of letters formed on the surface of a photonic paper by stamping with a silicone fluid (T11, $M_w=1250$). This paper was assembled from PS beads of 202 nm in diameter, and it exhibited a green color when viewed at normal incidence. Note that the pattern shown in (B) had an edge resolution better than 50 μm.

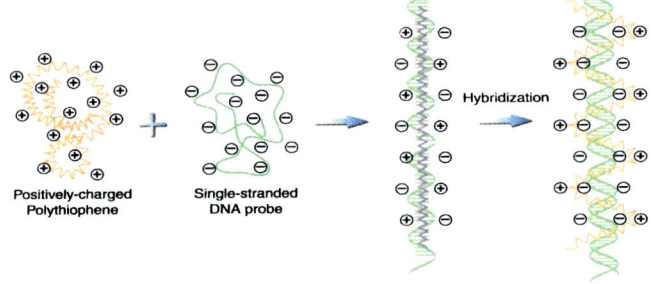

Plate 27.1. Schematic description of the formation of polythiophene/single stranded oligonucleotide duplex and polythiophene/hybridized oligonucleotide triplex.

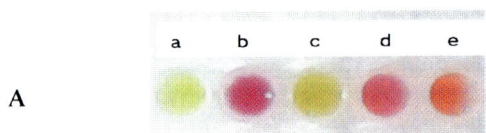

Plate 27.2. A) Photographs of 7.9×10^{-5} M (on a monomeric unit basis) solutions of a) polymer 1, b) polymer 1 / X1 duplex, c) polymer 1 / X1 / Y1 triplex, d) polymer 1 / X1 / Y2 mixture, and e) polymer 1 / X1 / Y3 mixture after 5 minutes of mixing at 55 C in 0.1 M NaCl/H_2O. B) UV-visible absorption spectrum corresponding to the different assays of photograph A.

Chapter 4

Novel Near-IR Electrochromic Ruthenium Complex Polymers

Pierre Desjardins and Zhi Yuan Wang*

Department of Chemistry, Carleton University, 1125 Colonel By Drive, Ottawa, Ontario K1S 5B6, Canada

A series of azodicarbonyl dinuclear ruthenium polymers were synthesized. Spectroscopic and electrochromic properties of these polymers were studied. These polymers were found to exhibit good thermal stability and electrochromic properties such as good coloration efficiency in the near infrared region. Furthermore, long-term switching trials were performed, which indicated good chemical stability of the material and potential application for attenuation of near infrared light.

Introduction

Electrochromic materials, those which undergo electrochemically stable and reversible colour change, have aroused a great deal of attention since the discovery of the phenomena in tungsten oxide materials over 30 years ago (*1*). Many materials have been explored, ranging from the transition metal oxides (*2-12*) and salts (*13-17*), as well as the conductive organic polymers such as polythiophene (*18-24*). Historically, much of the technological development has centered upon visible light attenuation devices of value in architectural glassing for energy conservation in comfort cooling (*25*) and the so-called NVS (*26*) (night vision safety) mirrors for the reduction glare from vehicle headlights. Of particular interest to some is the possibility of constructing flexible displays useful in variety of applications requiring non-planar geometry, ranging from electronic books or writing pads to glasses or window coatings. The newest area of endeavor is the development of materials that possess strong near infrared (NIR) absorbance bands within the telecom wavelength regions of 1250-1350 nm or 1500-1600 nm, which are useful in variable optical attenuator (VOA) (*27*) devices necessary for wavelength-division multiplexing.

It was our intention of constructing polymeric analogues to the dinuclear ruthenium dicarbonylhydrazine (DCH) or DCH-Ru complexes of the form $[\{Ru(bpy)_2\}_2 \, \mu\text{-DCH}]^{n+}$ (where bpy = 2,2'-bipyridine and n = 2, 3 or 4) previously studied by our group (*28*) and others (*29-31*). In addition to possessing visible electrochromism, these ruthenium complexes also possess strong NIR electrochromism (i.e. 1500 to 1600 nm). Metallic complexes, and particularly dinuclear systems in which metal-metal electronic communication exists, possess interesting electrochromism as a result of the interaction of metal and ligand (*32-34*). This chromism can be tuned therefore by the appropriate selection of both ligand and metal, permitting these systems to be tunable for a wide variety of applications. Unfortunately, the dinuclear complexes do not possess suitable physical properties to be spin coated onto transparent conductive substrates and are therefore only suitable for all liquid devices. Therefore, it is our intention to preserve as much as possible the dinuclear properties in the polymeric metal complex with good film forming properties. Having evaluated the work of Mohamed *et al* (*35*), in which metallized films were prepared from a series of wholly aromatic polyhydrazines at room temperature in N,N-dimethylacetamide (DMAc) by combination of the polymers with salts of Cu(II), Ag(I), Ni(II), Pb(II), Fe(III), and Cd(II). The films prepared by Mohamed were free standing, thermally stable and showed interesting electrical properties. The metal exchange reaction was performed via a simple combination of the polymer and metallic salt, suggesting that ruthenium analogues may be equally simple to prepare.

Aromatic and aliphatic polyhydrazines as polymeric ligands shown in Figure 1 were synthesized according to the method of Frazer *et al* (*36-37*).

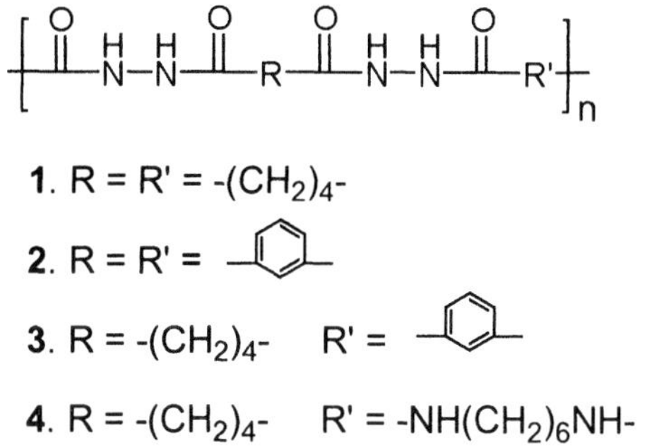

$$\left[\overset{O}{\underset{}{C}}-\overset{H}{\underset{}{N}}-\overset{H}{\underset{}{N}}-\overset{O}{\underset{}{C}}-R-\overset{O}{\underset{}{C}}-\overset{H}{\underset{}{N}}-\overset{H}{\underset{}{N}}-\overset{O}{\underset{}{C}}-R'\right]_n$$

1. $R = R' = -(CH_2)_4-$
2. $R = R' = $ —⌬—
3. $R = -(CH_2)_4-$ $R' = $ —⌬—
4. $R = -(CH_2)_4-$ $R' = -NH(CH_2)_6NH-$

Figure 1. Polyhydrazines as ligands.

Subsequent to isolation of the polymer, the metallization could be conducted in a similar manner as that of the complexes reported elsewhere for DCH-Ru complexes (*27-28*). The ruthenium complex polymers, shown in Figure 2, thus made would be expected to exhibit similar electrochromic properties as the dinuclear ruthenium complexes while possessing the film forming characteristics of the parent polymer. We report herein the synthesis and characterization of these polymers.

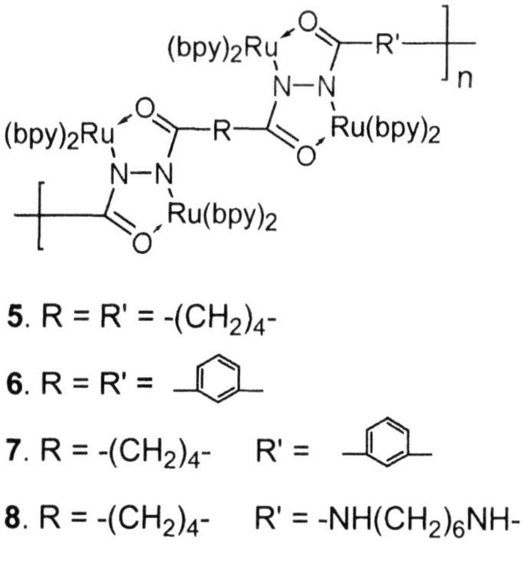

5. $R = R' = -(CH_2)_4-$
6. $R = R' = $ —⌬—
7. $R = -(CH_2)_4-$ $R' = $ —⌬—
8. $R = -(CH_2)_4-$ $R' = -NH(CH_2)_6NH-$

Figure 2. Ruthenium complex polymers.

Experimental Section

Measurements.

Electrochemical and spectroscopic measurements were taken in the solid state by casting thin films of the polymer on either platinum disk electrode (BAS 2013 MF, 1.6 mm diameter) or ITO coated glass. The ITO coated glass was pre-treated by first cleaning it by momentarily dipping it in concentrated H_2SO_4, followed by rinsing with deionized water and drying under a stream of Argon. The clean ITO surface was then treated with a 2% solution of 3-aminopropyltriethoxysilane (Aldrich) in toluene via spin coating, followed by air drying for 1 h, and finally coating with a solution of the polymer in DMF/$CHCl_3$. The polymer coating on ITO was then allowed to air dry for several hours prior to testing. Measurements were performed using an electrolyte composed of 0.1 M tetra-n-butylhexafluorophosphate (TBAH) in either $CHCl_3$ or tetrahydrofuran (THF). Ancillary electrodes used were a silver wire as a reference and platinum wire as a counter.

Spectroelectrochemical and switching experiments were performed with polymer coated on ITO and mounted within a 1 cm quartz cuvette. The cuvette was also fitted with a platinum wire counter electrode and silver wire pseudo-reference electrode. Sufficient electrolyte solution, THF or $CHCl_3$ with 0.1 M TBAH, was then added to the cell in order to make good contact with all electrodes and fully flood the spectrometer light path.

Reagents and Synthesis.

All ligands were synthesized from the appropriate hydrazide and either the appropriate diacid chloride, dihydrazide or diisocyanate. Reagents include phthaloyl dichloride, hydrazine hydrate, adipic dihydrazide, adipoyl chloride, 1,6-diisocyanohexane, *N*-methylpyrrolidone (NMP), and *N,N*-dimethylformamide (DMF) and all were obtained from Sigma-Aldrich Co.

Polymeric ruthenium complexes were synthesized in the same manner as that of the DCH-Ru complexes. The general synthetic procedure is to combine 0.38 mmol (200 mg) Ru(bpy)$_2$Cl$_2$·2H$_2$O with 0.19 mmol of the ligand polymer and 100 mg of Na_2CO_3 in 80 mL of 5:1 H_2O/EtOH. The mixture was then set to reflux for approximately 14 h under ambient atmosphere, after which it was cooled to room temperature, the product was then precipitated by the addition of excess NH_4PF_6, isolated by filtration, and dried under vacuum (5 mmHg) at room temperature.

Isophthalic dihydrazide

To 20 mL of hydrazine hydrate in a flask was added 1.0 g (4.9 mmol) of phthaloyl chloride. The mixture was stirred for 1 h, followed by precipitation of the product from methanol. The yield of the product was 0.9 g (94%). IR ν(C=O) 1665 cm^{-1}, ν(NH) 3290 cm^{-1}; ^{1}H NMR (200 MHz, d$_6$-DMSO) δ 4.60 (s, 4H), 7.56 (t, 1H), 7.97 (d, 2H), 8.31 (s, 1H), 9.88 (s, 2H); MS (EI, m/z, relative intensity) 194 (M$^+$, 30.2); mp 227-230 °C.

Ligand Polymer 1

To 30 mL of NMP in a flask were added 1.74 g (10 mmol) of adipic dihydrazide and 1.83 g (10 mmol) of adipoyl chloride. The mixture was allowed to stir overnight at room temperature, under N$_2$, followed by precipitation of the product in 300 mL of vigorously stirring methanol. The product was isolated by filtration and dried overnight under vacuum. Yield of the crude polymer was 1.7 g (61%). IR ν(C=O) 1600 cm^{-1}, ν(NH) 3216 cm^{-1}; ^{1}H NMR (200 MHz, d$_6$-DMSO) δ 1.5 (*s*, 4H), 2.1 (*s*, 4H), 9.7 (*s*, 2H).

Ligand Polymer 2

The synthesis was the same as for polymer 1, except with 15 mL of NMP, 0.99 g (5 mmol) of isophthalic dihydrazide, 1.02 g (5 mmol) of isophthaloyl chloride. Yield of the polymer was 1.4 g (69%). IR ν(C=O) 1649 cm^{-1}, ν(NH) 3235 cm^{-1}; ^{1}H NMR (200 MHz, d$_6$-DMSO) δ 7.7 (*t*, 1H), 8.2 (*d*, 2H), 8.5 (*s*, 1H), 10.8 (*s*, 2H).

Ligand Polymer 3

The synthesis was the same as for polymer 1, except with 1.74 g (10 mmol) of adipic dihydrazide and 2.03 g (10 mmol) of isophthaloyl chloride. The yield of the polymer was 2.6 g (69%). IR ν(C=O) 1654 cm^{-1}, ν(NH) 3241 cm^{-1}; ^{1}H NMR (200 MHz, d$_6$-DMSO) δ 1.6 (*s*, 4H), 2.4 (*s*, 4H), 7.6 (*t*, 1H), 8.1 (*d*, 2H), 8.2 (*s*, 1H), 9.9 (*s*, 2H), 10.4 (*s*, 2H).

Ligand Polymer 4

To 15 mL of DMF in a flask, were added 523 mg (3 mmol) of adipic hydrazide and 505 mg (3 mmol) of 1,6-diisocyanohexane. The polymer, which precipitated from the reaction mixture, was isolated by filtration and dried overnight under vacuum. Yield of the polymer was 0.8 g (78%). IR ν(C=O) 1654 cm^{-1}, ν(NH) 3353, 3224 cm^{-1}; ^{1}H NMR (200 MHz, d$_6$-DMSO) δ 1.4 (*m*, 12H), 2.1 (*s*, 4H), 3.0 (*m*, 4H), 6.3 (*s*, 2H), 7.6 (*s*, 2H), 9.4 (*s*, 2H).

DCH-Ru Polymer 5

The synthesis was performed in the manner described previously for dinuclear ruthenium complexes, using 200 mg (0.38 mmol) of Ru(bpy)$_2$Cl$_2$·2H$_2$O, 35 mg (0.25 mmol based on the repeat units) of ligand polymer **1**, and 100 mg of Na$_2$CO$_3$. Yield of the polymer was 215 mg (87% based on reacted polymer). IR ν(C=O) 1605 cm^{-1}, ν(P-F) 830 cm^{-1}.

DCH-Ru Polymer 6

The synthesis was performed in the manner described previously for dinuclear complexes, using 200 mg (0.38 mmol) of Ru(bpy)$_2$Cl$_2$·2H$_2$O, 35 mg (0.22 mmol) of ligand polymer **2**, and 100 mg of Na$_2$CO$_3$. Yield of the crude polymer was 218 mg (88% based on reacted polymer). IR ν(C=O) 1605 cm^{-1}, ν(P-F) 841 cm^{-1}.

DCH-Ru Polyme 7

The synthesis was performed in the manner described previously for dinuclear complexes, using 200 mg (0.38 mmol) of Ru(bpy)$_2$Cl$_2$·2H$_2$O, 30 mg (0.099 mmol) of ligand polymer **3**, and 100 mg of Na$_2$CO$_3$. Yield of the crude polymer was 230 mg (95% based on reacted polymer). IR ν(C=O) 1605 cm^{-1}, ν(P-F) 842 cm^{-1}.

DCH-Ru Polymer 8

The synthesis was performed in the manner described previously for dinuclear ruthenium complexes, using 200 mg (0.38 mmol) of

Ru(bpy)$_2$Cl$_2$·2H$_2$O, 33 mg (0.097 mmol) of ligand polymer **4**, and 100 mg of Na$_2$CO$_3$. Yield of the crude polymer was 137 mg (56% based on reacted polymer). IR ν(C=O) 1605 cm^{-1}, ν(N-H) 3326 cm^{-1}, ν(P-F) 841 cm^{-1}.

Results and Discussion

Synthesis

The ligand polymers **1-4** (Figure 1) for making ruthenium complex polymers **5-8** (Figure 2) were synthesized from equimolar quantities of each monomer in NMP. The ligand polymers were colorless and were found to be sparingly soluble in DMF, acetonitrile (CAN) and, to an extent, dimethylsulfoxide (DMSO). Only the fully aliphatic derivatives were sufficiently soluble in DMSO to permit viscosity measurements. Inherent viscosities were measured for DCH-Ru polymers **5** and **8** were found to be 0.139 and 0.262 dL/g, respectively, which when compared to those of Higashi *et al* (*38*), showed that they were of low molecular weight (possibly below 8000 Dalton).

The synthesis of DCH-Ru polymers **5-8** was performed in the same manner as that of the comparable dinuclear ruthenium complexes previously reported, since early experimentation revealed that the polyhydrazines were freely soluble in weak aqueous base solutions. The ruthenium-containing polymers were isolated by precipitation, but unfortunately could not be further purified by filtering through a neutral alumina-packed column as was used for purification of DCH-Ru complexes. The DCH-Ru polymers were absorbed strongly on the alumina gels and therefore were used as synthesized for all physical characterization reported herein.

Thermal Analysis

Thermal analysis was performed for ligand polymers **1-4** and ruthenium complex polymers **5-8**. The ligand polymers were found to be less thermally stable (100-150 °C) than the ruthenium complex polymers (>200 °C), except for polymer **4** (251 °C) vs. polymer **8** (209 °C), as assessed by thermogravimetry for 5% weight loss in nitrogen. Studies by Frazer *et al* (*36*), revealed that aliphatic polyhydrazines undergo dehydration above 150 °C to yield a cyclic oxydiazole, which could be accounted for the observed low onset temperatures for weight loss of polyhydrazines **1-4**.

DSC experiments were run on all polymers from 30 °C to ~10 °C at a heating rate of 10 °C/min below the thermal decomposition temperature; however, no peaks indicative of a phase transition were observed.

Electrochemistry

Electrochemistry was performed on ruthenium complex polymers as films on a Pt disk electrode. The cyclic voltammograms showed two quasi-reversible couples attributable to the two one-electron couples per dinuclear DCH-Ru fragment in the polymer. The results are listed in Table I.

The peak-to-peak separations for both couples ranged between 75 and 272 mV. The magnitude of this separation was generally higher versus the dinuclear complexes and can be attributed to lower kinetics due to the uptake and expulsion of electrolyte from the film. The movement of electrolyte, particularly the counterions, into and out of the film is necessary for charge balancing and generally controls the electrochemical rate. An anomalous result was derived for polymer **8**, which showed peak-peak separations of the same magnitude as that of the solution of DCH-Ru complexes (*28-31*), presumably due to a much thinner film.

Table I. **Electrochemical data for polymeric complexes in solid state on Pt disk electrode in chloroform (with 0.1 M TBAH).**

Polymer	$1^{st} E_{1/2}{}^a$ (p-p)^b	$2^{nd} E_{1/2}{}^a$ (p-p)^b	ΔE (mV)
5	310(210)	920(270)	610
6	450(180)	1040(222)	590
7	310(260)	930(240)	620
8	120(75)	730(100)	610

[a] All redox couples in mV versus silver reference electrode at 10 mV/s scan rate. [b] Peak to peak separation for each redox couple in mV.

In order to measure long term switching stability, the polymer films were subjected to repeated switching between the reduced or $-[Ru^{II}Ru^{II}]-$ and the mixed valence $-[Ru^{II}Ru^{III}]-$ state at a frequency ranging from 1.2 to 2.4 s. The choice of frequency was made based upon the relative speed at which the redox current minimized. Results for the trials are displayed in Table II, along with total charge per unit area of electrode taken from the integration, with respect to time of the current trace initially, at the middle and finally the end of the experiment. The results show a small degree of change over time for polymer **8** only, an increase of nearly a factor of two for polymer **5** and a significant drop in electrochemical response for polymers **6** and **7**. The increase in the redox charge of polymer **5** upon switching is not expected and might be due to other unknown redox processes upon cycling, while the latter likely being a result of film delamination.

Table II. Electrochemical switching of ruthenium complex polymers on Pt disk electrode (BAS 2013, 1.6 mm dia.) in chloroform (with 0.1 M TBAH).

Polymer	Switch Time (ms)	Total Time (h) (Cycles)	Initial Charge µC/cm²	Median Charge µC/cm²	Final Charge µC/cm²
5	1200	16.7 (50,000)	184	241	258
6	2400	22.7 (34,000)	241	80	63
7	2400	16 (24,000)	262	93	57
8	1200	21.7 (65,000)	78	76	69

Spectroscopic Study

Spectra for all polymers in the reduced, mixed valence and oxidized form were taken in-situ with the polymer coated on ITO glass, and exhibited similar peak patterns as shown in Figure 3 for polymer **5** as that observed for the dinuclear ruthenium complexes. This is consistent with the presence of isolated DCH fragments within a saturated polymer where no electronic interaction between the dinuclear DCH fragments exists. It is interesting to note that insertion of a secondary amine function into the polymer backbone resulted in a significant blue shift in the MMCT transition for polymer **8**. Although this material possessed an MMCT band maximum at 1188 nm, lower than the 1250-1350 nm transmission region, it does suggest that tuning of the band energy between both NIR transmission regions is possible.

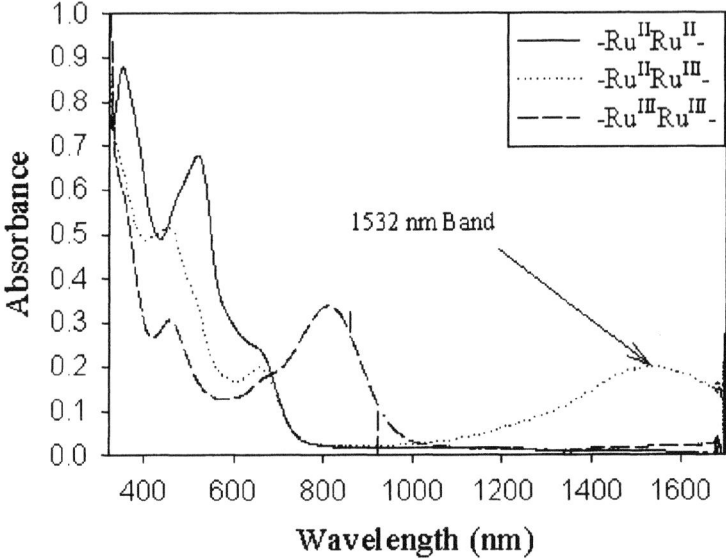

Figure 3. Spectroelectrochemical spectra for polymer 5 on ITO in chloroform with 0.1 M TBAH.

The data for all polymers, taken in a solution of 0.1 M TBAH in $CHCl_3$, are given in Table III. All polymers were found to be insoluble in $CHCl_3$, so the majority was evaluated spectroscopically using TBAH solutions of this solvent as the electrolyte. The data for polymer 5 showed only a slight solvatochromic shift in the major spectral bands and so it was concluded that data derived in either solvent would be equivalent. It was, however, necessary to evaluate some of the polymers in THF since, for long-term switching trials, delamination of the polymer layer occurred in $CHCl_3$.

Table III. Spectral data for polymeric complexes on ITO glass.

Polymer	λ_{max} (nm) for three different Ru/Ru states		
	Ru^{II}/Ru^{II}	Ru^{II}/Ru^{III}	Ru^{III}/Ru^{III}
5[a]	353, 518	451, 1521	315, 463, 830
5[b]	350, 520	454, 657, 1532	457, 810
6[a]	455, 516	452, 647, 1585	452, 805
7[b]	351, 463, 513	450, 1576	454, 809
8[b]	460	455, 579, 1188	460, 910

Electrolyte: [a] THF with 0.1 M TBAH, [b] $CHCl_3$ with 0.1 M TBAH.

A measure of the extinction coefficients may indicate the level of metal incorporation within the polymers. By choosing the bands characteristic of the DCH-Ru complex, such as the [Ru(II)]$d\pi \rightarrow$[bpy]π^* MLCT or bipyridyl $\pi \rightarrow \pi^*$ transitions, this may be accomplished by assuming the extinction coefficients are not significantly affected by the metal environment.

The extinction coefficients for the MLCT bands at ca. 450 to 520 nm were generally lower compared to those found for dinuclear complexes previously reported. In the case of polymers **7** and **8**, this is not as clear from the numbers presented in Table IV. However, as the complex polymers contain two adc fragments per repeat unit, the two have log(ε) values of 3.93 and 3.84 for polymers **7** and **8** respectively. The observed differential in extinction may be the result of differences in environment of each metal complex, however the differential is rather subtle when compared to that of the dinuclear complexes reported elsewhere (*28-31*).

Table IV. Spectral data for polymeric complexes in acetonitrile.

Polymer	λ_{max}(nm)	ε ($M^{-1}cm^{-1}$)	log(ε)
5	245	2.63 x 10^4	4.42
	289	5.46 x 10^4	4.74
	350	9.00 x 10^3	3.95
	516	7.50 x 10^3	3.88
6	244	2.99 x 10^4	4.48
	288	5.46 x 10^4	4.74
	454	7.60 x 10^3	3.88
7	244	6.80 x 10^4	4.83
	289	1.27 x 10^5	5.10
	457	1.69 x 10^4	4.23
8	244	6.37 x 10^4	4.80
	290	1.22 x 10^5	5.09
	455	1.37 x 10^4	4.14

A solid state, long term optical switching experiment was conducted using ITO coated glass as the substrate and THF/0.1M TBAH as the electrolyte (i.e., same experimental set-up as was used for spectral studies) in order to evaluate the stability of the films to withstand repeated bleach-colour cycling. Polymer **5** was the only material (studied here) to form films stable toward dissolution over the full length of the experiment. Given this, the film (~ 0.5 micron in thickness) of polymer **5** on ITO was cycled between –200 mV and 600 mV (i.e., switched electrochemically between the Ru^{II}/Ru^{II} and Ru^{II}/Ru^{III} states). The period of the switching was 4 s with each potential being held for 2 s and timebase absorbance at 1550 nm were taken for 120 s at intervals through the experiment initially, at 9 h and finally at 19 h. Figure 4 shows the optical switching data taken, and the results of the trial are given in Table V.

Table V. Optical switching data for polymer 5 in THF with 0.1 M TBAH.

Time (cycles)	$\%T_b/\%T_c$	$\Delta\%T$	Q_p (μC)	area (cm^2)	Q_d ($\mu C/cm^2$)	ΔOD ($x10^{-2}$)	CE (cm^2/C)
Initial	79/74	5	180	3	60	2.84	470
9 h (8100)	68/63	5	160	3	53	3.32	630
19 h (17100)	69/65	4	130	3	43	2.59	600

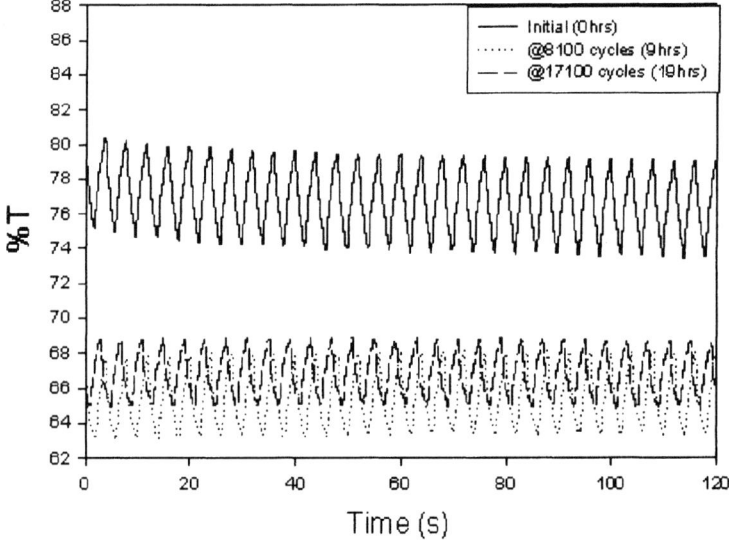

Figure 4. Optical attenuation of ITO/glass coated with polymer 5 in THF with 0.1 M TBAH as function of switching time.

The coloration efficiency (CE) was calculated according to the method of Sapp *et al* (*39*) by using equations 1 and 2. The results show that the CE value increases over the first half of the experiment and drops slightly over the second half.

$$CE(\lambda) = \frac{\Delta OD(\lambda)}{Q_d} \quad (1)$$

$$\Delta OD(\lambda) = \log[T_b(\lambda)/T_c(\lambda)] \quad (2)$$

It should be noted that the observed increase in the CE over the course of the experiment is likely linked with swelling or wetting of the film. This results in greater void space within the polymer matrix, augmenting ion charge transport, and thus decreasing resistance of the film and increasing CE. Over time, however, this swelling can result in a reduction in the electrical contact between the film and the ITO surface, which would account for the reduction in the CE in the later half of the experiment.

Conclusions

The ruthenium-containing polymers produced reasonably good films on ITO glass. Long-term switching trials performed using polymer **5** indicated good chemical stability of the material and comparable coloration efficiencies to other well-established electrochromic materials. Moreover, the significant blue shift in the NIR MMCT transition of polymer **8** versus the other metallic polymer complexes of this study provides positive confirmation of the tunability of these systems.

References

(1) Deb, S. K. *Appl. Optics, Suppl.* **1969**, *3*, 192.
(2) Cogan, S. F.; Plante, T. D.; Parker, M. A.; Rauh, R. D. *J. Appl. Phys.* **1986**, *60*, 2735.
(3) Habib, M. A.; Maheswari, S. P. *Solar Energy Mater., Solar Cells* **1992**, *25*, 195.
(4) Deb, S. K. *Solar Energy Mater., Solar Cells* **1992**, *25*, 327.
(5) Granqvist, C. G. *Mater. Sci. Eng.* **1993**, *A168*, 209.
(6) Greenberg, C. B. *J. Electrochem. Soc.* **1993**, *140*, 3332.
(7) Passerini, S.; Scrosati, B. *J. Electrochem. Soc.* **1994**, *141*, 889.
(8) Deb, S. K. *Solar Energy Mater., Solar Cells* **1995**, *39*, 191.
(9) Gravqvist, C. G. *Handbook of Inorganic Electrochromic Materials*; Elsevier: Amsterdam, 1995.

(10) Monk, P. M. S.; Duffy, J. A.; Ingram, M. D. *Electrochim. Acta* **1998**, *43*, 2349.
(11) Granqvist, C. G. *Solar Energy Mater., Solar Cells* **2000**, *60*, 201.
(12) Franke, E. B.; Trimble, C. L.; Hale, J. S.; Schubert, M.; Woolam, J. A. *J. Appl. Phys.* **2000**, *88*, 5777.
(13) Sharpe, A. G. *The Chemistry of Cyano Complexes of the Transition Metals,* Academic Press: New York, 1976.
(14) Mortimer, R. J.; Rosseinsky, D. R. *J. Chem. Soc., Dalton Trans.* **1984**, 2059.
(15) Mortimer, R. J. *Chem. Soc. Rev.* **1997**, *26*, 147.
(16) Monk, P. M. S.; Delage, F.; Vieira, S. M. C. *Electrochim. Acta* **2001**, *46*, 2195.
(17) Mortimer, R. J.; Warren, C. P. *J. Electroanal. Chem.* **1999**, *460*, 263.
(18) Verghese, M. M.; Ram, M. K.; Vardhan, H.; Malhotra, B. D.; Ashraf, S. M. *Polymer* **1997**, *38*, 1625.
(19) Mortimer, R. J. *Electrochim. Acta* **1999**, *44*, 2971.
(20) Shibata, M.; Kawashita, K.-I.; Yosomiya, R.; Gonzheng, Z. *Eur. Poly. J.* **2001**, *37*, 915.
(21) Schottland, P.; Zong, K.; Gaupp, C. L.; Thompson, B. C.; Thomas, C. A.; Giurgiu, I.; Hickman, R.; Abboud, K. A.; Reynolds, J. R. *Macromolecules* **2000**, *33*, 7051.
(22) Gaupp, C. L.; Zong, K.; Schottland, P.; Thompson, B. C.; Thomas, C. A.; Reynolds, J. R. *Macromolecules* **2000**, *33*, 1132.
(23) Pomerantz, M.; Cheng, Y.; Kasim, R. K.; Elsenbumer, R. L. *J. Mater. Chem.* **1999**, *9*, 2155.
(24) Thompson, B. C.; Schottland, P.; Zong, K.; Reynolds, J. R. *Chem. Mater.* **2000**, *12*, 1563.
(25) Byker, H. J.; Cammenga, D. L.; Poll, D. L. US Patent 5805330 (Gentex Corporation, Zeeland, Mich. USA, 1998).
(26). Bauer, F. T.; Byker, H. J.; Cammenga, D. J.; Roberts, J. K. US Patent 5808778 (Gentex Corp., Zeeland, Mich. USA, 1998).
(27) (a) Qi, Y.; Desjardins, P.; Wang, Z. Y. *J. Opt. A: Pure Appl. Opt.* **2002**, *4*, S273. (b) Qi, Y.; Wang, Z.Y. *Macromolecules*, **2003**, *36*, 3146.
(28) (a) Qi, Y.; Desjardins, P.; Birau, M.; Wu, X.; Wang, Z. Y. *Chin. J. Polym. Sci.* **2003**, *21*, 147. (b) Qi, Y.; Desjardins, P.; Meng, X. S.; Wang, Z. Y. *Opt. Mater.* **2003**, *21*, 255.
(29) Kasack, V.; Kaim, W.; Binder, H.; Jordanov, J.; Roth, E. *Inorg. Chem.* **1995**, *34*, 1924.
(30) Kaim, W.; Kasack, V.; Binder, H.; Roth, E.; Jordanov, J. *Angew. Chem. Int. Ed. Engl.* **1988**, *27*, 1174.
(31) Kaim, W.; Kasack, V. *Inorg. Chem.* **1990**, *29*, 4696.

(32) Crutchley, R. *Adv. Inorg. Chem.* **1994**, *41*, 273.
(33) Creutz, C.; Taube, H. *J. Am. Chem. Soc.* **1973**, *95*, 1086.
(34) Creutz, C.; Taube, H. *J. Am. Chem. Soc.* **1969**, *91*, 3988.
(35) Mohamed, N. A. *Eur. Polym. J.* **1998**, *34*, 387.
(36) Frazer, A. H.; Wallenberger, F. T. *J. Polym. Sci. Part A* **1964**, *2*, 1137.
(37) Frazer, A. H.; Wallenberger, F. T. *J. Polym. Sci. Part A* **1964**, *2*, 1147.
(38) Higashi, F.; Lee, Y.-N. *J. Polym. Sci.: Part A: Polym. Chem.* **1993**, *31*, 1453.
(39) Sapp, S. A.; Sotzing, G. A.; Reynolds, J. R. *Chem. Mater.* **1998**, *10*, 2101.

Chapter 5

Far-IR-through-Visible Electrochromics Based on Conducting Polymers for Spacecraft Thermal Control and Military Uses

Application in NASA's ST5 Microsatellite Mission and in Military Camouflage

P. Chandrasekhar[1], B. J. Zay[1], D. Ross[1], T. McQueeney[1], G. C. Birur[2], T. Swanson[3], L. Kauder[3], and D. Douglas[3]

[1]Ashwin-Ushas Corporation, Inc, 500 James Street, Unit 7, Lakewood, NJ 08701
[2]Jet Propulsion Laboratory, 4800 Oak Grove Drive, Pasadena, CA 91109
[3]NASA-Goddard Space Flight Center, Greenbelt, MD 20771

The largest known, dynamic infrared signature variation in any material, to our knowledge -- > 50% Reflectance variation in the 2 to 25 µm region and an emittance variation > 0.5 -- is reported for Conducting Polymer-based electrochromic flat panels which are to be flown on NASA's ST5 microsatellite mission. The very thin (< 0.5 mm), flexible, lightweight, variable area (1 cm² to 0.5 m²), entirely solid state flat panels show durability in a space environment, switching times < 5 s, and concomitant Visible-region electrochromism. Applications in military camouflage, e.g. against IR-homing missiles, are also briefly described.

Introduction

Electrochromics and Conducting Polymers: Electrochromics are materials or devices that change color when a voltage, typically a DC voltage $< \pm 5$ V, is applied to them. They may work in the Visible region, and, less commonly, in the IR region, 2 to 45 μm, or other spectral regions. They may be, self-evidently, transmission-mode, as in electrochromic sunglasses or automobile rear-view mirros, or reflectance mode, as in most IR-region devices *(1)*. They may be 2-electrode mode (working, counter electrodes) or 3-electrode mode, (i.e. with an additional reference electrode). Conducting Polymers (CPs) have been well known to display varied electrochromism *(1)*, whose basis is "doping/de-doping", i.e. redox, of the CPs. Electrochromism may be characterized by reflectance, ρ, and, in the IR region, by emissivity, ϵ. Although the relation of these two parameters involves spatial integrals *(2)*, very crudely, $\epsilon = 1 - \rho$. A parameter of greater interest in the aerospace industry is the *emittance*, which is the integrated emissivity, typically in the 2 to 45 μm region. In essence, thus, IR-electrochromics are *variable emittance materials*.

Need for Visible-to-far-IR Region Electrochromics: An urgent need currently exists *(3)* for materials capable of showing large signature variation in the Visible through far-infrared (IR) in a dynamic (i.e. switchable, controllable) way for two very specific applications: 1) In spacecraft thermal control; 2) As countermeasures against IR sensors in the battlefield.

The spacecraft requirement stems from the need for spacecraft to conserve heat, and thus battery power, when not exposed to sunlight ("darkside"), to reflect heat when facing the Sun, and, on occasion, to emit excess heat on the darkside. The two major extant technologies, mechanical louvers and various variations of heat pipes (e.g. loop heat pipes), are expensive and increasingly inept in handling the higher heat load requirements of modern electronics *(3)*. Furthermore, their size and weight make them unusable for microspacecraft (typically of weight < 50 kg), which are the trend in future spacecraft; launch costs, i.e. weight, represent a major cost component in spacecraft. They are also unusable for specialized military applications such as space-based radars. With the US alone launching > 100 spacecraft per year, the need thus remains urgent. The key requirement is materials with a large, dynamic variation in emittance, $\Delta\epsilon > 0.4$, coupled with a low solar absorptance, $\alpha(s) < 0.4$. Additional requirements are of course durability in a space environment, e.g. 10^{-6} Torr, ± 75 °C, radiation, Solar Wind (charged particles), micrometeoroids, etc..

The military requirement *(3)* stems from the need for countermeasures against common missiles, e.g. anti-tank and air-to-surface missiles, nearly all of which use IR-homing sensors operating at 3 to 5 μm built into their radomes for final target selection. There is currently no countermeasure against these. These sensors are to be distinguished from common night-vision devices, which are merely image intensifiers operating in the near-IR (to 1.5 μm). The requirements for military uses include a reflectance (specular and diffuse) variation of > 30% in the 3 to 5 μm and 8 to 12 μm regions, switching times < 3 s, and individual control of devices ("pixels") in arrays of up to 100.

Conducting Polymer (CP) Electrochromics: Following the first report of significant, dynamic IR electrochromism in CPs *(4)*, we reported recently briefly on its applications *(5)*. Here, we report in detail on the use of these CP IR-electrochromics in a thermal control panel on NASA's ST5 microsatellite-constellation mission, to be launched in early 2004, and in anti-missile IR-countermeasures.

Features of the Technology: The patented *(6)* technology features electrochromic flat panels which are thin (< 0.5 mm), flexible, lightweight (160 mg/cm^2), entirely solid state, of variable area (1 cm^2 to 0.5 m^2), low power (< 40 μW/cm^2 in normal operation), high physical durability, and very low cost (US$10 K/m^2 vs., e.g., US$200 K/m^2 for mechanical louvers).

Experimental

Materials and Device Assembly: These have been described in detail elsewhere *(5-6)*. The electrochromic device essentially comprised a layer of solid electrolyte sandwiched between two electrodes. Each electrode comprised the CP electro-deposited on a Au/microporous membrane substrate, with the thickness of CP on the bottom, counter electrode about 5 X that on the top, working electrode, thus giving the system high Faradaic reversibility (**Figure 1**). The CP was a diphenyl amine-aniline copolymer, the dopant a pendant-sulfate-group containing polymer, and the microporous membrane typically poly(vinylidene fluoride) of 0.4 μm poresize *(5-6)*. For spacecraft use, a CsI window (transparent 0.5 to 40 μm), coated with indium tin oxide (ITO) for ESD protection, was used, with the devices hermetically sealed in vacuo at 60 °C *(5-6)*. This vacuum hermetic seal was a painstaking procedure requiring 9 months to refine, since the solid electrolyte required some residual water content to function *(5-6)*. For military use, no vacuum hermetic seal was required and the CsI window was replaced with a heat-laminated polyethylene window.

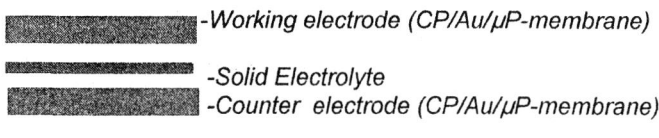
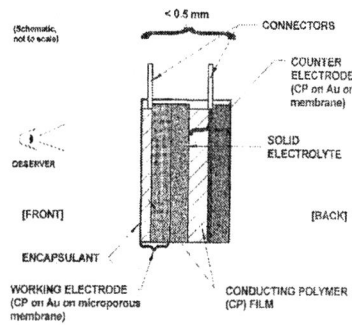

Figure 1. Schematics of electrochromic device. The ITO-coated CsI window in the spacecraft devices is not shown.

Electrochemical Control and Reflectance, Emittance and Solar Absorptance Measurements: A Princeton Applied Research (PARC) Model 263 potentiostat with PARC's 270/250 software was used for preliminary voltammetric characterization and to control devices for spectral measurements. *In-situ* (i.e. as a function of applied potential) Specular IR (16° incidence), Diffuse IR, and Diffuse UV-Vis-NIR (0.2 to 1.1 µm) reflectance measurements were carried out, respectively, on a Perkin-Elmer (P-E) Model 1615 FTIR, a Bio-Rad FTS 6000 FTIR and a P-E Model Lambda 12. Vendor-supplied mirrors or Au surfaces, as appropriate, were used as references. *In-situ* emittance and solar absorptance measurements were carried out, respectively, on an A-Z Tek Model Temp 1000A emissometer (2.5 to 45 µm range) and a Gier-Dunkle Emissometer Solar Absorptometer (0.3 to 2.5 µm range).

Thermal Cycling, "Thermal Vacuum" and Calorimetric Measurements: As a first qualification, spacecraft devices were cycled between -70 and + 75 °C at 10^{-6} Torr with gradients of 2 °C/min and dwell times of 10 mins for 200 cycles. Calorimetric measurements of modulation of the heat transfer under high vacuum (10^{-6} Torr) from a heater placed behind a device to a cold plate in front of the device (cf. **Figure 5**) under a thermal shroud were carried out in a dedicated, specialized apparatus at JPL.

Radiation, Solar Wind (charged particles) Exposure, Outgassing and Other Space Durability Tests: A Co(60) γ-radiation source and the JPL measurement facility were used. For Solar Wind, the facility at NASA-Goddard was used, with fluxes of 1 X 10^{16} p/cm^2 @ 5 KeV and 6.5 X 10^{16} e/cm^2 @ 10 KeV, and IR/Visible reflectance and emittance monitored. For outgassing tests, an effusion chamber facing four quartz crystal microbalances (QCMs) and incorporating a residual gas analyzer having a quadrupole mass spectrometer with a range of 1 to 511 AMU and a detection limit of 10^{-15} Torr (for N_2) was used. Based upon mass impingement rate onto the QCMs and the transport (view) factor, outgassing rates and identities of outgassing species at various temperatures could be determined. Other space durability tests included 500 h UV exposure, shelf life (6 months), vibration, EMI/EMP, and similar tests per standard NASA procedures *(3)*.

Terrestrial IR:: A FLIR Systems FLIR Systems Model MilCam XP (3 to 5 µm) was used. Data were collected under a variety of backgrounds, e.g. foliage, sand, sky, water, dir/earth, and scenarios, e.g. no heat (total darkness), heat from behind (e.g. warm tank), cool from behind (e.g. cold tank), heat from front (e.g. flare), and day/night.

Results and Discussion

Features of Technology and Principle of Operation

The features of the technology have been described briefly in the Introduction above. The extreme physical durability of the devices -- imperviousness to

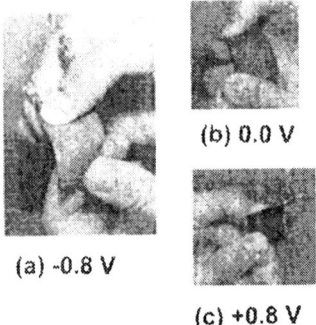

Figure 2. Photos of terrestrial device showing imperviousness to bending/distortion in actual operation, and also Visible color changes (as function of applied voltage indicated).

flexing/bending while in active operation -- are seen in the photos of a terrestrial device **Figure 2**. This figure also shows the typical Visible-region color changes.

The principle of operation of the IR electrochromism comprises modulation of the reflectance of the underlying Au layer on the working electrode (Au is the most IR-reflective material known) by the overlying CP layer as a function of applied potential. However, as we noted in an earlier communication *(5)*, within the purview of this modulation by the CP/dopant, scattering effects due to morphological changes on doping/de-doping (i.e. in particle size, approximating the IR wavelength) are also involved.

Summary of Spectral, Thermal, Space Durability and Switching Data

Figures 3 a, b, c summarize the typical spectral IR and Visible region spectral properties of devices. The large light/dark variation ("dynamic range") is clearly seen therein as well. An interesting property, discussed at length in conjunction with electrochemical data in one of our earlier communications *(4)*, is that the Visible- and IR-region reflectances vary in tandem between applied potentials of -1.1 V and 0.0 V, but behave in an opposite fashion between 0.0 and +0.8 V. As we discussed earlier *(5a)*, this is due to transitions between the various poly(aniline) states: reduced non-conductive leuco-emeraldine, partially oxidized conductive doped-emeraldine, and highly oxidized and again non-conductive pernigraniline.

Table I shows typical emittance data for two devices. The emittance data may be compared with the variation observed in extant mechanical louvers, ca. 0.16 to 0.56 ($\Delta\epsilon = 0.4$) *(3)*. *We also note that, to the best of our knowledge, these dynamic range (i.e. light/dark electrochromic contrast) values, e.g. > 50% Reflectance at 3 to 12 µm (Figure 3) and $\Delta\epsilon > 0.5$ (Table I), represent the largest dynamic (i.e. switchable) IR signature variation in any material.*

Table I: Typical Emittance Data

Device Designation	Applied Voltage (V)	ϵ	$\Delta\epsilon$
Ka_007ad	-0.7	0.21	
	+0.2	0.69	**0.48**
Kb_046ad	-1.2	0.25	
	0.0	0.79	**0.54**

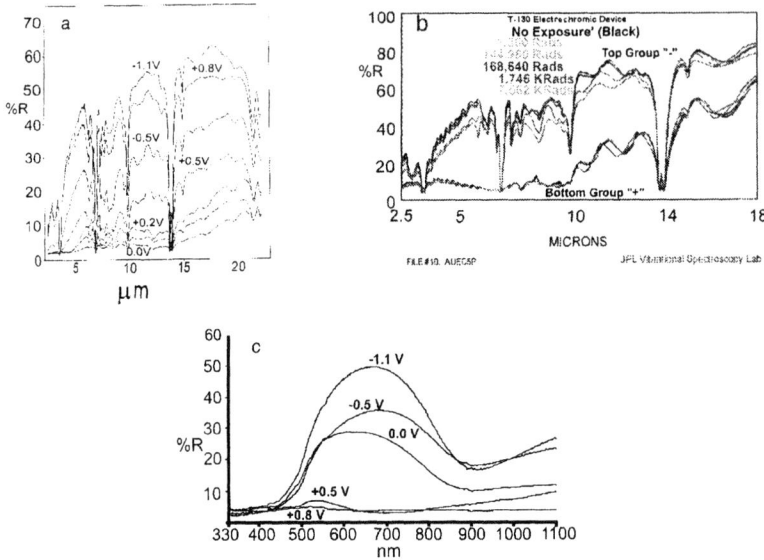

Figures 3. Typical IR specular reflectance (a), IR diffuse reflectance (b) and UV-Vis-NIR reflectance (c) electrochromic data, applied potentials as shown. The diffuse reflectance data also show the effect of γ-radiation, at exposures indicated ("top group"= light state, "bottom group"= IR dark state).

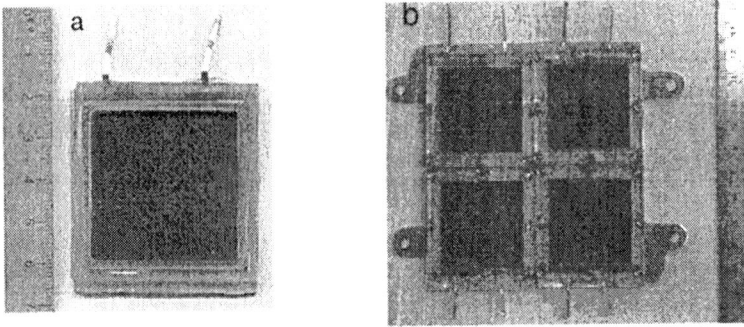

Figure 4. (a): Typical spacecraft device. (b) 4-device panel to be flown on the ST5.

The spacecraft devices passed all space durability tests outlined in the Experimental section above. In the case of the Solar Wind, devices passed "static" tests; "dynamic" tests, i.e. switching the devices actively and monitoring their reflectance while under Solar Wind irradiation, are in progress. Calorimetric "Thermal Vacuum" tests, monitoring the modulation of heat transfer from a heater placed behind a device to a heat sink in front of it (**Figures 5**) showed the calculated emittance behavior in vacuo and under test was identical to that measured with the A-Z Tek instrument outside the test setup. Six electrochromic panels, of the type depicted in **Figure 4b**, together with two rad-hard Controllers, are being delivered to NASA. Each of these will undergo the full gamut of space durability tests outlined in the Experimental section above. They will also undergo additional vibration, EMI/EMP, micrometeoroid, and 500 h of thermal cycling tests prior to their being declared flightworthy. Although the space viability of the technology will be firmly established with these tests, the 2004 spaceflight remains an important, psychological technology barrier. In this respect, it is noted that the ST5 mission is purely experimental, with, e.g., the thermal control panel not being mission-critical.

Room temperature switching times, per the standard literature definition *(1)*, were ca. 30 s for the space devices and < 2 s for the terrestrial devices, the longer times for the former being due to the more desiccated state of the solid electrolyte in them. At -35 °C switching times were ca. 20 minutes. Cyclabilities *(1)*, were > 10^4 cycles for all devices.

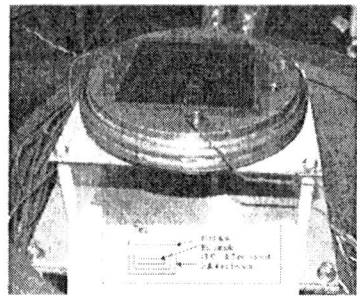 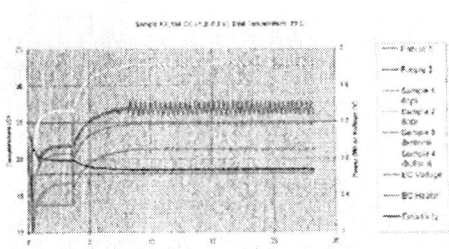

Figure 5. (a) Test setup for Thermal Vacuum and calorimetric tests. (b) Typical results. Black curve labeled "Emissivity" is calculated Emittance based on thermal data.

Controller

The design of a proper Controller for control of arrays (e.g. up to 100 devices each or area 5 cm X 5 cm) is critical to both terrestrial and space applications. The use of a continously applied DC power source is to be avoided, since the electrochemical "wear and tear" on the active electrochromic (the CP) in such a case is large, greatly reducing the device lifetime. Our Controller periodically "interrogated" each device in an array for its Open Circuit Potential (OCP), compared it to the value desired, and then applied a short (typically 50 ms) overpulse to bring the device closer to its desired OCP; it then re-interrogated it, etc., with each device in an array being addressed serially in this fashion. Further, the space-use Controller needed to be constructed of radiation-hard (rad-hard) parts. In our work, the Controller was a modular unit separate from the electrochromic device panels and connected thereto with a simple rad-hard harness. A non-rad-hard version was first constructed and tested. The rad-hard version used a rad-hard 80C51 microprocessor programmed in C++ (due to the length of time in space qualifying microprocessors, these versions are typically three years behind the commercial market -- e.g. even the updated Hubble telescope still uses a 386 processor). The Controller was interfaceable to both a PC, for testing, and the spacecraft computer. **Figure 6** shows a Controller under test with an ST5 4-device electrochromic panel.

Figure 6. Non-rad-hard Controller under test with ST5 electrochromic panel.

The ST5 Mission

NASA's ST5 mission comprises three identical *micro-spacecraft* flying *in constellation* in a 3-month mission. Each spacecraft is 20 kg, and of 45 cm diameter X 20 cm height. It is a purely experimental mission, the *first* microspacecraft constellation mission ever and is thus a proof-of-concept demonstration. Launch is scheduled for early 2004. The flight path will include frequent traversing of the van Allen radiation belts. The specifications for our electrochromic panels a panel dimension of 9 cm X 10 cm, maximum weight of 350 g (of which 190 g for the panel and 160 g for the separate Controller), power of 217 mW (normal operation) and 365 mW (peak transients), and of course passing all space durability tests. The basic performance specs are $\Delta\epsilon > 0.4$, $\alpha(s) < 0.4$, switching times < 30 s, temperature durability -70 to + 85 °C, operating temperature - 40 to + 60 °C. **Figure 7** shows pictures of one of the microsatellites, which, as seen, is just the size of a birthday cake; it also depicts the three satellites flying in constellation. The space labeled "Thermal Control System" in the first photo is where our electrochromic panel will be placed on each satellite. It is important to note that this is a technology demonstrator rather than a mission-critical thermal control panel. Nine panels (one flight article, and two backups) and two rad-hard Controllers (one flight article and one backup) are being delivered to NASA. Each item will undergo again all the space durability tests outlined in the Experimental section above, including. Additional tests will include further vibration, EMI/EMP, and extended (500 h) thermal cycling and calorimetric measurements. Although the technology has essentially already been pre-qualified for space use on the ground, the 2004 demonstrator spaceflight remains an important, psychological technology barrier to be passed. Other issues we continue to address on the ground meanwhile include further optimization of the $\Delta\epsilon$, and, with a view to commercialization, reduction of the "attrition" (loss) rate of devices during production and qualification, which still runs > 50%.

Military Applications

One must needs start the discussion of the military application with a small caveat: We are not talking about the night vision cameras and devices of the type typically seen in newsreels. Those are image-intensifiers, working in the near-IR (to 1.5 μm) and costing as little as US$500. Quite simply, they cannot work in complete darkness. Rather, we are addressing the large stockpile of common, cheap missiles (i.e. not the intercontinental kind). These include common anti-tank and air-to-surface missiles, of which the current US stockpile alone is estimated at > US$3 billion. These missiles typically have a 3 to 5 μm IR sensor built into their radomes (which use sapphire or ALON windows), and use it for *final* target selection. The targets needing protection include *all* craft (ships, tanks, trucks, etc.) and personnel (e.g. soldiers' textile uniforms). There is currently *no* technology available to address this IR-sensor countermeasure.

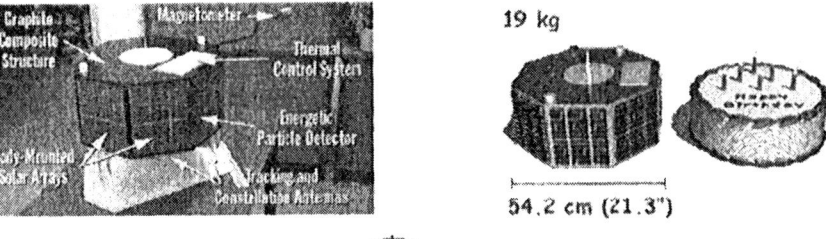
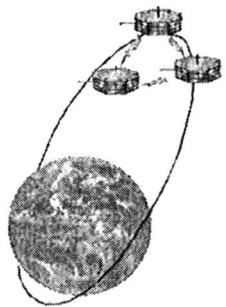

Figure 7. (Left, top): One of the ST5 microsatellites ("Thermal Control Area" is where our electrochromic panel will sit). (Right, top): Size of satellite. (Bottom): Depiction of satellite constellation orbit.

To be meaningful, IR countermeasure data must be collected under a variety of *backgrounds*: Foliage, sand, sky, water, dir/earth. And a variety of *scenarios*: no heat (total darkness), heat from behind (e.g. warm tank); cooling from behing (e.g. cold tank); heat from front (e.g. flare); and, lastly, under both day and night.

Figures 8-9 show sample military camouflage data under many of the above backgrounds and scenarios, as labeled. These were taken with an IR camera operating in the 3 to 5 µm region, and a standard Visible-region digital camera (for the Visible analogs). The device shown exhibited a *ΔR* (specular reflectance) at 5 µm of about 50%. What is to be noted is the matching of the device to the surround background (on the periphery of each photo). It can be seen that this is truly excellent.

Other Electrochromic and Thermal Control Technologies

Subsequent to our original work in 1995 *(4)*, Topart's group *(7)* reported on the IR electrochromism of a P(ANi)/(CSA)/liquid electrolyte (1.0 M $HClO_4$) system. They observed a dynamic range for Specular reflectance of ca. 35% at 10 µm. These appeared to be static measurements, taken after the active electrochromic film was first brought to a specific potential and then introduced into the spectrometer. More recently, the Reynolds group presented some IR Reflectance data for "PEDOT" CPs in the 0.2 to 5.0 µm region *(8)*, using an archaic, liquid-electrolyte, device design borrowed, with due acknowledgment, from our patent *(4)*. The type of Reflectance, i.e. Specular or Diffuse, hemispherical or not, was not specified. They reported dynamic ranges of ca. 40% Importantly, however, they stated that they subtracted the absorption of the window assembly they used, which would greatly increase the dynamic range. In contrast, our data are direct, raw data and have no such "correction"

Some of the earliest attempts at IR electrochromism incorporated the use of phase change materials, e.g. poly(vinyl stearate), poly(dodecene) and poly(octadecene) and an IR emitter (i.e. heater). More recent work from the Daimler-Benz Dornier Aerospace GmbH group used a specular mirror approach with WO_3 as the active electrochromic and Ge windows, and Li-ion based gel electrolytes.[6-7] A group at EIC Laboratories fabricated WO_3 devices based on the earlier Dornier/Daimler work *(9)*. Notably, these devices yielded a $\Delta\epsilon$ of < 0.15, were heavy, had cyclabilities of not more than a few hundred cycles, and burst in vacuum durability tests. Innovative approaches for spacecraft application, such as the Johns Hopkins Applied Physics Lab's *(10)* µm-sized "micro-louvers", suffer from problems of fragility and dust-failure. A more detailed discussion of these alternative technologies was presented by us elsewhere *(5)*. Suffice it to say that none of them achieve $\Delta\epsilon$'s of > 0.3, while also being space-incompatible.

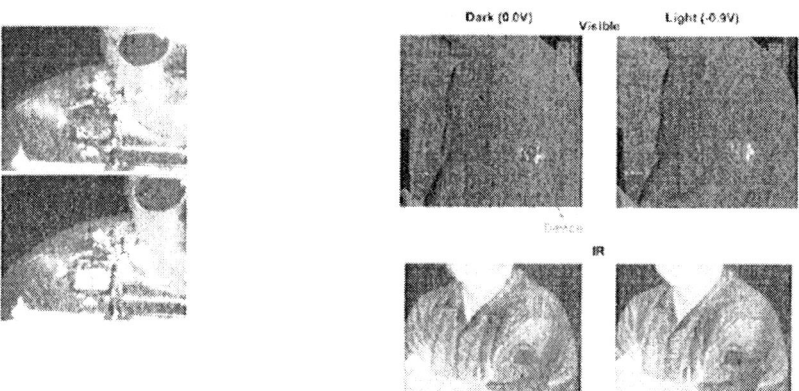

Figure 8. Sample military camouflage data: IR pictures (3 to 5 µm region) showing camouflage of selected targets (top right Visible for reference). Note excellent match of device (circled) to background in one of its states.

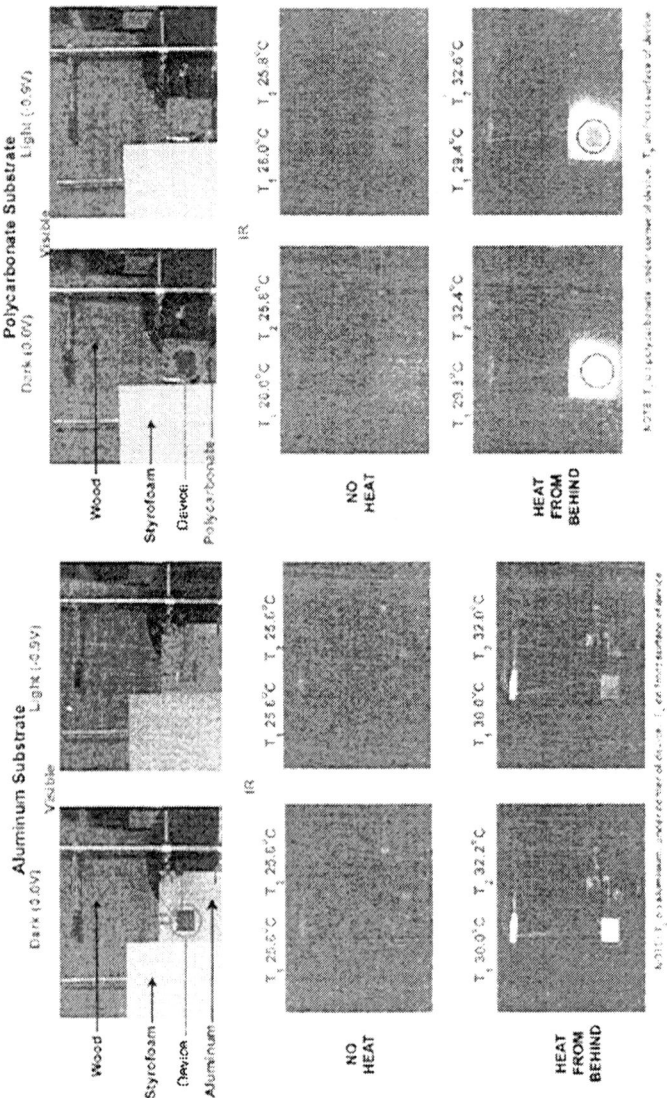

Figure 9. Additional IR camouflage data, as in Fig. 2. Again, note excellent match of device (circled) to background in one of its states. (Top picture in each is Visible, for reference.)

Conclusion

In conclusion, we have demonstrated the largest dynamic IR and thermal signature variations in any material, to our knowledge. Actual applications in a spacecraft mission (launch 2004) and in military camouflage are presented.

References

(1) a) Chandrasekhar, P., *Conducting Polymers, Fundamentals and Applications: A Practical Approach.*, Kluwer Academic Publishers, Dordrecht, The Netherlands and Norwell, MA, USA,**1999**. Chapters 3, 20.

(2) Duffie, J. A.; Beckman, W. A., *Solar Engineering of Thermal Processes*, J. Wiley, New York, NY, USA, **1991**, Chap. 4, Sec. 4.5, pp. 194 ff.

(3) See, e.g., a) *Proc. Twelfth Annual Spacecraft Thermal Control Technology Workshop,* 28 February - 2 March **2001**, El Segundo, CA, USA; b) *Proc. Space Technology and Applications International Forum (STAIF-2001)*, 11-14 February **2001**, Albuquerque, NM, USA.

(4) a) Chandrasekhar, P.; Dooley, T. J. *Proc. SPIE*, **1995**, *2528*, 169. b) Masulaitis, A. M.; Chandrasekhar, P., *Proc. SPIE*, **1995**, *2528*, 190.

(5) a) Chandrasekhar, P.; Zay, B.; Birur, G. C.; Rawal, S.; Pierson, E. A.; Kauder, L.; Swanson, T.; *Adv. Func. Mater.*, **2002**, *12*, 95-103. b) Chandrasekhar, P. Birur, G. C.; Stevens, P.; Rawal, S.; Pierson, E. A.; Miller, K. L.; *Synth. Met.*, **2001**, *119*, 293.

(6) a) Chandrasekhar, P., "Electrochromic Display Device", U.S. Patent No. 5,995,273 (30 Nov. **1999**) b) Chandrasekhar, P., "Electrolytes", U.S. Patent No. 6,033,592 (7 March **2000**).

(7) Topart, P.; Hourquebie, P., *Thin Solid Films*, **1999**, *352*, 243-248.

(8) Schwendemann, I.; Hwang, J.; Welsh, D. M.; Tanner, D. B.; Reynolds, J. R., *Adv. Mater.*, **2001**, *13*, 634-637.

(9) a) Huchler, M.; Natusch, A.; Rothmund, W. , *SAE Ser. (Proc. 25th Intl. Conf. Env. Syst.*, San Diego, USA, **1995**, *7/1995*, 951674. b) Braig, A.; Meisei, T.; Rothmund, W.; Braun, R., *J. Aerospace*, **1994**, *1229,* 941465.

(10) a) Osiander, R.; Champion, J. L.; Darrin, M. A. G.; Douglas, D. M.; Swanson, T. D.; Allen, J. J., in Ref. 1a) above. b) Swanson, T., in Ref. 1a) above. **2001**.

Photochromic and Related Stimuli-Responsive Polymers

Chapter 6

Chromic Transitions and Nanomechanical Properties of Poly(diacetylene) Molecular Films

Robert W. Carpick[1], Alan R. Burns[2], Darryl Y. Sasaki[2], M. A. Eriksson[3], and Matthew S. Marcus[3]

[1]Engineering Physics Department and [3]Physics Department, University of Wisconsin, Madison, WI 53704
[2]Biomolecular Materials and Interfaces Department, Sandia National Laboratories, Albuquerque, NM 87185

Polymerization of ultrathin films containing the diacetylene group has produced a variety of robust, highly oriented, and environmentally responsive films with unique chromatic properties. We present recent developments in the preparation and analysis of ultrathin poly(diacetylene) layers on solid substrates, one to three molecular layers thick. This chapter reviews the structural properties, mechanochromism, and in-plane mechanical anisotropy of these films. Atomic force microscopy (AFM) and fluorescence microscopy confirm that the films are organized into highly ordered domains, with the conjugated backbones parallel to the surface. The number of stable layers is affected by the head group functionality. Local mechanical stress applied by AFM and near-field optical probes induces a transition in the film at the nanometer scale involving substantial optical and structural changes. In addition, we show that AFM reveals the relation between the highly anisotropic character of the chromatic polymer backbone and the associated mechanical properties. In particular, we observe that friction depends dramatically upon the angle between the polymer backbone and the sliding direction, with the maximum found when sliding perpendicular to the backbones. The observed threefold friction and associated structural anisotropy also leads to contrast in the phase response of intermittent-contact AFM, indicating for the first time that in-plane anisotropy of polymeric systems in general can be investigated using this technique.

Introduction

Ultrathin organic films, prepared through methods such as Langmuir deposition or self-assembly (*1,2*), offer the possibility of tailoring the optical, mechanical, and chemical properties of surfaces at the molecular scale. Such control of surface properties is required to implement micro- and nano-scale sensors, actuators, and computational devices. Materials that change in response to external stimuli are especially important for such applications. Poly(diacetylene)s (PDAs) (*3*) merit particular interest as these molecules exhibit strong optical absorption and fluorescence emission that change dramatically with various stimuli, namely optical exposure (*photochromism*) (*4-7*), heat (*thermochromism*) (*8-12*), applied stress (*mechanochromism*) (*7,13-15*), changes in chemical environment (*16,17*), and binding of specific chemical or biological targets to functionalized PDA side chains (*affinochromism/biochromism*) (*18-20*). These transitions, along with other properties such as high third-order nonlinear susceptibility, interesting photo-conduction characteristics, and strong nanometer-scale friction anisotropy(*21*), render PDA a uniquely interesting material.

Optical absorption in PDAs occurs via a π-to-π^* absorption within the linear π-conjugated polymer backbone (*3*). Frequently the first chromic state of the PDA appears blue in color. The chromic transitions described above all involve a significant shift in absorption from low to high energy bands of the visible spectrum, so the PDA transforms from a blue to a red color. The mechanism behind these transitions is not fully established. It is believed that conformational changes such as side chain packing, ordering, and orientation, impart stresses to the polymer backbone that alter its conformation, thus changing the electronic states and the corresponding optical absorption (*3,9*).

In this chapter, we discuss the Langmuir deposition of ultrathin PDA films and the subsequent measurement of their structural, optical, and mechanical properties at the nanometer scale. By altering the head group functionality, we can choose between mono- and tri-layer PDA film structures. We then show that we can use the tip of an atomic force microscope (AFM) or a near field scanning optical microscope (NSOM) tip to locally convert the PDA from the blue form to the red form via applied stress. This represents the first time that mechanochromism has been observed at the nanometer scale. Dramatic structural changes are associated with this mechanochromic transition.

AFM measurements also reveal strongly anisotropic friction properties that are correlated with the orientation of the conjugated polymer backbone. The threefold contrast in friction and the associated mechanical anisotropy produces unexpected contrast in intermittent-contact atomic force microscopy (IC-AFM). In IC-AFM, the cantilever tilt breaks the tip-sample rotational symmetry and enables measurements of in-plane anisotropic forces. The anisotropic forces result in varying energy dissipation depending on the cantilever-sample orientation, yielding phase contrast. The unique anisotropic properties of PDA have therefore allowed us to demonstrate very generally that in-plane properties, as opposed to the commonly discussed out-of-plane properties, can be measured with IC-AFM.

Experimental

Film Preparation

Details of our materials and sample preparation are described elsewhere *(22)*. Briefly, diacetylene molecules with two distinct head groups were made into separate films (Fig. 1). The first, 10,12-pentacosadiynoic acid (PCDA) (**I**) (Farchan/GFS Chemicals) was received as a bluish powder which was purified to remove polymer content. The second molecule, N-(2-ethanol)-10,12-pentacosadiynamide (PCEA) (**II**) was prepared by coupling ethanolamine with 10,12-pentacosadiynoyl chloride in tetrahydrofuran and triethylamine. The acid chloride was prepared from the PCDA using oxalyl chloride in methylene chloride. PCEA was isolated by flash column chromatography on silica gel (25% ethylacetate/hexanes, $R_f = 0.23$).

Langmuir film preparations were performed on a Langmuir trough (Nima) which was situated on a vibration isolation table inside a class 100 clean room. The pure water subphase was kept at 15 ± 0.2 °C. Diacetylene monomers were

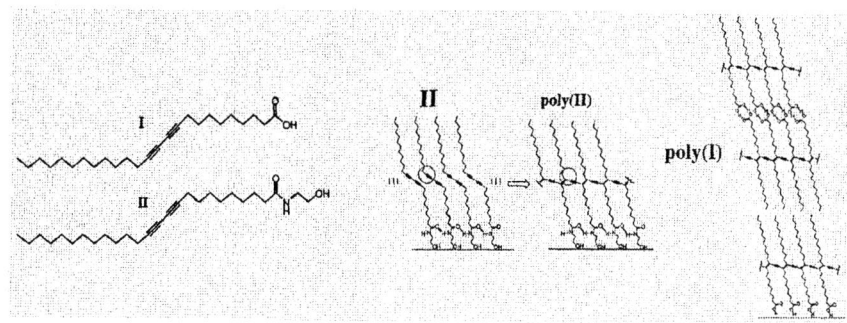

Figure 1. Left: PCDA (I) and PCEA (II) molecules. Middle: schematic of molecular orientation of II and its subsequent conversion to poly(II) upon UV irradiation. A hydrogen bonded network at the headgroup position in the monolayer is drawn. Right: poly(I) in its trilayer form. Hydrogen-bonded carboxylate dimers bind the top two layers to each other, and the lowest layer is bonded to the substrate through hydrogen bonds. Van der Waals' interactions bond the lowest and middle layers.

spread on the water surface in a 50% chloroform/benzene solution. All films were incubated for 10 – 15 minutes at zero pressure prior to compression.

For polymerization at the air-water interface, the films were compressed to a surface pressure of 20 mN/m, then equilibrated for 20 – 30 minutes. UV irradiation of the compressed films was performed with a pair of pen lamps (Oriel). UV exposure was controlled by setting the lamp height above the air-water interface and choosing specific exposure times as described elsewhere (*22*). A few minutes after UV exposure, the water was slowly drained off by aspiration. The films were laid down on mica (freshly cleaved) or silicon (piranha-cleaned) substrates that were pre-submerged horizontally in the aqueous subphase before monolayer spreading. The substrate was then removed and dried in clean room air. This horizontal transfer method proved to be the most effective for producing high quality films, as polymerization creates a degree of rigidity in the film on the water surface. This rigidity renders vertical transfer methods unreliable as the films would not uniformly compress during vertical transfer.

Instrumentation

A Nanoscope IIIA AFM (Digital Instruments) operating in contact mode was used to obtain topographic and friction force images. The same type of AFM was used to obtain IC AFM measurements. AFM data were acquired under laboratory ambient conditions. Silicon nitride cantilevers (Digital Instruments) with a nominal normal force constant of 0.06 N/m were used for all contact-mode measurements. For IC-AFM images, Si cantilevers were used. A novel home-built NSOM (23) was used to simultaneously observe sample fluorescence with sub-wavelength resolution as well as normal forces and shear forces. The tips used were Al-coated etched optical fibers.

Results and Discussion

Film Structure

Pressure-area isotherms indicate the amphiphiles of **I** and **II** on pure water both had identical take-off areas of 25 $Å^2$/molecule, corresponding to the molecular cross-section of the hydrocarbon-diacetylene structure. The film of **I** collapses at low pressure (~12 mN/m), but upon over-compression reaches a stable solid phase with a limiting molecular area of ~ 8 $Å^2$/molecule. This over-compressed state corresponds to a stable trilayer structure. The film of **II** was stable as a monolayer with a collapse pressure of ca. 35 mN/m and an extrapolated molecular area at zero pressure of 25 $Å^2$/molecule. After equilibration, films were polymerized to the blue-phase by exposure to incidence powers of 40 $\mu W/cm^2$ for **I** and 23 $\mu W/cm^2$ for **II** over a period of 30 sec. Red-phase films were produced by exposing the trilayer of **I** to 500 $\mu W/cm^2$ and the monolayer of **II** to 40 $\mu W/cm^2$ for 5 min.

AFM images of the blue- and red-phase forms of poly(**I**) and poly(**II**) on mica or silicon substrates confirm that the coverage for all films was nearly uniform for the entire substrate. Over 95% of the transferred film was flat to within ±0.5 nm, with up to 100 μm crystalline domains observed. AFM measurements confirmed that films of **I** and **II** formed trilayers and monolayers respectively. There were distinct height differences between the blue- and red-phase films of both **I** and **II**. The heights of the blue- and red-phase trilayers of poly(**I**) were measured at 7.4±0.8 and 9.0±0.9 nm, respectively. The blue- and red-phase poly(**II**) monolayer films had similarly proportional height differences of 2.7±0.3 and 3.1±0.3 nm, respectively. The films possess highly aligned striations corresponding to small height variations of ~2Å discussed further below, and similar to previous reports (9). These striations appear to be small variations in density or side chain tilt angle and are aligned with the polymer backbone direction, as confirmed with polarized fluorescence microscopy.

These results provide insight into the stabilization of diacetylene films. The headgroup interactions and alkyldiyne chain stacking should dominate the film structure of the monomeric diacetylene Langmuir films. The ability of the amide headgroup of **II** to form lateral intermolecular hydrogen bonded structures (Fig. 1, center), similar to β-sheets in proteins, may explain the stability of this monolayer film on pure water. In contrast, **I** films on pure water are unstable as monolayers but stack favorably into trilayers. Carboxylic acid dimer formation aids in stabilizing this structure. Indeed, stable bilayer islands are commonly observed on top of the **I** trilayer. Thus, by altering the head group, we can control whether the resulting film will be structured as a monolayer or a trilayer. Further details of the film preparation and structure are published elsewhere (*22*).

Mechanochromism

The blue-to-red transition can be activated at the nanometer scale using NSOM or AFM tips on both the trilayer poly(**I**) and monolayer poly(**II**) (*7,24*). Fig. 2 shows simultaneous NSOM topography and fluorescence images on a blue poly(**I**) film. In the first scan (left pair), no fluorescence is seen over the flat PCDA region. In the subsequent scan (right pair), topographic changes are created, and localized fluorescence emission is produced. A fluorescence spectrum obtained over this region reveals the spectral fingerprint of red PCDA. These observations were reproducible. In general, when this transition is observed, the fluorescent regions grow in size with each image acquired.

The blue-to-red transition has also been produced using AFM tips with both trilayer poly(**I**) and monolayer poly(**II**) blue films (Fig. 3). With AFM, local topographic changes, discussed below, are observed *in-situ*. These changes indicate the transition is taking place. By creating a large (>1 μm) red region, *ex-situ* fluorescence microscopy is used to confirm that a red region has been created by the AFM tip. With both AFM and NSOM, normal forces alone are

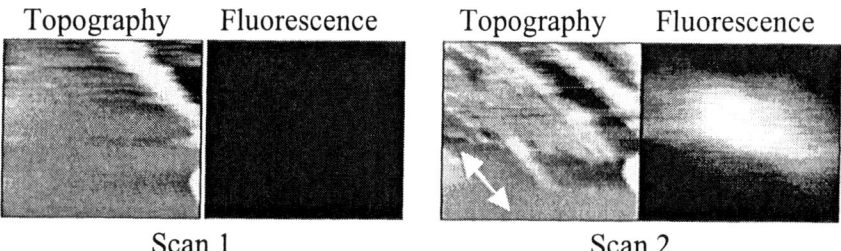

Figure 2. NSOM shear force topography and simultaneous fluorescence images (2.4×2.4 μm^2) showing tip-induced mechanochromism.

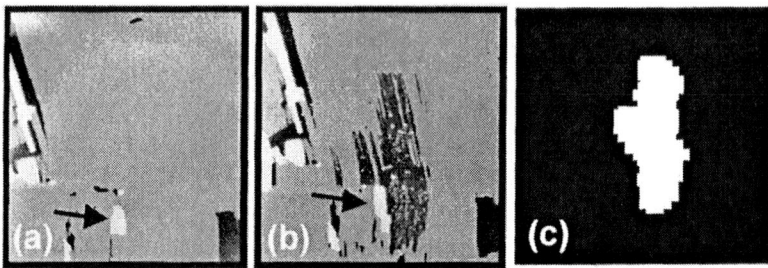

Figure 3. 10×10 µm² topographic AFM images showing tip-induced patterning of red PCDA domains. (a) Initial image of a blue film. (b) Final image with a patterned red region. The backbones are oriented in roughly the vertical direction. The patterning was formed by multiple, high-load scans within the patterned region. The black arrow indicates a bilayer island that has grown in size after patterning. (c) Far-field fluorescence image of the same region. Characteristic red PCDA fluorescence is localized within the patterned region.

not sufficient to cause the transition. Shear forces must also be applied, *i.e.* during the scanning process, to produce the blue-to-red transition. In all cases, the observed transitions are irreversible up to at least several months.

The transformed regions consistently exhibit higher friction, increased roughness, and a surprising height reduction of typically 40-50% of the original film height. High resolution images of the transformed regions, however, consistently reveal backbone-related striations. This indicates a substantial degree of preservation of the conjugated backbone despite the dramatic height reduction. No such structural change for other types of PDA have been previously reported. While this height reduction could be explained by a removal of one or more layers in the trilayer poly(I) film, this cannot explain the comparable height reduction for the monolayer. One possibility is that some molecules are removed during the transformation process, and the remaining molecules substantially increase the tilt angle of the hydrocarbon side chains. The all-*trans* nature of the side chains may also be strongly disturbed, but the backbone structure remains. This increased tilt angle and conformational changes allow stress within the backbone and side chains to be relieved as discussed below. This picture is consistent with the observation via AFM that the backbones remain in tact, and the film is greatly compressed. It is also consistent with the observation that friction force between the tip and PDA sample is higher over the transformed region, since highly tilted and defective side chains would expose more methylene groups to the tip, as opposed to the terminal methyl groups with have a lower surface energy.

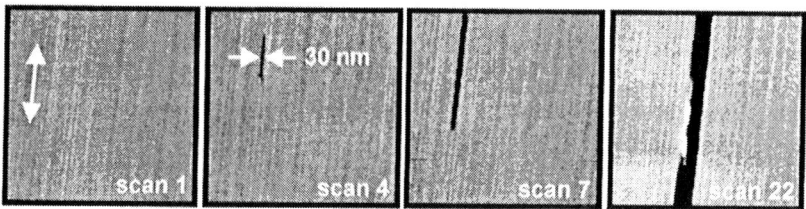

Figure 4. Series of 1×1 µm² topographic AFM images of blue PCDA showing the progressive growth of the tip-induced red domains. In the first scan, striations indicative of the polymer backbone direction are observed. By the fourth scan, a topographically distinct (i.e. lower) region, only 30 nm wide, appears. This region continues to grow in subsequent scans.

The blue phase therefore appears to be a metastable phase. On the Langmuir trough, the high registry of the diacetylene packing permits rapid topochemical polymerization of the diyne monomers to the ene-yne conjugation upon UV illumination resulting in the blue-phase polydiacetylene. Little change in the amphiphile packing, and thus little reorientation of the alkyl side chains occurs. However, the hybridization change from sp to sp^2 for the terminal alkyne carbons creates a stress on the polymer as a result of the 180° to 120° bond angle conversion (see Fig. 1). With initial UV illumination to create the blue form, significant molecular stress is built into the film. At higher degrees of polymerization, or with the application of mechanical stress or heat, the film's original structure breaks down as the alkyl chains of the blue-phase polymer reorganize to accommodate the bond angle conversion. In the case of UV polymerization or heating, this yields a closer packing (film contraction) and reorientation (vertical height increase) of the alkyl chains. In the case of mechanochromism, this leads to a totally different collapsed film structure.

These reorganizations, although thermodynamically more stable, produce a loss of π-conjugation and results in the red form of the polydiacetylene. These results are consistent with recent NMR investigations by Lee et al. (12) which show that the blue-to-red thermochromic transition in other PDA bulk samples involves a release of mechanical strain on the backbone and reorganization of the side groups. Furthermore, FTIR data of Lio et al. (9) and ^{13}C NMR data of Tanaka et al. (25) suggest that some of the tilted side chains rotate toward the surface normal in the red phase for thermochromic films. Theoretical calculations indicate that a rotation of only a few degrees about this bond dramatically changes the π-orbital overlap (26), causing a significant blue-shift of the absorption spectrum. Recent molecular modeling studies (24) of PDA oligomers also show that a loss of backbone planarity leads to shifts in absorption spectra corresponding to the blue-to-red transition.

Friction Anisotropy

AFM measurements demonstrate that the films possess strong friction anisotropy (*21*). For example, measurements on the red poly(II) monolayer (Fig. 5(a) and 5(b)) reveal a domain structure. The friction force varies substantially from one domain to the next, and is nearly uniform within each domain. The topographic image reveals an essentially flat film. As mentioned above, topographic images within a single domain reveal parallel striations of varying width and uniform direction (Fig. 4). These striations are associated with the direction of the underlying polymer backbone, and allow us to determine the relative angle between the sliding direction and the backbone direction.

By measuring the friction force at the same load for different orientations, we find that friction is lowest when sliding parallel to the backbones, and 2.9 times larger when sliding perpendicular (Fig. 5(c)). This dramatic effect may be due to anisotropic film stiffness caused by anisotropic packing and/or ordering of the alkyl side chains, as well as the anisotropic stiffness of the polymer backbone structure itself. Along the backbone direction, the conjugated polymer bonds provide a rigid link between alkyl chains (Fig. 1). However, the spacing between alkyl chains linked to *neighboring* backbones is determined by weaker interchain van der Waals' forces and head group-substrate interactions. In other words, the lack of covalent bonding between neighboring polymer chains allows some freedom in their spacing, consistent with previous studies of a similar PDA film (*9*). Variations in film density would also explain the typical film height contrast of ~ 2Å due to the striations observed in Fig. 4 (*27*).

A simple model for a scalar in-plane anisotropic tip-sample interaction force $F_{in-plane}$ is an isotropic dissipative force F_1, plus an anisotropic term that varies as $\sin(\theta)$ with maximum value F_2:

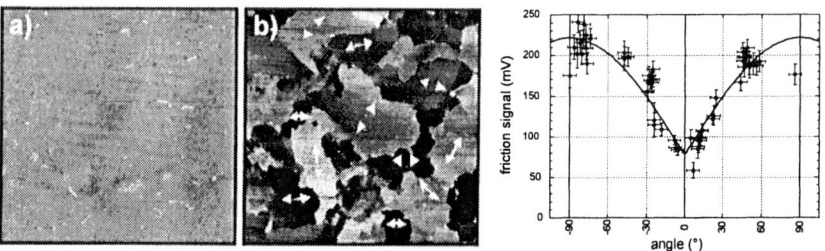

Figure 5. (a) 50×50 µm² AFM topography image of a red poly(I) monolayer. (b) simultaneous friction image. The friction image reveals the different domains. White arrows indicate the domain orientation. (c) Friction force (raw signal units) vs. angle. 0° indicates sliding parallel to the backbone direction. The standard deviation is used for the friction error bars. The solid line represents the fit of Eq. (1) to the data.

$$F_{in-plane} = F_1 + F_2|\sin(\theta)| \quad (1)$$

where θ represents the domain orientation. The anisotropic term is consistent with the notion that the frictional work done is equal to the vector dot product of the distance traveled and a force that only acts perpendicular to the backbones. The absolute value is used to ensure that this contribution is positive. Eq. (1) provides a consistent fit to the data as shown in Fig. 5(c), giving $F_1 = 77$ mV and $F_2 = 144$ mV (uncalibrated raw signal units). Thus, according to the fit, the total friction anisotropy is $F_{max} = \dfrac{F_1 + F_2}{F_1} = 2.9$.

The anisotropic contribution F_2 may have several sources. Lower stiffness along the perpendicular direction may lead to larger molecular deformation when sliding in that direction, and thus a larger contact area, more gauche defect creation, and more bending of the hydrocarbon chains. These would all contribute to larger friction forces (*28*).

Imaging In-Plane Anisotropy with Intermittent Contact AFM

In IC AFM, the AFM cantilever is driven at or near its resonance frequency so that the tip oscillates with respect to the substrate. The tip makes contact with the sample for a small portion of its cycle, and the reduced amplitude that results is used as a feedback signal to map out the topography of the sample. The corresponding phase shift between the drive and response is monitored simultaneously, and is generally considered to be a map of dissipation during compression of the sample along the sample normal.

Fig. 6(a) shows an IC AFM topographic image of a PCEA monolayer film with large domains (*29*). Islands of extra PCEA layers are also visible. As with contact-mode AFM, each domain can be identified in the phase image by the orientation of the striations along which the PDA backbones lie (*9,22*). The typical phase ϕ in Fig. 6(b) is approximately 116° (*30*). Given that the properties of PDA films *normal* to the substrate are highly uniform between domains, it is surprising that the phase ϕ differs from domain to domain by up to 2° in Fig. 6(b). The maximum phase ϕ_{max} occurs when the long axis of the cantilever is parallel to the striations ($\theta = 0°$).

Phase shifts in IC-AFM indicate energy loss (*31*). When the tip's motion is sinusoidal, the power dissipated due to the tip-sample interaction is (*31,32*):

$$\overline{P}_{tip} = \frac{1}{2}\frac{kA^2\omega_0}{Q}\left(\frac{A_0}{A}\sin(\phi) - 1\right) \quad (2)$$

where ϕ is the phase of the oscillation relative to the drive, k is the spring constant of the cantilever, ω_0 is the cantilever's resonance frequency, Q is the quality factor of the cantilever, A_0 is the free oscillation amplitude of the lever,

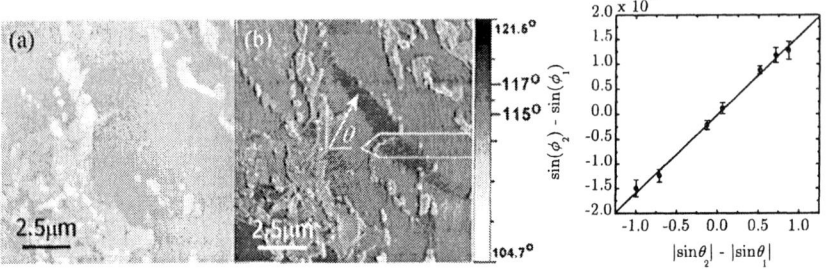

Figure 6. Topographic (a) and phase (b) images of a PDA monolayer thin film on mica. θ is the angle between the local PDA backbone striations and the long axis of the cantilever. The orientation of the cantilever is sketched at the right. (c) The difference in the sines of the phase angles ϕ, proportional to the difference in energy loss between domains, versus the difference in the absolute values of the sines of the angles θ, proportional to the difference in the in-plane tip-sample dissipative forces.

and A is the reduced amplitude during measurement. We have shown that, consistent with many IC-AFM measurements, the tip motion is very nearly sinusoidal in our experiments (29).

From Fig. 6(b) and Eq. (2), we find that the power dissipated is smallest (*i.e.* phase shift largest) when the striations are parallel to the long axis of the cantilever. In fact, the cantilever loses an extra amount of energy $\Delta E \approx 2.4$ eV per cycle in domains where the striations are perpendicular, rather than parallel to the long axis of the cantilever. The effect observed in Fig. 6(b) can now be explained by considering the fact that the cantilever is tilted along its long axis (11° in our case). Therefore, there will be a small but significant component of tip motion parallel to the sample during each oscillation cycle. The direction of larger dissipation corresponds, as we would expect, with the direction of high friction for this component of in-plane sliding. The amount of extra energy dissipated is roughly 10% of the total energy dissipated through the tip-sample interaction. That this level of energy loss should occur due to in-plane forces is reasonable, given that the tip moves in the plane of the sample a distance that is ~20% of the total tip displacement.

Fig. 6(c) is a plot of $\Delta\sin(\phi)$ vs. $\Delta|\sin\theta|$ for the data in Fig. 6(b) (29). Remarkably, we find that $\Delta\sin(\phi)$ is proportional to $\Delta|\sin\theta|$, with proportionality constant $\alpha = (1.58 \pm 0.05) \times 10^{-2}$. This linear proportionality is discussed in detail in reference (29). If Eq. (2) describes the power dissipation, the observed proportionality is expected. From Eq. (2), the difference in $F_{in-plane}$ between two domains *1* and *2* is simply proportional to

$\Delta|\sin(\theta)| \equiv |\sin(\theta_2)| - |\sin(\theta_1)|$. The difference in power dissipated between two domains is proportional to $\Delta\sin(\phi) \equiv \sin(\phi_2) - \sin(\phi_1)$, if we assume that Eq. (1) correctly describes the in-plane dissipation. We therefore conclude that the anisotropic forces in our experiment arise from friction, or in addition, inelastic shear deformation.

We have constructed a model which, unlike previous models of IC-AFM, takes the tilt of the cantilever into account. The model assumes Hertzian tip-sample contact, with both in-plane and out-of-plane dissipative components (*29*). The key result is that components of motion both normal and parallel to the sample occur, and therefore in-plane dissipative processes can cause phase shifts. Using parameters appropriate for our system, we solve for the steady state motion of the tip. The model indicates a maximum tip-sample in-plane tip motion of δ=49.9±0.1 pm parallel to the sample. The distance δ is extremely small, and it is difficult to make firm distinctions between friction and shear deformation at such a small scale, as discussed below. The important result is that δ is virtually independent of the in-plane damping. Furthermore, the model produces a nearly sinusoidal tip motion, indicating that Eq. (2) remains valid for the tilted-cantilever geometry.

In principle, modeling can be used to quantitatively associate the measured phase shifts with the dissipative in-plane properties of the material being imaged. These properties are friction (as quantified by the interfacial shear strength τ between the tip and sample) (*28*) and dissipative shear deformation (due to viscoelasticity of the sample, as quantified by the loss tangent of the material, $\tan\Delta$) (*33*). Both of these mechanisms contribute to the observed dissipation and so we cannot explicitly separate them in our data. However, we can use our data to determine the upper limits of τ and $\tan\Delta$ by finding the values that result when attributing all the dissipation to each mechanism respectively. The current difficulty with this approach is that the phase shifts predicted by our Hertzian model have the opposite sign to the phase shifts we observe. The reason for this discrepancy is that we have ignored adhesion in our model. Preliminary results from modeling that includes adhesion show that the phase shift changes sign and becomes consistent with our data. A description of this adhesive model is in progress (*34*).

Conclusions

We have produced high-quality ultrathin PDA films using a horizontal Langmuir deposition technique. The number of stable layers in the film is controlled by altering the head group functionality. The films exhibit strong friction anisotropy that is correlated with the direction of the polymer backbone structure. Shear forces applied by AFM or NSOM tips locally induce the blue-to-red chromatic transition in the PDA films.

Monolayer films of PCEA exhibit strong threefold friction anisotropy. Friction is highest when scanning perpendicular to the polymer backbone direction. We propose that this effect results from anisotropic film deformation modes.

The highly anisotropic nature of PDA films allows us to show that in-plane properties of materials can be observed using IC-AFM. This is due to the tilt of the AFM cantilever which produces a small but significant in-plane component to the tip's motion. In the case of PDA monolayers, in-plane friction and shear deformation anisotropy leads to contrast in the IC-AFM phase image. The results can be explained using a simple model that incorporates Hertzian contact mechanics with in-plane dissipation, and may be generalized to the study of other anisotropic materials.

Acknowledgement

We acknowledge funding from the NSF CAREER program, grant #DMR0094063 (MAE) and grant #CMS0134571 (RWC), the Research Corporation (MAE), and the NSF MRSEC program, grant #DMR0079983, and the University of Wisconsin-Madison. Some images in this paper were prepared using WSxM freeware from Nanotech Electronica. Sandia is a multiprogram laboratory operated by Sandia Corporation, a Lockheed Martin Company, for the United States Department of Energy under Contract DE-AC04-94AL85000.

References

(1) Ulman, A. *Introduction to Ultrathin Organic Films from Langmuir-Blodgett to Self-Assembly*; Academic Press: New York, 1991.
(2) Schwartz, D. K. *Surf. Sci. Rep.* **1997**, *27*, 245.
(3) Bloor, D.; Chance, R. R. *Polydiacetylenes: Synthesis, Structure, and Electronic Properties*; Martinus Nijhoff: Dordrecht, 1985.
(4) Day, D.; Hub, H. H.; Ringsdorf, H. *Isr. J. Chem.* **1979**, *18*, 325.
(5) Tieke, B.; Lieser, G.; Wegner, G. *J. Polym. Sci., A, Polym. Chem.* **1979**, *17*, 1631.
(6) Olmsted, J.; Strand, M. *J. Phys. Chem.* **1983**, *87*, 4790.
(7) Carpick, R. W.; Sasaki, D. Y.; Burns, A. R. *Langmuir* **2000**, *16*, 1270.
(8) Wenzel, M.; Atkinson, G. H. *J. Am. Chem. Soc.* **1989**, *111*, 6123.
(9) Lio, A.; Reichert, A.; Ahn, D. J.; Nagy, J. O.; Salmeron, M.; Charych, D. H. *Langmuir* **1997**, *13*, 6524.

(10) Chance, R. R.; Baughman, R. H.; Muller, H.; Eckhardt, C. J. *J. Chem. Phys.* **1977**, *67*, 3616.

(11) Carpick, R. W.; Mayer, T. M.; Sasaki, D. Y.; Burns, A. R. *Langmuir* **2000**, *16*, 4639.

(12) Lee, D. C.; Sahoo, S. K.; Cholli, A. L.; Sandman, D. J. *Macromolecules* **2002**, *35*, 4347.

(13) Muller, H.; Eckhardt, C. J. *Mol. Cryst. Liq. Cryst.* **1978**, *45*, 313.

(14) Nallicheri, R. A.; Rubner, M. F. *Macromolecules* **1991**, *24*, 517.

(15) Tomioka, Y.; Tanaka, N.; Imazeki, S. *J. Chem. Phys.* **1989**, *91*, 5694.

(16) Cheng, Q.; Stevens, R. C. *Langmuir* **1998**, *14*, 1974.

(17) Jonas, U.; Shah, K.; Norvez, S.; Charych, D. H. *J. Am. Chem. Soc.* **1999**, *121*, 4580.

(18) Charych, D. H.; Nagy, J. O.; Spevak, W.; Bednarski, M. D. *Science* **1993**, *261*, 585.

(19) Reichert, A.; Nagy, J. O.; Spevak, W.; Charych, D. *J. Am. Chem. Soc.* **1995**, *117*, 829.

(20) Charych, D.; Cheng, Q.; Reichert, A.; Kuziemko, G.; Stroh, M.; Nagy, J. O.; Spevak, W.; Stevens, R. C. *Chemistry and Biology* **1996**, *3*, 113.

(21) Carpick, R. W.; Sasaki, D. Y.; Burns, A. R. *Trib. Lett.* **1999**, *7*, 79.

(22) Sasaki, D. Y.; Carpick, R. W.; Burns, A. R. *J. Colloid Interface Sci.* **2000**.

(23) Burns, A. R.; Houston, J. E.; Carpick, R. W.; Michalske, T. A. *Langmuir* **1999**, *15*, 2922.

(24) Burns, A. R.; Carpick, R. W.; Sasaki, D. Y.; Shelnutt, J. A.; Haddad, R. *Trib. Lett.* **2001**, *10*, 89.

(25) Tanaka, H.; Gomez, M. A.; Tonelli, A. E.; Thakur, M. *Macromolecules* **1989**, *22*, 1208.

(26) Orchard, B. J.; Tripathy, S. K. *Macromolecules* **1986**, *19*, 1844.

(27) Fenter, P.; Eisenberger, P.; Liang, K. S. *Phys. Rev. Lett.* **1993**, *70*, 2447.

(28) Carpick, R. W.; Salmeron, M. *Chem. Rev.* **1997**, *97*, 1163.

(29) Marcus, M. S.; Carpick, R. W.; Sasaki, D. Y.; Eriksson, M. A. *Phys. Rev. Lett.* **2002**, *88*, 226103.

(30) The reported phase shifts are true phase shifts with respect to the drive signal. The phase shifts reported by the instrument are not properly scaled and are shifted by 90°.

(31) Cleveland, J. P.; Anczykowski, B.; Schmid, A. E.; Elings, V. B. *Appl. Phys. Lett.* **1998**, *72*, 2613.

(32) Tamayo, J.; Garcia, R. *Appl. Phys. Lett.* **1997**, *71*, 2394.

(33) Lakes, R. S. *Viscoelastic solids*; CRC Press: Boca Raton, 1999.

(34) D'Amato, M. J.; Marcus, M. S.; Eriksson, M.A.; Carpick, R.W.; in preparation.

Chapter 7

Functional Amphiphilic and Bolaamphiphilic Poly(diacetylene) Assemblies with Controlled Optical and Morphological Properties

Jie Song[1], Raymond C. Stevens[2], and Quan Cheng[3,*]

[1]Materials Sciences Division, Lawrence Berkeley National Laboratory, Berkeley, CA 94720
[2]Department of Molecular Biology, The Scripps Research Institute, La Jolla, CA 92037
[3]Department of Chemistry, University of California, Riverside, CA 92521

Amino acid-terminated amphiphilic and bolaamphiphilic diacetylene lipids were synthesized and assembled to form microstructures of varied morphologies. UV irradiation of the assemblies leads to conjugated polymers with unique optical properties. Chromatic transition of polydiacetylene materials in response to pH and thermal effect and their morphological transformation upon lipid doping are discussed.

Conjugated polymers capable of responding to external stimuli by changes in optical, electrical or electrochemical properties are of great interest for the design of various sensors (*1-3*). They are attractive materials for constructing direct sensing devices because the signal transducer element (the conjugation system of the polymer) and molecular recognition moiety can potentially be built within a single unit, rendering them amenable to microfabrication. Polydiacetylenes (PDAs) represent one of the most promising chemical platforms for sensors, especially with colorimetric detection of analytes. The delocalized electronic structure allows strong absorption in the UV-visible range, giving the material a blue appearance. The optical properties of PDA can be dramatically altered from blue to red by external stimuli such as heat, organic solvent, pH and mechanical stress (see scheme below). A particularly interesting stimulus is the binding of biological analytes at the polymer-media interface.

<chemical scheme: diacetylene monomers with R$_1$, R$_2$ substituents undergo hv polymerization to form blue phase ene-yne polymer, which undergoes chromatic transition to red phase>

blue phase → red phase

The recognition of a ligand by a membrane-associated receptor or an enzyme (covalently or non-covalently incorporated into the PDA scaffold) provides the needed driving force to induce chromatic transition of PDA upon the occurrence of the interfacial binding event (i.e., biochromism), leading to the birth of colorimetric biosensors for influenza virus, bactcrial toxins and *E. coli* (*4-6*).

While the mechanistic detail of chromic shifts of PDAs in response to various environmental perturbations has been extensively investigated (*7-9*), the design of functional PDAs with controlled supramolecular structures and customized optical properties is still in its infancy. PDA as a sensing material has limitations, particularly in detection sensitivity, processability and durability. New chemistries allowing for modification of the PDA systems should focus on improvements in these areas. For instance, headgroup derivatization can affect the original color of the polymer in the coplanar conformation of the ene-yne conjugation backbone, and therefore determine the type of chromic transition occurring upon the departure from coplanarity (*10*). Attachment of an ionic molecule as headgroup could provide a useful means to alter the surface charge distribution and hydrophobicity around the recognition interface, striking a delicate balance between repulsive steric interactions and attractive *van der Waals* interactions, and thereby offering the possibility to optimize the chromatic transition properties (*11,12*). More effective approaches involve alteration of hydrophobic lipid core length and the replacement of amphiphilic diacetylene lipids with a bolaamphiphilic lipid. Strengthening of hydrogen bonding interactions on both faces of the transmembranic structures could enhance the crystallinity of the lipid packing arrangement and possibly lead to the formation of novel microstructures (*13,14*).

Much remains to be learned before the true rational design of functional PDAs can be realized, as either the transducer of chemical and biological colorimetric sensors or the molecular template of functional composites and devices with structural control ranging from nanoscopic to microscopic levels. In this chapter, we will discuss the design and preparation of mono-functional (amphiphilic) and bis-functional (bolaamphiphilic) PDAs that adopt exotic morphologies but retain characteristic optical properties that are commonly observed with conventional Langmuir-Blodgett film *(15,16)* and vesicular *(17,18)* PDA assemblies. Emphasis will be placed on the derivatization of PDA templates with various amino acids. We will demonstrate that morphological transformations from extended helical ribbons to organized nanofibers, along with chromatic transition, can be manipulated via pH control. Transformations from ribbons to vesicular structures by doping with ganglioside G_{M1} and lipophilic cholesterol in a controlled manner will also be discussed.

I. Amino Acid-Terminated Amphiphilic Polydiacetylenes

Monomer synthesis, microstructure formation and characterization

Modification of amphiphilic diacetylene lipids with a series of naturally occurring amino acids is straightforward. 10,12-Pentacosadiynoic acid was converted to a succinimidyl ester in the presence of N-hydroxysuccinimide and EDC, and then coupled with the N-terminus of corresponding amino acids in a THF/H_2O mixed solvent to yield the derivatized lipids through an amide linkage (Figure 1). The choice of amino acids as headgroup is to create a compatible surface on microstructures for protein-related sensing applications. In addition,

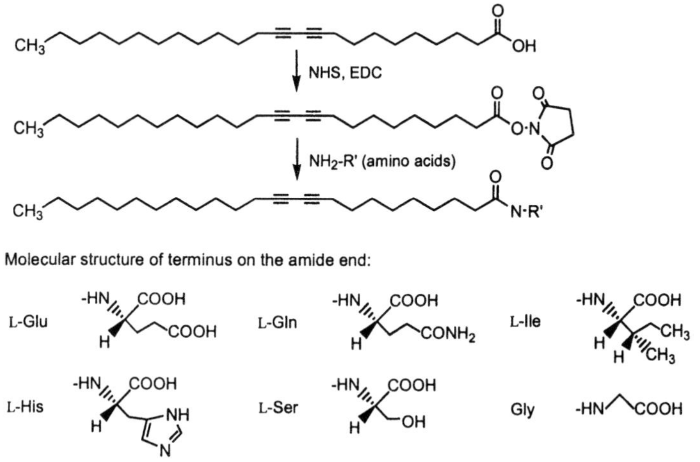

Figure 1. Synthetic scheme and molecular structures of amino acid terminated diacetylene lipids.

these optically pure amino acids vary in polarity, allowing surface charge and hydrophilicity to be manipulated in a broad range and controllable manner. To obtain lipid microstructures, dried lipid was dissolved in methanol to which warm deionized water was added dropwise under vigorous stirring. Prior to polymerization and colorimetric characterization, the sample was dialyzed against water using a Spectra/Por membrane tubing to remove residual methanol. Lipid bilayer vesicles were obtained by hydration of diacetylenic lipids using probe sonication. Photopolymerization of diacetylene microstructures and bilayer vesicles was realized by UV irradiation at 254 nm.

Figure 2 shows the TEM images of microstructures made from amino acid terminated diacetylene lipids. For L-Glu-PDA (Fig. 2A), the aggregate consists of twisted and untwisted ribbons and fibers, with their lengths varying from

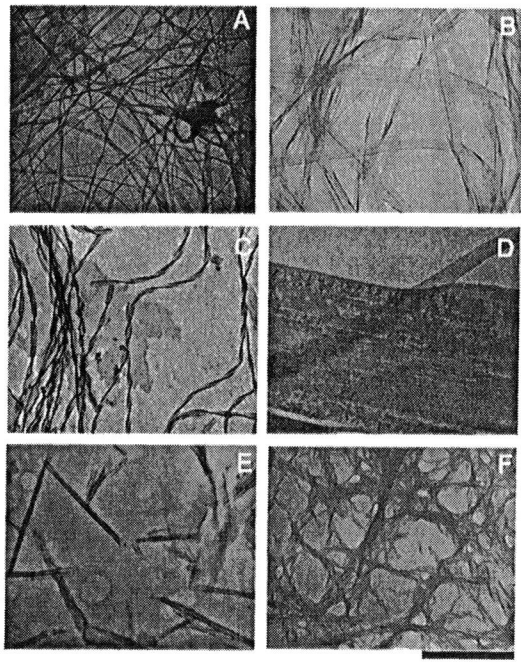

Figure 2. TEM images of microstructures made from amino acid terminated diacetylene lipids. (A) L-Glu-PDA, (B) L-Gln-PDA, (C) L-His-PDA, (D) Gly-PDA, (E) L-Ser-PDA and (F) L-Ile-PDA. The bar is 0.6 μm for (A), (B) and (F); 0.8 μm for (C) & (D); 2 μm for (E).

several to hundreds of microns. Under UV irradiation, L-Glu-PDA microstructures readily polymerize to give a dark blue color. L-Gln-PDA lipid forms similar ribbon shaped microstructures (Fig. 2B). However, the ribbon assemblies are more uniform in size. The typical ribbon width and thickness is around 150 nm and 8 nm, respectively. For L-His-PDA, the formation of extended (up to several microns) helical assemblies was readily observed (Fig.

2C), with right-handed helical twists averaging 60 nm in diameter. A large amount of planar platelets coexist with the helices. To verify the headgroup chirality effect on helical microstructure formation, achiral Gly-PDA lipid was used as a negative control. As expected, large amount of flat sheets and platelets without apparent twisting or curvature was obtained (Fig. 2D). L-Ser-PDA forms open tubular assemblies (Fig. 2E). The average diameter of the tubules is around 0.2 μm, with an estimated tubular wall thickness around 20 nm. The coexistence of a significant amount of sheets in the sample, especially the layers wrapped around the open end of the tubules, suggests that the formation of L-Ser-PDA tubules is through a rolling up mechanism (*18*). L-Ile-PDA lipids, on the other hand, form networks of highly twisted braided ribbons (Fig. 2F). It is worth mentioning that twisted ribbons are the exclusive morphology observed with the L-Ile-PDA assembly.

The relationship between molecular structure and the preferential morphology of a supramolecular assembly is poorly understood. Extensive experimental studies indicated that chirality, conformation and hydrogen bonding forming ability of the polar headgroup of a lipid amphiphile are determining factors. For all the microstructures studied here, hydrogen bonding exists extensively, with an especially high degree in L-Glu-PDA assembly. The TEM results here seem to concur that headgroup chirality and the balance of headgroup size, polarity, and the degree of favorable interactions (e.g. H-bonding), in addition to the balance of dipolar forces (*19*), are critical to the rise of lipid bilayer curvatures.

Colorimetric properties of amphiphilic PDA microstructures

The polymerization of the organized diacetylene lipid assemblies is a topochemical process, requiring optimal packing of the diacetylenic segments to allow propagation of an extended ene-yne conjugation backbone. The conjugated polymer absorbs light strongly around 650 nm, giving the material a blue appearance. Considerable investigations have been recently reported on the chromism of PDA bilayer vesicles and LB thin films (*10*), in the light of using PDA assemblies for colorimetric sensors. Diacetylene lipid microstructures are structurally similar to their bilayer or monolayer counterparts and possess the intrinsic features as organized assemblies that should allow topochemical reaction and polymerization.

It is worth noting that all the amino acid-diacetylene lipid microstructures studied here could be polymerizable to form blue colored PDAs. However, only hydrophilic amino acid lipids can readily form bilayer vesicles and allow polymerization (*11*). The intensity of the initial blue color, however, varies with headgroups. Amino acids with hydrophilic segments give the darkest blue appearance, while hydrophobic amino acids (Ile-) produce barely noticeable blue appearance.

The colorimetric properties of amino acid terminated PDAs were investigated by thermochromism and solution pH induced chromism. Quantitative analysis of chromatic transition was conducted by analyzing the colorimetric response (CR) as a function of solution pH (*16*). The CR is defined as the percent change in the maximum adsorption at 646 nm with respect to the total absorption at 542 nm and 646 nm. Figure 3 shows the CR vs. pH for the amino acid terminated PDA microstructures. Sigmoidal curves were obtained

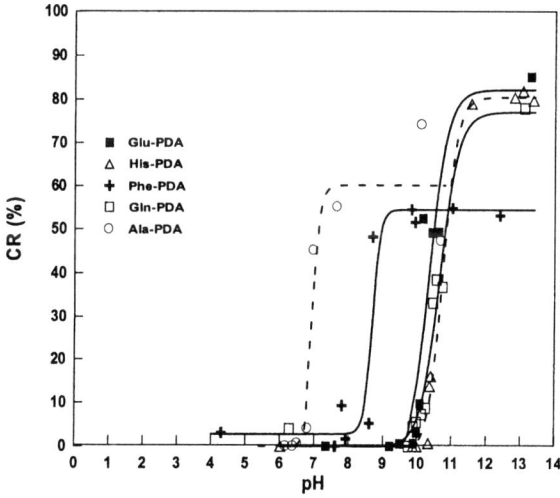

Figure 3. Colorimetric response for the polymerized microstructures as a function of solution pH.

for all microstructures studied, indicating sharp transitions from blue to red color upon pH increase. The transition point is defined by the CR_{50} values, namely the pH required to achieve 50% of the maximal color transition (*11*). The CR_{50} values for "hydrophilic" amino acids (Glu, Gln and His) all fall between pH 10 to 11. For comparison, response curves for two "hydrophobic" amino acid lipids, Phe-PDA and Ala-PDA, are shown in Figure 3. The values for Ala and Phe are much lower (7.1 and 8.6, respectively). The "base-resistant" nature for "hydrophilic" amino acids differs from that obtained with the amino acid terminated PDA bilayer vesicles, where L-Glu-PDA was found to be the most base-sensitive. The color change of PDA microstructures can also be achieved by thermal treatment (thermochromism), as well documented in literature. Similar sigmoidal curves were obtained for amino acid terminated PDA lipids where L-Gln-PDA microstructure is the most heat resistant. A temperature as high as 71°C is needed for the assembly to convert 50% of its color. Thermochromism of bilayer vesicles formed by amino acid terminated PDA lipids was also studied. Contrary to microstructures, the trend for vesicles seems totally reversed. L-Gln-PDA vesicles are the most thermal sensitive while L-Glu-PDA is the least.

It has to be pointed out that thermochromism and pH induced chromism of PDA microstructures appear to proceed through different mechanisms. The temperature-induced side chain conformation transition is responsible for the thermochromism of the PDA microstructures. For pH induced chromatic transition, the headgroup undergoes significant reorganization as a result of ionization, causing a new conformational adjustment (staggered packing) that thereby imposes strains to the backbone. This has been further confirmed by FTIR studies on PDA microstructures (*12*).

II. Bolaamphiphilic Polydiacetylenes

Amino acid-terminated bolaamphiphilic diacetylene lipid

The formation of a robust supramolecular assembly can be achieved through the deliberate installation of various functionalities throughout the molecular architecture that enforce the intermolecular association between assembling units. We designed an *L*-glutamic acid derivatized wedge-shaped bolaamphiphilic diacetylene lipid *L*-Glu-Bis-3 (structure seen below) as the

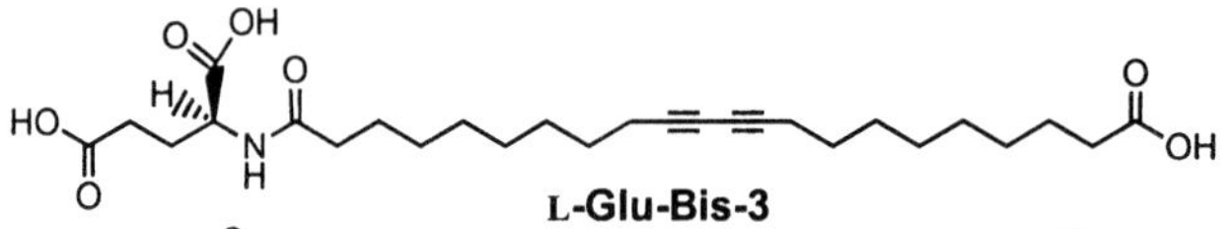

L-Glu-Bis-3

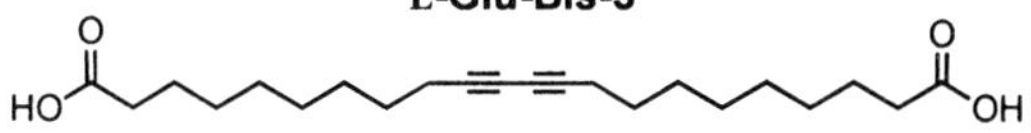

Bis-1

self-assembling unit of a highly organized molecular architecture. Compared to their amphiphilic lipid counterparts, bolaamphiphiles tend to form well-organized systems under very mild conditions. They mimic transmembranic lipids that some microorganisms synthesize for stabilizing membrane structures in response to extreme pH and temperature (*20*). *L*-Glutamic acid residue attached to one end of 10,12-docosadiynedioic acid (Bis-1), along with the free carboxylate on the other end of the lipid, is designed to enhance favorable H-bonding interactions on the polar faces of the assembly. The diacetylene unit was placed at the center of the molecule to maximize the chance of proper alignment of polymerization units in different packing arrangements.

The synthesis of *L*-Glu-Bis-3 was reported elsewhere (*13*). One terminal of Bis-1 was activated with N-hydroxysuccinimide before it was coupled with *L*-glutamic acid through an amide linkage, giving an overall 61% yield. Alternatively, activation of both carboxylate groups before the attachment of a glutamate residue and the hydrolysis of the unreacted ester terminal could lead to an improved overall yield.

The self-assembling of L-Glu-Bis-3 occurred rapidly under mild conditions. Instead of probe sonication and subsequent low temperature incubation that are commonly required for amphiphilic lipids, vortexing and room temperature incubation was sufficient to ensure the formation of a stable for L-Glu-Bis-3 supramolecular assembly in aqueous media. UV-irradiation of the assembled material resulted in instantaneous polymerization of L-Glu-Bis-3, affording the material an intense blue appearance. The rapid polymerization indicates a highly ordered packing arrangement and the good alignment of diacetylene units.

Morphology and surface packing arrangement of bolaamphiphilic PDA

The morphology and surface packing arrangement of the polymer was characterized by transmission electron microscopy (TEM) and atomic force microscopy (AFM). TEM micrographs revealed the formation of ribbons tens of microns long (Figure 4). These ribbons are either flat or twisted with various

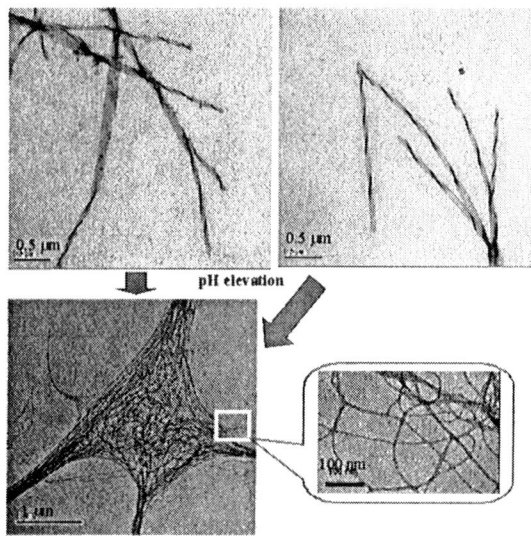

Figure 4. TEM images of poly-L-Glu-Bis-3 microstructures. The top images were obtained at pH 5.8 (D.I. water) showing rupture that is indicative of the origination of helices (top, left) and right-handedness of the twist (top, right). Lower images show structural transformation into nanofibers upon treatment of above microstructures with pH 7.5 Tris buffer.

degrees of right-handed helicity, consistent with the observation of helical ribbons formed by L-amino acid terminated amphiphilic PDAs. Strips of parallel domains were clearly visible on wider ribbons, apparently parallel with the propagation of the polymer backbone. These ribbons are 5 to 10 nm thick, corresponding to either a monolayer or a double layer lipid stacking. The widths of the ribbons vary from tens to hundreds of nanometers, with generally wider dimension for flat structures.

Among the many experimental discussions (21,22) and theoretical treatments (23,24) aiming at the explanation of tubular or helical lipid assembly formation, the chiral packing theory has been the cornerstone. It has been postulated that when bilayer chiral lipid amphiphiles aggregate, they first form wide sheets with sharply separated domains (21), which would then break up along the domain edge to form narrower ribbons that are free to twist into helices, driven by chiral packing effect. Helical ribbons may further fuse into tubular structures to reduce edge energy. Our TEM data provides direct evidence to support this theory in the context of chiral bolaamphiphiles. The micrographs shown in Figure 4 captured the initiation of the transition from flat strips to helical ribbons through rupturing of wider flat ribbons along the parallel domain edges. The narrower strips could then continue to twist into helical structures as a result of chiral bolaamphiphiles' cumulative tilt away from the local surface normal. Formation of tubular structures, as observed at certain regions, is evidence of further winding of the helical ribbons to reduce the edge energy.

Contact mode AFM was used to characterize the surface packing of bolaamphiphilic PDA ribbons on atomic level. The 2-D fast Fourier Transformation (2-D FFT) of scans over a flat ribbon surface suggests the formation of highly compact hexagonal packing arrangement of the polymer, with an approximate cell area of 20 $Å^2$, which is characteristic for tightly packed hydrocarbon chains. In contrast, earlier thermochromic studies on monofunctional PDA films using AFM showed that pseudo-rectangular packing arrangement was predominantly observed at room temperature for the blue phase film even when it was over-compressed during the preparation (25). Our results demonstrate that the bolaamphiphilic lipid is able to form more stable and better-organized assemblies at ambient conditions. However, using this technique the distinction of the terminal carboxylate on the glutamate end from the one on the single carboxylate end would be difficult.

pH-induced optical and structural transformation in bolaamphiphilic PDA

A sharp blue-to-red color change was observed with Poly-L-Glu-Bis-3 upon the increase of pH. As expected, the existence of multiple base-sensitive carboxylic acid residues in the molecule resulted in a chromatic transition at a lower pH region compared to the poly-L-Glu-PDA assemblies discussed earlier. At pH 7.5, the blue polymer turned completely red as a result of significantly shortened conjugation length induced by the side chain disorder arising from increased electrostatic repulsion between deprotonated surface carboxylates.

Dramatic morphological changes accompanied the pH-induced colorimetric response of Poly-*L*-Glu-Bis-3 (Fig. 4). Extended ribbons were frayed into oriented nanofibers less than 10 nm in diameter upon the increase of pH. By exposing the polymer to more basic conditions for a longer time, more randomly coiled fibers were obtained. Figure 5 illustrates a proposed model of the transformation. Increased surface electrostatic repulsion upon the addition of base disrupts favorable H-bonding networks at the polar surface and overcomes

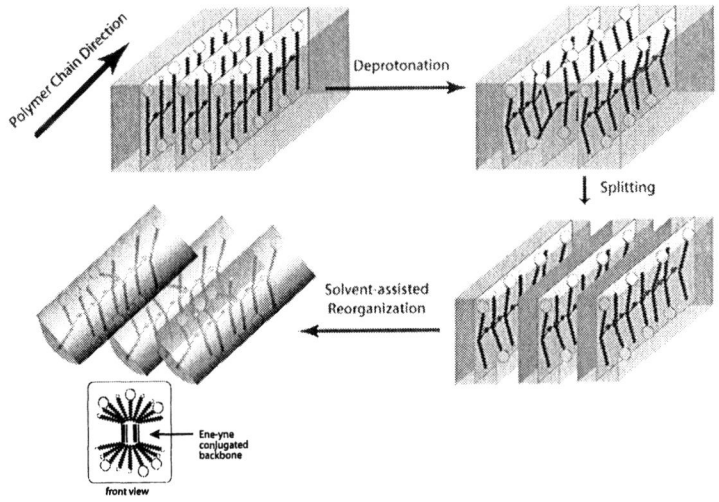

Figure 5. A cartoon illustration of pH-triggered morphological transformation of poly-L-Glu-Bis-3 from ribbons to nanofibers.

the attractive hydrophobic interactions between lipid cores, effectively splitting closely packed polymer chains into aligned fibers. The pH induced morphological transformation also reaffirms the linear propagation as the predominant format of polymerization of diacetylene units.

III. Lipid Doping-Induced Structural Transformation in Bolaamphiphilic PDA Assemblies

One fundamental consideration in designing biosensors is to establish an effective signal transduction pathway upon the interaction of incorporated receptors with analytes at the detection interface. For PDA-based colorimetric biosensors, this requirement transforms into a balance between the rigidity (which determines the extent of polymerization as well as the signal transduction efficiency) and flexibility (which is necessary for effective binding of analytes with surface receptor as well as the lowering of transition energy barrier for the conformational change of the conjugation backbone) of the sensor scaffold. To strike a balance in the fluidity of the bolaamphiphilic PDA sensor scaffold, controlled doping with either the receptor as the only additive or by incorporating additional lipid dopants would be necessary. The diverse chemical structures of many naturally occurring lipids provide abundant possibilities to fine-tune the fluidity and morphological properties of bolaamphiphilic PDA-based biosensors.

Specifically, we are investigating the effect of lipid doping on the microstructural morphology of L-Glu-Bis-3 assemblies with the addition of G_{M1} ganglioside (structure shown below), a known receptor of cholera toxin, and/or cholesterol. Gangliosides are a family of glycosphingolipids localized to the outer leaflet of the plasma membrane of vertebrate cells. When inserted into artificial membranes, the oligosaccharide motif of gangliosides is exposed at the membrane surface and functions as a recognition group for a number of bacterial toxins. Polycyclic cholesterol is known to directly participate in the formation of lipid rafts with glycosphingolipids. Both lipids have been shown to modulate domain structure and phase separation in model membrane systems (26).

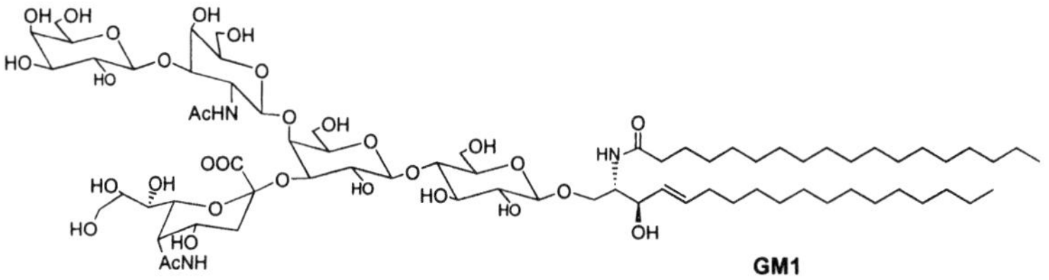

Ribbon-to-vesicle microstructural transformation

When 5% G_{M1} ganglioside was introduced into the L-Glu-Bis-3 system, vesicles were formed along with ribbons (Fig. 6B). A significant number of vesicles, varied from less than 100 nm to greater than 500 nm in diameter, appeared to be attached to the ribbon structures, typically at the junction of several entangled ribbons. Incorporation of cholesterol at a low concentration (5%) along with G_{M1} led to the formation of an aggregate that allowed the development of a uniform blue color upon UV irradiation. However, the ternary system with high cholesterol content (20%) only led to the formation of turbid suspensions even after prolonged vortexing or probe sonication, leaving its photo-polymerizability at a minimum. TEM micrographs revealed that with increased cholesterol content, more vesicles were formed with continued coexistence of the ribbon structures (Fig. 6A). Apparently, addition of cholesterol further facilitates and stabilizes the formation of vesicles.

Cholesterol has long been known to stabilize membrane structures. Sphingolipids were thought to associate laterally with one another through interactions between their headgroups, whereas cholesterol molecules function as spacers, filling the voids at the hydrophobic regions between associating sphingolipids. Such preferential packing was believed to lead to the formation

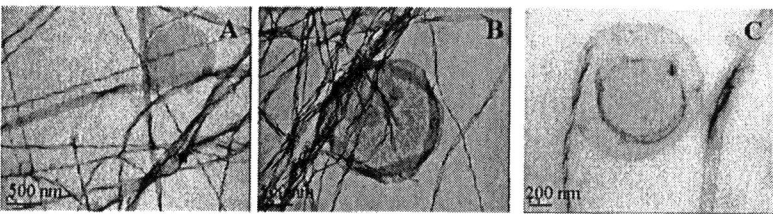

*Figure 6. Transmission electron micrographs of L-Glu-Bis-3 doped with GM1 and cholesterol. (A) **L-Glu-Bis-3** doped with 5% G_{M1} and 5% cholesterol; (B) doped with 5% G_{M1}; (C) doped with 5% G_{M1} and 10% cholesterol. (Reproduced with permission from reference 14. Copyright 2002 Elsevier.)*

of rafts within the membrane bilayers. In the three-component systems, we speculate that cholesterol molecules are inserted in the outer-surface of the vesicles, filling the voids at the hydrophobic region between aggregated gangliosides and membrane spanning lipids. Given the fact that none of G_{M1} ganglioside, cholesterol, or L-Glu-Bis-3 assembles by themselves to form vesicles, it is apparent that inserting the proper dopants between membrane spanning lipids is essential to inducing surface curvature and vesicle formation.

Some intriguing morphological details of doped L-Glu-Bis-3 assemblies captured by TEM (Fig. 6B,C) provided an opportunity to examine intermediate states of microstructural transformation between ribbons and vesicles. These micrographs clearly suggest that ribbons (relatively rigid) and vesicles (relatively fluid) were physically interrelated during the formation of different microstructures in these multi-component systems.

Morphological details of interconnected microstructures shown in Figures 6B (doped with 5% G_{M1}) and 6C (doped with 5% G_{M1} and 10% cholesterol), where the edges of vesicles or vesicle domains were outlined in the shape of ribbons, suggest a vesicle-to-ribbon transition mechanism at the periphery of vesicles. The growth of a ribbon and its extension away from a vesicular microstructure is most clearly seen in the image shown in Figure 6C. A vesicle-to-ribbon transition is a probable process during domain reorganizations within less crosslinked and more fluid vesicles. Lateral reorganization of lipids within these areas may have resulted in phases or domains with particularly low unpolymerizable dopant concentrations, thus a higher continuity of chirally packed matrix lipid L-Glu-Bis-3.

IV. Conclusions

Optically pure amino acid-terminated amphiphilic and bolaamphiphilic diacetylene lipids have been synthesized and assembled into tubes or ribbons with various dimensions and right-handed helicity. These microstructures can be photo-polymerized upon UV irradiation to form conjugated polymers with retained morphology and intense blue color, suggesting intrinsic highly ordered lipid packing arrangement and good alignment of diacetylene units that allow for topochemical polymerization. These conjugated polymer ribbons and tubes respond to external stimuli such as pH and heat via characteristic blue-to-red color change that is commonly observed with conventional thin film or vesicular PDA assemblies.

Careful chemical modifications made throughout the diacetylene lipid, including the installation of particular polar headgroup at either one end or both ends of the lipid, the selection of headgroup size, the manipulation of surface charge density as well as the positioning of the polymerization unit allow for the optimization of favorable inter-lipid interactions (e.g. H-bonding and *van der Waals* interaction) and the balance of dipolar forces that are critical to the formation of highly ordered supramolecular assemblies with unique microstructural morphology. Such rational design also brings control over the type and extent of optical and microstructural transitions of the material in response to specific external perturbation. In the case of Poly-*L*-Glu-Bis-3, the conjugated polymer responds to pH increase with a sharp colorimetric response as well as dramatic morphological changes from helical ribbons to aligned nanofibers.

In addition to physical environmental cues such as heat and pH, the incorporation of different lipid dopants into the bolaamphiphilic diacetylene assemblies also effectively induces microstructural transformations. Specifically, controlled doping of *L*-Glu-Bis3 with naturally occurring glycosphigolipid G_{M1} and cholesterol, which are relevant receptors for a number of potential biosensors, triggers the formation of fluid vesicles along with more crystalline ribbons. The multi-component system also allows for the observation of unique phase separation and microstructural transformaiton intermediates, resulting from dynamic clustering of the unpolymerizable lipid dopants and the reorganization of the polymerizable lipids in a relatively fluid environment.

Lessons learned from these studies provide valuable guidance to the rational design of future generations of PDA-based colorimetric sensors as well as other advanced nanomachinery where the microscopic morphology and optical properties of the material are crucial to its function.

References:
(1) Leclerc, M. *Adv. Mater.* **1999**, *11*, 1491-1498.

(2) Englebienne, P. *J. Mater. Chem.* **1999**, *9*, 1043-1054.
(3) McQuade, D. T.; Pullen, A. E.; Swager, T. M. *Chem. Rev.* **2000**, *100*, 2537-2574.
(4) Huo, Q.; Russell, K. C.; Leblanc, R. M. *Langmuir* **1999**, *15*, 3972-3980.
(5) Kuriyama, K.; Kikuchi, H.; Kajiyama, T. *Langmuir* **1998**, *14*, 1130-1138.
(6) Foley, J. L.; Li, L.; Sandman, D. J.; Vela, M. J.; Foxman, B. M.; Albro, R.; Eckhardt, C. J. *J. Am. Chem. Soc.* **1999**, *121*, 7262-7263.
(7) Charych, D.; Nagy, J. O. *Chemtech* **1996**, *26*, 24-28.
(8) Jelinek, R.; Kolusheva, S. *Biotech. Adv.* **2001**, *19*, 109-118.
(9) Song, J.; Cheng, Q.; Zhu, S. M.; Stevens, R. C. *Biomed. Microdevices* **2002**, *4*, 213-221.
(10) Okada, S.; Peng, S.; Spevak, W.; Charych, D. *Acc. Chem. Res.* **1998**, *31*, 229-239.
(11) Cheng, Q.; Stevens, R. C. *Langmuir* **1998**, *14*, 1974-1976.
(12) Cheng, Q.; Yamamoto, M.; Stevens, R. C. *Langmuir* **2000**, *16*, 5333-5342.
(13) Song, J.; Cheng, Q.; Kopta, S.; Stevens, R. C. *J. Am. Chem. Soc.* **2001**, *123*, 3205-3213.
(14) Song, J.; Cheng, Q.; Stevens, R. C. *Chem. Phys. Lipids* **2002**, *114*, 203-214.
(15) Charych, D. H.; Nagy, J. O.; Spevak, W.; Bednarski, M. D. *Science* **1993**, *261*, 585-588.
(16) Charych, D.; Cheng, Q.; Reichert, A.; Kuziemko, G.; Stroh, M.; Nagy, J. O.; Spevak, W.; Stevens, R. C. *Chem. Biol.* **1996**, *3*, 113-120.
(17) Spevak, W.; Nagy, J. O.; Charych, D. H.; Schaefer, M. E.; Gilbert, J. H.; Bednarski, M. D. *J. Am. Chem. Soc.* **1993**, *115*, 1146-1147.
(18) Fuhrhop, J.-H.; Schnieder, P.; Boekema, E.; Helfrich, W. *J. Am. Chem. Soc.* **1988**, *110*, 2861-2867.
(19) Frankel, D. A.; O'Brien, D. F. *J. Am. Chem. Soc.* **1994**, *116*, 10057-10069.
(20) Jung, S.; Zeikus, J. G.; Hollingsworth, R. I. *J. Lipid Res.* **1994**, *35*, 1057-1065.
(21) Schnur, J. M. *Science* **1993**, *262*, 1669-1676.
(22) Fuhrhop, J.-H.; Schnieder, P.; Rosenberg, J.; Boekema, E. *J. Am. Chem. Soc.* **1987**, *1987*, 3387-3390.
(23) Selinger, J. V.; MacKintosh, F. C.; Schnur, J. M. *Phys. Rev. E* **1996**, *53*, 3804-3818.
(24) Thomas, B. N.; Lindermann, C. M.; Clark, N. A. *Phys. Rev. E* **1999**, *59*, 3040-3047.
(25) Lio, A.; Reichert, A.; Ahn, D. J.; Nagy, J. O.; Salmeron, M.; Charych, D. *Langmuir* **1997**, *13*, 6524-6532.
(26) Hwang, J.; Tamm, L. K.; Bohm, C.; Ramalingam, T. S.; Betzig, E.; Edidin, M. *Science* **1995**, *270*, 610-614.

Chapter 8

Chromogenic Polymer Gels for Reversible Transparency and Color Control

Arno Seeboth, Jörg Kriwanek, André Patzak, and Detlef Lötzsch

Fraunhofer Institute for Applied Polymer Research, Richard-Willstätter-Strasse 12, 12489 Berlin, Germany

A current overview about the preparation and characterization of novel chromogenic polymer materials is given. The topic of chromogenic materials has developed extremely rapidly in the last few years. Among them, thermotropic and thermochromic polymer gel networks have met with growing interest, because of their advanced properties. These novel polymer gels exhibit pronounced changes in transparency and/or color over a modest temperature range. Some of them possess a temperature independent volume, which is an essential condition for many future technical applications.

Introduction

The preparation and characterization of chromogenic polymer gels changing their optical properties in response to temperature has met with growing interest in the last decade. Two classes of such gels are generally distinguished in literature: thermotropic polymer gels, which switch in dependence on temperature reversible between a transparent and a translucent state and thermochromic polymer gels which change their color or color strength upon changes in temperature. However, so-called chromogenic polymer gels combining thermotropic and thermochromic properties in one material have been developed.

From the practical point of view, temperature sensitive chromogenic polymer gels seem to be promising not only as temperature indicators, but also for use in smart windows, temperature tunable light filters and large area displays. Especially their potential application as thermally self-adjusting light and heat filters in the external glazings of buildings has motivated their development in this field. For most of the desired applications a constant volume of the chromogenic material is required. Thus gels whose optical switching is accompanied by considerable shrinking or swelling will not be considered in this chapter.

Thermotropic Gel Networks for Reversible Transparency Control with Temperature

Thermotropic gel networks transform at a certain temperature from a highly transparent into a light scattering state. Such an optical effect can be either caused by a phase separation process or by a phase transition between an isotropic and an anisotropic lyotropic liquid crystalline state.

With the presentation of advanced prototypes of sun-protective glazing by Charoudi (*1*), Watanabe (*2*) and Seeboth (*3*) the interest in thermotropic gel networks has rapidly increased in recent years. A detailed description of the material development in this field was recently reviewed (*4*). In this chapter we will focus on new results on the preparation of thermotropic hydrogels based on biopolymers, the usefulness of calorimetric measurements for the characterization of thermotropic gel networks, the influence of the addition of salts on the material properties and the construction of hybrid solar and electrically controlled light filters.

Preparation of Thermotropic Polymer Gel Networks Based on Biopolymers

For commercial use of hydrogels, non-toxic and inexpensive raw materials are required. Biopolymers, like polysaccharides, offer the possibility to fulfil all these requirements and, compared with synthetic polymers, they have the benefit of their environmental compatibility. Therefore, the suitability of biopolymers for the preparation of novel thermotropic hydrogels was extensively studied in recent years. Thermotropic hydrogels based on cellulose derivatives in combination with an amphiphilic component were developed by Watanabe (2). By using one of these thermotropic hydrogels a window of 1m^2 size was constructed and successfully tested under practical conditions over a period of two years. Thermochromic hydrogels, which are also based on cellulose derivatives have been reported (5). It was shown, that no amphiphilic component is necessary to prevent an irreversible flocculation of suspended hydroxy propylcellulose, if hydroxy ethylcellulose is added. Although the biological decomposition is often mentioned as an advantage of biopolymers, it is also the most important hindrance for their commercial use because contact of the biopolymers with microorganisms must be prevented during production and throughout the lifetime of the product.

Calorimetric Measurements on Aqueous Polymer Gel Networks

In order to characterize the properties of polymer gel networks optical and dielectric spectroscopy, rheological investigations and calorimetric measurements are commonly used. Calorimetric measurements are well known to provide information on the water binding properties of the polymeric systems. Recent published results show that differential scanning calorimetry (DSC) is also suitable to detect phase transitions and phase separation processes of hydrogels (6-8). Whereas optical techniques can only determine phase transition temperatures, DSC measurements also provide the transition enthalpy data.

Alexandridis et al. investigated the influence of the addition of salts on the phase separation and on the unimer-to-micelle transition temperatures in aqueous solutions of a poly(ethylene oxide)-block-poly(propylene oxide)-block-poly(ethylene oxide) copolymer (6). In this system the detection of both transitions by optical techniques is hindered by the presence of a hydrophobic impurity which causes turbidity of the solutions even below their respective phase separation temperatures. Therefore, DSC measurements were employed to detect the phase transitions. Moreover, the enthalpy data obtained for the unimer-to-micelle transition were used to calculate the unimer concentration above the critical micellization temperature.

The influence of the composition on morphology and phase transition temperatures of a polyalkoxide/water/LiCl system has been reported (7). For the determination of the phase transition temperatures a combination of DSC measurements and optical techniques was used, whereby both methods were found to give similar results. With increasing water content the formation of an anisotropic phase as well as the occurrence of two different phase separation processes takes place in the investigated polyalkoxide/water system. In Figure 1 the DSC curves of three polyalkoxide/water mixtures, with mixing ratios of 5:1, 4:1 and 3:1 by mass are shown. The appearance of two additional DSC peaks at a polyalkoxide/water mixing ratio of 3:1 clearly displays the change of the polymorphy in dependence on the water content. In all three mixtures a broad DSC peak can be seen, which corresponds to a separation of a water phase. Whereas optical techniques can only detect the beginning of this process, the DSC measurements show that the phase separation takes place over a wide temperature range.

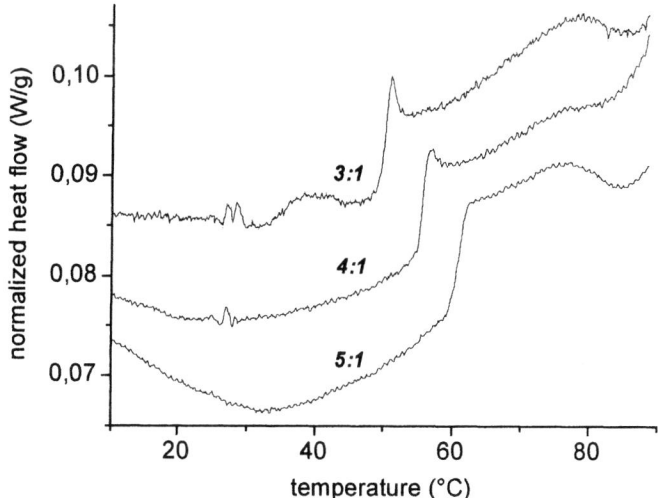

Figure 1. Differential scanning calorimetry curves of three different polyalkoxide/water samples with mixing ratios of 5:1, 4:1 and 3:1 by mass.

The determination of phase separation temperatures of aqueous polymer solutions by a combination of DSC measurements and optical techniques is also reported in reference (8). Again, the DSC data were found to be in good agreement with the transition temperatures obtained by optical techniques.

In summary, the results discussed prove the usefulness of DSC measurements for the determination of the thermodynamic properties of aqueous polymeric systems.

Incorporation of Salts into Aqueous Polymer Systems

It is well known that salts can strongly influence the material properties of aqueous polymer systems. Their addition often leads to a physical cross-linking of the polymer chains, whereby three-dimensional networks are built up. Systematic investigations of the effect of different salts on polymorphy, phase transition temperatures, water binding capability and macroscopic properties of aqueous polymeric systems have been investigated. At a given concentration different salts were found to shift the phase separation temperatures of thermotropic aqueous polymer systems according to their "salting-in" or "salting-out" strength (*6, 9, 10*), which is described by the so-called Hofmeister series. Salts with a salting-in phenomena cause an increase of the phase separation temperature, while salts with a salting-out phenomena have the opposite effect. The influence of the addition of LiCl on the water binding properties of an aqueous polyalkoxide system was investigated (*7*). Proportional to the LiCl content an increase of the non-freezing-bound-water capacity of the polyalkoxide and an increase of the binding enthalpy of freezing-bound-water were observed. Both results indicate that the interaction between water and polymeric system becomes stronger with increasing LiCl content.

Hybrid Solar and Electrically Controlled Light Filters

A hybrid solar and electrically controlled transmission changing light filter based on thermotropic hydrogels was recently described (*11*). A 2-3 mm thick thermotropic hydrogel layer was placed between two indium tin oxide (ITO) coated glass-substrates, whereby the ITO layers were placed either inside or outside the double glazing item. Such an arrangement can be switched on demand either passive by solar energy or active by electrical energy through heating of the ITO layers. It was shown that an increase of the ITO layer thickness reduces the required wattage to achieve the same switching time. On the other hand the transparency of the glazing item is also reduced. For one of the investigated glazing items a transmission change from about 62% to $\leq 1\%$ and a switching time of 5 min was achieved by applying a wattage of 0.246 W/cm^2.

It can be expected that a further optimization of all components of the glazing item and especially of the layer thickness as well as the material properties of the incorporated thermotropic hydrogel will lead to a further significant reduction of the required wattage.

Thermochromic Gel Networks for Reversible Color Control with Temperature

Polymer gel networks with thermochromic properties are only rarely described in the literature, whereby most of the systems are composed of a polymer and an organic solvent. Fujimatsu et al. reported that a gel consisting of poly(1-butene) and tetrachloroethylene exhibits reversible color changes with temperature between the melting point of the solvent and the sol-gel transition temperature (*12*). The thermochromism of this gel is caused by light scattering effects. A similar behavior was observed for a gel composed of 3 wt.% isotactic polypropylene in benzene which is blue at room temperature and becomes yellow at about 70-80°C (*13*). An example of thermochromism at the gel-sol transition was found in polydiacetylene gelled in o-dichlorobenzene or other gel forming solvents (*14*). These transparent gels show, upon heating above their respective gel-sol transition temperatures, pronounced reversible color changes. Colors and transition temperatures could be varied by using different solvents as well as by changing the composition. Another type of thermochromic effect in a gel was reported by Gelinck et al. (15). After a few days in a refrigerator a poly(2-(3,7-dimethyloctoxy)-5-methoxy-1,4-phenylenevinylene) / benzene system forms a clear red gel phase. On heating this gel shows at approximately 35°C a gradually red to yellow color change which is not fully reversible with temperature. The authors explained the thermochromic behavior of the gel by a reduction of interchain π-π interaction with increasing temperature.

An example of a thermochromic hydrogel system was reported by Asher et al. (*16*). By embedding a crystalline colloidal array of polystyrene spheres in a poly(N-isopropyl-acrylamid)-hydrogel, which swells or shrinks depending on temperature, tunable thermochromic hydrogel films were created. As long as the layer thickness of the films is below 500 µm they are transparent. Above 500 µm the films become translucent, because of light scattering on the colloidal particles. The color effect in this system is caused by Bragg reflection from the periodic structure of the crystalline colloidal array. A variation of the temperature leads to volume changes of the hydrogel matrix; the crystalline colloidal array of polystyrene spheres follows, changing the lattice spacing and thus the diffracted wavelength. A 125 µm thick hydrogel film was presented which changes the diffracted wavelength continuously from 704 nm at 11.7°C to 460 nm at 34.9°C.

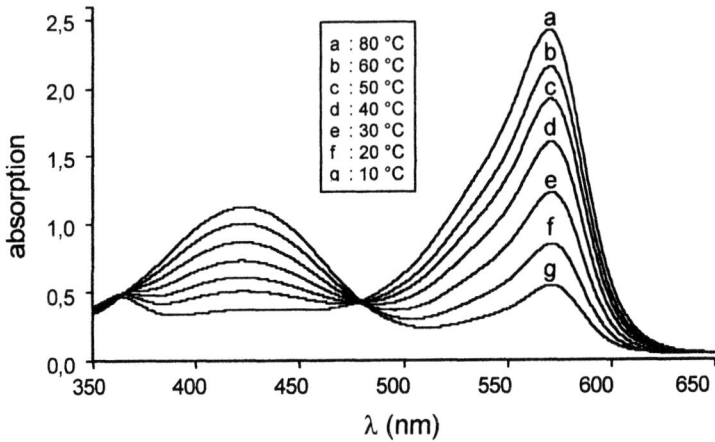

Figure 2. UV/Vis absorption spectra of Phenol Red in a (PVA)/borax/surfactant gel network at different temperatures.

The first report of a thermochromic effect of dyes embedded in a transparent hydrogel was given by Seeboth et al. (*17*). The gel network described in this paper is composed of a definite polyvinyl alcohol (PVA)/borax/surfactant mixture doped with suitable pH-sensitive indicator dyes with pKa-values between 7.0 and 9.4. With increasing temperature the phenol-phenolate equilibrium of the indicator dyes was found to be shifted in the hydrogel matrix towards their deprotonated phenolate form. Depending on the indicator dye(s) used switching between a colorless and a colored state or between two or even more different colored states was observed. For example, by using the so-called Reichard betaine dye 2,6-diphenyl-4-(2,4,6-triphenyl-1-pyridinio)phenolate (DTPP) a hydrogel was obtained which changes color gradually from colorless at 10°C to deep violet at 80°C. A Phenol Red containing PVA/borax/surfactant hydrogel on the other hand switches from yellow at 10°C to wine-red at 80°C and a Bromothymol Blue and Cresol Red containing PVA/borax/surfactant hydrogel from yellow below 5°C to green between about 15-25°C and further to violet above about 60°C. As an example the UV/Vis absorption spectra of the Phenol Red containing hydrogel at different temperatures are shown in Figure 2. With increasing temperature the absorption band at λ_{max} = 424 nm decreases while simultaneously the absorption band at λ_{max} = 570 nm increases. All spectra meet at an isosbestic point at λ = 479 nm, supporting thereby the suggested model of a temperature dependent equilibrium between two different forms of the indicator dye. An important advantage of these hydrogels is that the thermochromic effect is not accompanied by a shrinking or swelling of the gel networks. Therefore these materials are suitable not only as temperature indicators but also for the construction of smart windows, large area displays and tunable color filters.

Chromogenic Gel Networks for Reversible Transparency and Color Control with Temperature

A novel class of polymer gels changing both their color and transparency reversibly with temperature was recently reported (*18*). These gels were obtained by adding suitable pH-sensitive indicator dyes to a thermotropic hydrogel consisting of a polyalkoxide and an aqueous LiCl containing buffer solution. The thermochromic effect of these hydrogels was attributed to temperature induced pH-changes in the gel network and the thermotropic effect to a phase separation process, which is typical for such polyalkoxide/LiCl/water gel systems. It was found that the addition of indicator dyes only slightly influences the thermotropic properties of the hydrogel matrix.

As an example the temperature dependence of UV/Vis absorption spectra and transparency of a Bromothymol Blue containing sample are shown in Figures 3 and 4, respectively. In the temperature range from –5°C to about 33°C the hydrogel is green and highly transparent. Above about 33°C a color change from green to yellow takes place whereby the hydrogel remains transparent. On further heating the transparency is reduced and above about 36°C the hydrogel becomes yellow translucent.

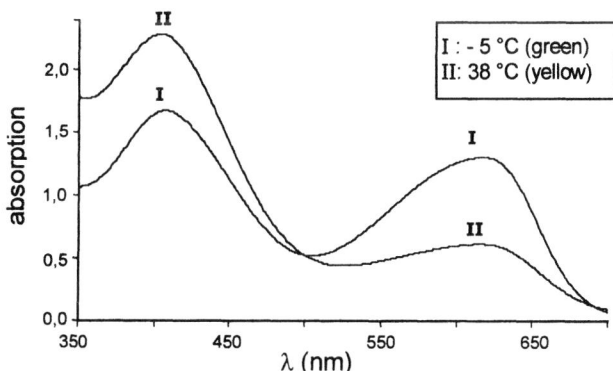

Figure 3. UV/Vis absorption spectra of Bromothymol Blue in a polyalkoxide/ salt/water gel network at two different temperatures.

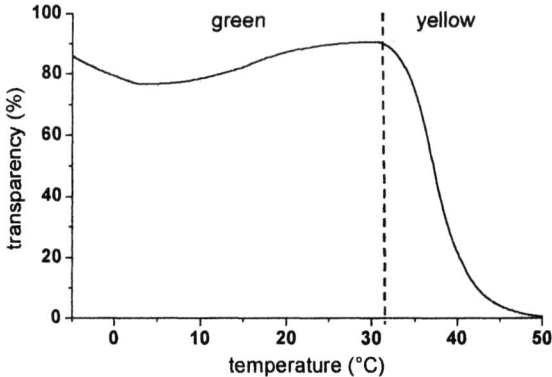

Figure 4. Temperature dependence of the transparency of a Bromothymol Blue containing polyalkoxide/salt /water gel network.

With increasing temperature the phenol-phenolate equilibrium of Bromothymol Blue is shifted in the polyalkoxide/salt/water system from the green colored phenolate form to the yellow colored phenol form. This result is in contrast to the behavior of phenol substituted indicator dyes in a PVA/borax/surfactant gel network for which the opposite shift of the phenol-phenolate equilibrium with temperature was observed. Obviously, the reversible color change of indicator dyes in gel systems depends on the specific composition of the gel networks. However, the origin of this effect on a molecular level is still under discussion.

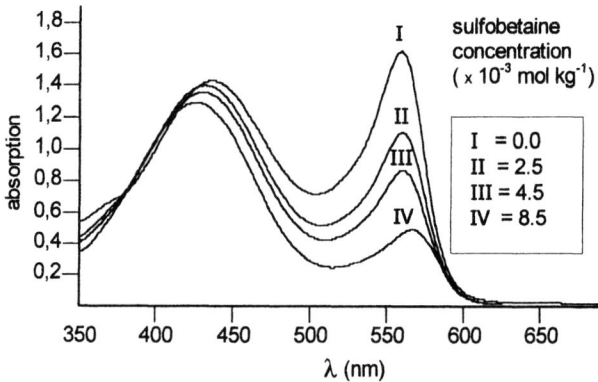

Figure 5. UV/Vis absorption spectra of Phenol Red in PVA/polyalkoxide/borax/ sulfobetaine gel networks with various sulfobetaine contents.

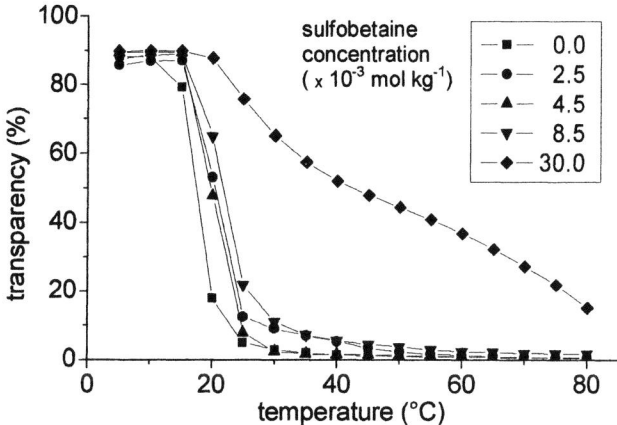

Figure 6. Temperature dependence of the transparency of Phenol Red containing PVA/polyalkoxide/borax/sulfobetaine gel networks with various sulfobetaine contents.

Another example of a chromogenic gel network was obtained by adding small amounts of a polyalkoxide to a thermochromic Phenol Red containing PVA/borax/surfactant hydrogel (19). A thermotropic behavior based on a phase separation process was found to appear at a polyalkoxide content of 0.8 wt.%. Moreover, in this paper the concentration dependence of the zwitterionic sulfobetaine surfactant on the thermochromic and thermotropic behavior was investigated. For this purpose a hydrogel containing 1.1 wt.% polyalkoxide was chosen to which sulfobetaine concentrations below and above the critical micelle concentration (CMC = 3.8×10^{-3} mol kg^{-1}) were added. The UV/Vis absorption spectra of these hydrogels are displayed in Figure 5. Two absorption bands are detected. The first one with a $\lambda_{max} \approx 440$ nm which corresponds to the phenol form of Phenol Red and the second one with a $\lambda_{max} \approx 563$ nm which corresponds to the phenolate form of Phenol Red. With increasing sulfobetaine concentration a decrease of the intensity of both UV/Vis absorption bands occurs. It is well known that above the CMC surfactants can influence the UV/Vis absorption behavior of water soluble dyes. However, here this effect takes place already at the lowest sulfobetaine concentration of 2.5×10^{-3} mol kg^{-1}, which is significant below the CMC. To characterize the thermotropic behavior the transparency of the gels were measured as a function of temperature (see Figure 6). Again even below the CMC a significant influence of the surfactant concentration on the thermotropic properties is observed. To explain this behavior the authors suggested the formation of complexes between dye molecules and aggregates of sulfobetaine, but discussed also an interaction of the dye with single sulfobetaine molecules as an alternative mechanism.

Outlook

The efforts in material development in recent years have led to novel thermotropic polymer gels with advanced properties. Furthermore, thermochromic as well as chromogenic polymer gels where developed. Some of the chromogenic polymer materials exhibit a practically temperature independent volume, which makes them promising for a series of future applications like smart windows, large area displays and tunable color filters. The first prototypes of such applications, that works in a practical temperature range, were already presented.

It can be assumed, that in a few years intelligent sun protecting glazing based on chromogenic polymer gels will be a common constituent of modern building architecture. Furthermore, thermochromic gels will find their application as temperature indicators and as operating layers in electro-optical modules like large area displays and devices with a high information density.

References

(1) Charoudi, D. U.S.-Patent No. 5404245, 1995.
(2) Watanabe, H. *Sol. Energy Mater. Sol. Cells* **1998**, *54*, 203.
(3) Seeboth, A.; Holzbauer, H.-R. *International Journal for Restoration of Buildings and Monuments* **1998**, *4*, 507.
(4) Seeboth, A.; Schneider, J.; Patzak, A. *Sol. Energy Mater. Sol. Cells* **2000**, *60*, 263.
(5) Schneider, J.; Seeboth, A. *Materials Science and Engineering Technology.* **2000**, *31*, 1.
(6) Alexandridis, P.; Holzwarth, J. F. *Langmuir* **1997**, *13*, 6074.
(7) Seeboth, A.; Lötzsch, D.; Potechius, E. *Colloid Polym. Sci.* **2001**, *279*, 696.
(8) Cai, W. S.; Gan, L. H.; Tam, K. C. *Colloid Polym. Sci.* **2001**, *279*, 793.
(9) Suwa, K.; Yamamoto, K. Akashi, M.; Takano, K.; Tanaka, N.; Kunugi, S. *Colloid Polym. Sci.* **2001**, *276*, 529.
(10) Okamura, H.; Masuda, S.; Minagawa, K.; Mori, T.; Tanaka, M. *European Polymer Journal* **2002**, *38*, 639.

(11) Fischer, Th.; Lange, R.; Seeboth, A. *Sol. Energy Mater. Sol. Cells* **2000**, *64*, 321.
(12) Fujimatsu, H.; Ogasawara, S.; Ihara, H.; Takashima, T.; Toyaba, K.; Kuroiwa, S. *Colloid Polym. Sci.* **1988**, *266*, 688.
(13) Fujimatsu, H.; Kuroiwa, S. *Colloid Polym. Sci.* **1987**, *265*, 938.
(14) Patel, G. N.; Ivory, D. M. U.S.-Patent No. 4,439,346, 1984.
(15) Gelinck, G. H.; Warman, J. M.; Staring, E. G. J. *J.Phys. Chem.* **1996**, *100*, 5485.
(16) Weissman, J. M.; Sunkara, H. B.; Tse, A. S.; Asher, S. A. *Science* **1996**, *274*, 959.
(17) Seeboth, A.; Kriwanek, J.; Vetter, R. *J. Mat. Chem.* **1999**, *9*, 2277.
(18) Seeboth, A.; Kriwanek, J.; Vetter, R. *Adv. Mater.* **2000**, *12*, 1424.
(19) Kriwanek, J.; Vetter, R.; Lötzsch, D.; Seeboth, A. *Polymers for Advanced Technologies* **2003**, *14*, 79.

Chapter 9

Modulation of Optical Properties of New Photosensitive Polymers: 3-D Optical Data Storage Media

Kevin D. Belfield[1,2], Katherine J. Schafer[1], and Stephen Andrasik[1]

[1]Department of Chemistry and [2]School of Optics/CREOL, University of Central Florida, Orlando, FL 32816-2366

We report the modulation of absorption and emission properties via single and two-photon photoinduced changes in polymeric media with two-photon fluorescence readout of multilayer structures. Photoinduced acid generation in the presence of a two-photon fluorescent dye possessing strongly basic functional groups underwent protonation upon exposure with UV or near-IR (740 nm fs pulses) irradiation. Solution studies demonstrate formation of monoprotonated and diprotonated species upon irradiation, each resulting in distinctly different absorption and fluorescence properties. Hence, two-channel, two-photon fluorescence imaging provides "positive" or "negative" image readout capability. Further, a poly-styrene-*co*-malaic anhydride copolymer containing a two-photon absorbing fluorophore was used to demonstrate near-IR two-photon based image formation, followed by two-photon based image readout. Results of solution and solid polymer thin films experiments are presented.

Introduction

Over the past 50 years, the field of organic photochemistry has produced a wealth of information, from reaction mechanisms to useful methodology for synthetic transformations. Many technological innovations have been realized during this time due to the exploits of this knowledge, including photoresists and lithography for the production of integrated circuits, photocharge generation for xerography, multidimensional fluorescence imaging, photodynamic therapy for cancer treatment, photoinitiated polymerization, UV protection of plastics and humans through the development of UV absorbing compounds and sunscreens, and fluorescence imaging, to name a few. The scientific basis of many of these processes continues to be utilized today, particularly in the field of organic three-dimensional optical data storage media and processes.

With the ever-pressing demand for higher storage densities, researchers are pursuing a number of strategies to develop three-dimensional capabilities for optical data storage in organic-based systems. Among the various strategies reported are holographic data storage using photopolymerizable media (*1*), including efforts by companies such as DuPont, IBM, Lucent, and Imation, photorefractive polymers (*2*), multilayer fluorescence-based techniques such as those developed by C3D and Call/Recall, and two-photon induced photochromism (*3*), to mention a few. It is known that fluorescent properties of certain fluorophores may be changed (quenched) upon protonation by photogeneration of acid (*4,5*). We have reported two-photon induced photoacid generation using short pulse near-IR lasers in the presence of a polymerizable medium, resulting in two-photon photoinitiated cationic polymerization and microfabrication (*6*). This inherent three-dimensional features associated with two-photon absorption provides an intriguing basis upon which to combine spatially-resolved two-photon induced photoacid generation and fluorescence quenching with two-photon fluorescence imaging.

The quadratic, or nonlinear, dependence of two-photon absorption on the intensity of the incident light has substantial implications ($dw/dt \propto I^2$). For example, in a medium containing one-photon absorbing chromophores, significant absorption occurs all along the path of a focused beam of suitable wavelength light. This can lead to out-of focus excitation. In a two-photon process, negligible absorption occurs except in the immediate vicinity of the focal volume of a light beam of appropriate energy. This allows spatial resolution about the beam axis as well as radially, which circumvents out-of-focus absorption and is the principle reason for two-photon fluorescence imaging (*7*). Particular molecules can undergo upconverted fluorescence through nonresonant two-photon absorption using near-IR radiation, resulting in an energy emission greater than that of the individual photons involved (upconversion). The use of a longer wavelength excitation source for fluorescence emission affords advantages not feasible using conventional UV or visible fluorescence techniques, e.g., deeper penetration of the excitation beam and reduction of photobleaching, and is particularly well-suited for fluorescence detection in multilayer coatings.

Rentzepis et al. reported two-photon induced photochromism of spiropyran derivatives at 1064 nm (8,9). Analogous to single-photon absorption facilitated isomerization, the spiropyran underwent ring-opening isomerization to the zwitterionic colored merocyanine isomer. The merocyanine isomer underwent two-photon absorption at 1064 nm, resulting in upconverted fluorescence. Spiropyrans are known to undergo photobleaching and photodegradation upon prolonged exposure, hence are not suitable for long term use. Nonetheless, an intriguing model for 3-D optical storage memory was proposed. An intriguing bacteriorhodopsin-based holographic recording media and process, using two-photon excitation, has been reported by Birge et al. (10).

We previously reported the synthesis and characterization of organic fluorescent dyes with high two-photon absorptivity (6,11,12). Several of these dyes also undergo substantial changes in the absorption and fluorescence spectral properties in the presence of strong acid, i.e., they undergo protonation changing, among other things, their polarizability, absorption and emission maxima, and fluorescence quantum yields (13). We wish to report results of the photoinduced

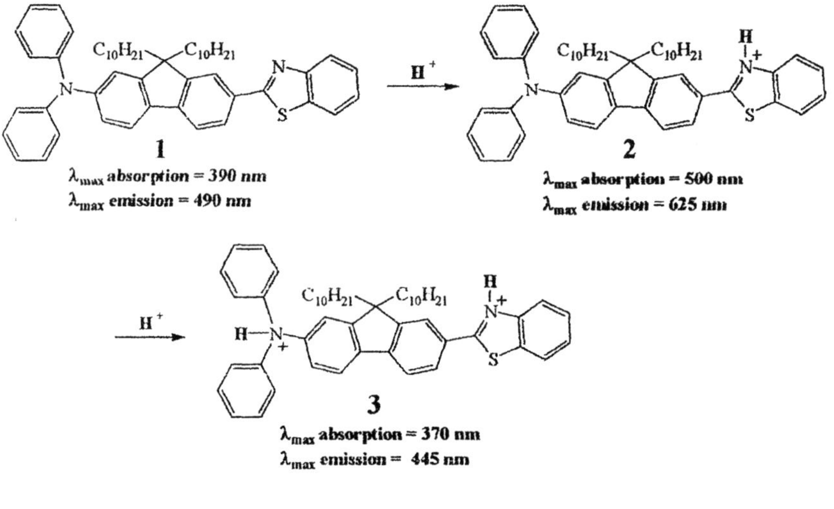

Figure 1. Reaction of fluorene 1 with acid (monoprotonated 2 and diprotonated 3 products).

protonation of fluorene dye **1** in liquid solution and polymer thin films (Figure 1). Further, image formation and two-photon induced fluorescence readout within a novel photosensitive polymer **4**, containing a two-photon absorbing fluorophore are also presented (*5*).

Results and Disscussion

Fluorene **1** was previously shown to undergo two-photon absorption and upconverted fluorescence on exposure to near-IR fs laser irradiation (*6,11*). The two-photon absorbing dye **1**, contains basic nitrogen-containing benzothiazolyl and triarylamino groups that are sensitive to the presence of acids. Pohers *et al.* have demonstrated the absorption spectrum of an acid-sensitive dye containing the benzothiazole group red shifts upon protonation in the presence of a photoacid generator (PAG) (*14*). Due to differences in basicity (pK_b), fluorene **1** undergoes selective, stepwise protonation, first by protonation of the benzothiazolyl nitrogen ($pK_b \cong 13$) (*15*) then the triarylamino nitrogen ($pK_b \cong 19$) (*16*). This leads to a mixture of three species in Figure 1 (**1**, **2**, and **3**), each with distinct UV-visible absorption and fluorescence emission properties.

To understand the behavior of the two-photon absorbing fluorophore and predict results expected in solid thin film studies, solution studies were performed in CH_2Cl_2. Time-dependent UV-visible absorption spectra for a solution containing **1** and the photoacid generator CD1010 (a triarylsulfonium salt) illustrate this nicely, as shown in Figure 2. Upon irradiation with broadband UV light (300-400 nm, 0.57 mW/cm^2), **1** undergoes protonation, resulting in formation of **2** whose absorption spectrum is red shifted by about 100 nm relative to that of **1**. The conversion of the neutral fluorophore **1** at early irradiation times (10 s) results in decreasing absorbance at its maximum at 390 nm, and increasing absorbance at 500 nm upon generation of the protonated form, **2**. The red shift was expected since fluorene **1** is of an electron donor-π-acceptor construct and protonation of the benzothiazolyl acceptor increases the electron deficiency of this group, affording a greater dipole moment and polarizability. When **2** undergoes protonation, a new absorption that is blue shifted relative to both **1** and **2** was observed, due to the fact that the once electron-donating diphenylamino group in **1** and **2** has been converted to an electron accepting moiety (quaternary ammonium salt) in **3**. The absorption due to the triarylsulfonium salt (λ_{max} = 310 nm) also decreases with time as expected but, for clarity, is not displayed in Figure 2. No evidence for charge transfer complex formation was observed between fluorine **1** and the PAG, however, more detailed investigations would be needed to fully explore this possibility.

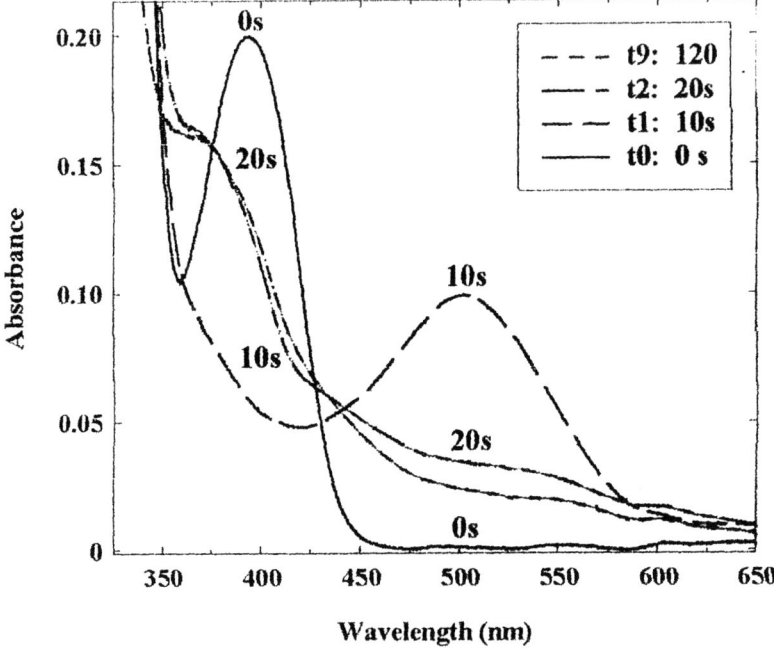

Figure 2. Time-dependent UV-visible absorption spectra of the irradiation of 1 and photoacid generator in CH_2Cl_2 at irradiation times from 0 to 120 s.

Changes in the fluorescence emission spectra corresponded with the observed changes in the absorption spectra. Protonation of **1** also resulted in a reduction of its fluorescence emission, while emission at longer wavelengths was observed due to excitation of the longer wavelength absorbing monoprotonated **2** (Figure 3). As can be seen, the fluorescence emission intensity at ca. 490 nm (390 nm excitation wavelength) decreases with irradiation while, at early exposure times, emission at ca. 625 nm appears, which then blue shifts upon further protonation to **3**. The emission at 625 nm is from monoprotonated **2** upon excitation at 500 nm. Eventually, diprotonation results in a relatively weak, blue shifted emission at ca. 445 nm (from **3**). Thus, in addition to observing fluorescence quenching at ca. 490 nm, fluorescence enhancement (creation) at longer wavelengths (ca. 625 nm) is observed upon short irradiation times. As demonstrated in the following section, this behavior facilitates two-channel fluorescence imaging, resulting in contrast due to fluorescence quenching at the shorter wavelengths ($\lambda_{emission}$ of **1** from 425 – 620 nm) and fluorescence enhancement at longer wavelengths ($\lambda_{emission}$ of **2** from 520 – 700 nm).

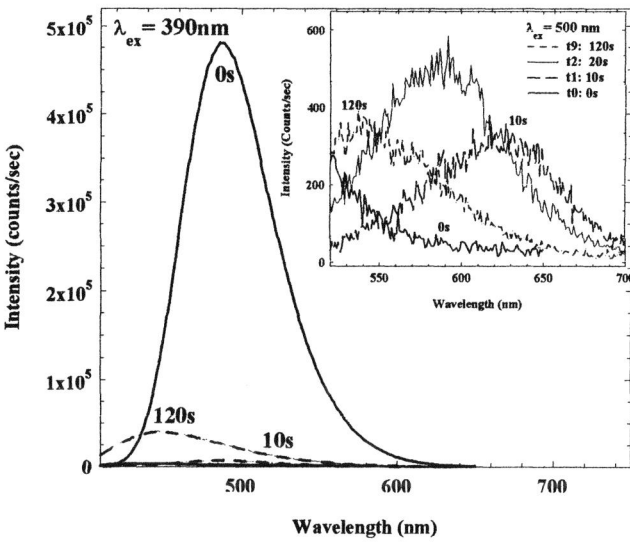

Figure 3. Time-dependent fluorescence emission spectra for the irradiation of 1 and photoacid generator in CH_2Cl_2 at irradiation times from 0 to 120 s (excitation at 390 nm). Inset shows fluorescence at longer wavelength with excitation at 500 nm.

To demonstrate the ability of fluorene **1** to exhibit two-photon upconverted fluorescence emission, fluorescence spectra were recorded upon excitation at a number of wavelengths using a CPA-2001 laser system from Clark-MXR. Femtosecond pulses from a frequency-doubled erbium-doped fiber ring oscillator were stretched to about 200 ps, then passed through a Ti: Sapphire regenerative amplifier and compressed down to 160 fs. The energy of the output single pulse (centered at λ = 775 nm) was 137 nJ at a 1 kHz repetition rate. This pumped a Quantronix OPO/OPA, producing fs pulsed output, tunable from 550 nm to 1.6 µm. Two-photon upconverted fluorescence spectra is illustrated in Figure 4a. From Figure 4a, it is readily apparent that fluorene **1** exhibits maximum two-photon upconverted fluoresence intensity when pumped at 800 nm. To further confirm that **1** undergoes two-photon absorption, the total integrated fluorescence intensity was determined as a function of incident intensity (pump power). Fluorescence from a two-photon absorption process will exhibit a quadratic dependence on incident intensity. Figure 4b indeed confirms that fluorene **1** underwent two-photon absorption as evidenced by the quadratic relationship between fluorescence emission intensity at several pump powers at two different pump wavelengths.

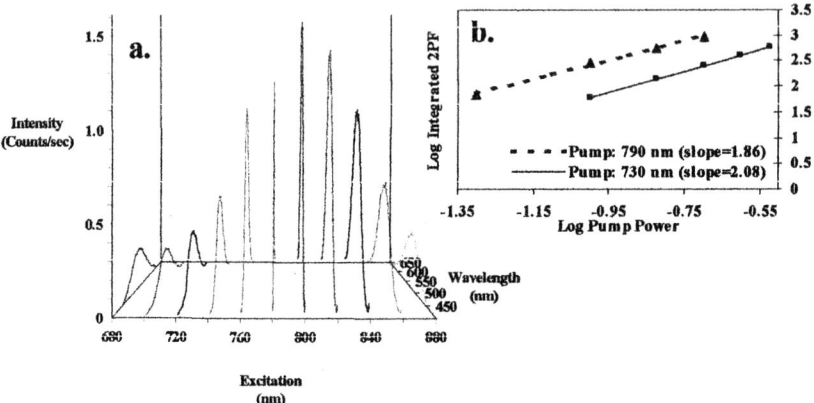

Figure 4. Two-photon upconverted fluorescence emission of 1 at several fs pulse pump wavelengths (a). Plot of total integrated fluorescence intensity as a function of pump power at two wavelengths (b).

Next, thin polymer films (ca. 2-3 μm film thickness) were prepared by spin coating (on 0.17 mm thick glass cover slips) a mixture of fluorene **1**, the photoacid generator, and polystyrene (or alternatively poly(methyl methacrylate) in a 1:3 v/v solution of acetonitrile/dioxane (*17*). Films were exposed to UV light through a number of different masks, including TEM grids, Air Force resolution targets, and photolithographic waveguide masks. For demonstration purposes, multilayer assemblies were constructed by placing an uncoated glass cover slip between two coated cover slips with the coated sides against the middle cover slip.

Three-dimensional two-photon fluorescence imaging was performed on the multilayer structures using a modified Olympus Fluoview laser scanning confocal microscopy system equipped with a broadband, tunable Coherent Mira Ti:sapphire laser pumped by a 10 W Coherent Verdi frequency doubled Nd:YAG laser (tuned to 800 nm, 115 fs pulsewidth, 76 MHz repetition rate). Two photomultiplier tube detectors with band pass filters, 510-550 nm (channel 1) and 585-610 nm (channel 2) were used for two channel fluorescence imaging. Results analogous to those obtained in solution studies were observed but, quite fortuitously, the slower acid generation/protonation rate resulted in formation and stabilization of monoprotonated fluorene **2**. With the beam focused in the plane of one of the fluorphore-containing layers, both channel 1 (green) and channel 2 (red) can be recorded. The contrast in the "green" channel was due to the fluorescence quenching of fluorene **1** (whose concentration decreases with irradiation). Contrast in the "red" channel was due to the fluorescence of monoprotonated **2** (whose concentration increases with irradiation). Time-dependent studies were performed by irradiating the films at various times to determine exposure times that led to optimum contrast in each detection channel.

Plate 1a and b shows films exposed using an Air Force image resolution target with images recorded by both channels. The fluorescence intensity profile as a function of position across one set of the elements for each image is shown in Plate 1c. The large differences in fluorescence intensity in exposed and unexposed regions can be clearly seen in the graph as well as the reverse parity of the images in the two channels, i.e. "positive" and "negative" image formation.

For demonstrative purposes, multilayer assemblies were constructed by placing an uncoated glass cover slip between two cover slips coated with patterned photosensitive films, with the coated sides against the middle cover slip (Plate 2). Three-dimensional two-photon fluorescence imaging was performed on the multilayer structures.

Two-photon fluorescent images of the photosensitive films constructed in a multi-layer configuration (developed via UV exposure through TEM square and hexagonal grid masks) are displayed in Plate 3a. An xy planar scan of each film (hexagonal grid image on the top and square grid image on the bottom) within the multi-layer, by focusing and scanning within the plane of the films, clearly shows the photo-patterned image resulting from formation of the protonated species in exposed areas. A cross-sectional scan, where an xy line scan is stepped in the z dimension (multi-layered image between the grid images in Plate 3a), clearly displays the separate film layers and demonstrates the three dimensional nature of image formation possible within layered assemblies, and the nondestructive optimal sectioning ability of two-photon fluorescent imaging. The signal readout (Plate 3b and 3c) establishes the possibility for a WORM binary optical data storage medium, where the valleys can be designated as a "0" and the peaks a "1".

Poly(styrene-co-maleic anhydride) was modified by condensation with a fluorenylamine (18), affording the modified copolymer 4 in 36% yield after purification. The fluorophore labeled copolymer structure and its linear absorption and fluorescence emission spectra are presented in Figure 5. A Stoke's shift of >100 nm was observed with a quantum yield of 0.61 (DMF). Shown in Figure 6 are the two-photon upconverted fluorescence spectra at excitation wavelengths of 660, 790, and 870 nm. Two components were observed. One component with the shorter wavelength emission centered at 450 nm, is attributable to the fluorenylimide fluorophore, while an additional component produced emission at longer wavelengths via longer wavelength excitation. This is likely a charge-transfer complex, and is the subject of further investigation. Emission lifetime measurements were consistent with the steady state spectra, revealing a short-lived species (< 100 ps), and a longer lived species with a lifetime of ca. 3 ns.

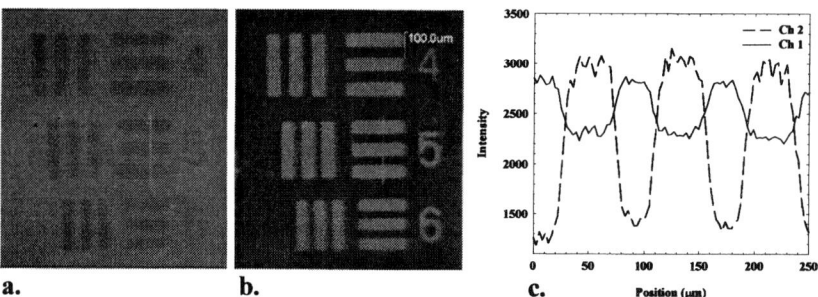

Plate 1. Two-photon fluorescence images of photosensitive films developed (via 350 nm broadband exposure, 4.4 mW/cm^2) through Air Force targets. Image recorded by channel 1 (a), image recorded by channel 2 (b), and fluorescence intensity by scanning an x,y line across one set of three-member elements (c).

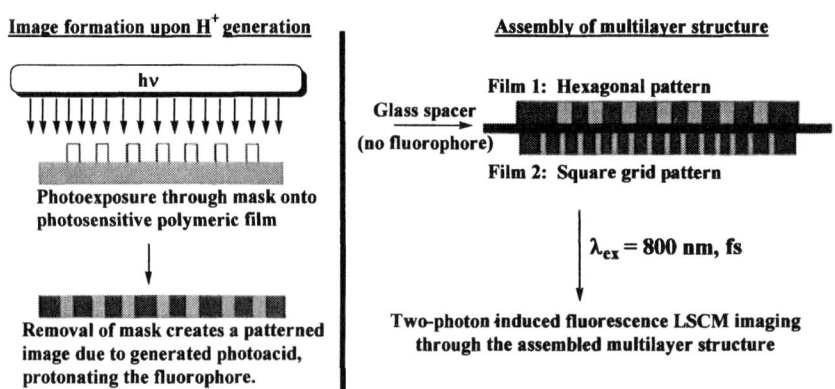

Plate 2. Image formation (upon photoacid generation) within photosensitive polymer films for assembly of multi-layered structures. Two-photon fluorescence LSCM imaging using fs pulsed near-IR pump allows for 3-D volumetric imaging of the layered structure.

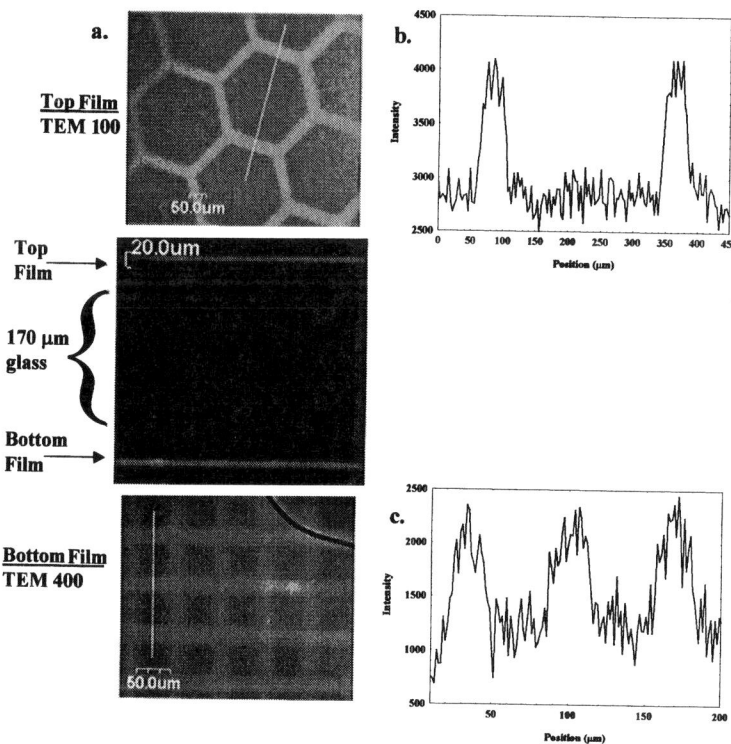

Plate 3. Two-photon fluorescent images of multi-layered films developed via 350 nm, broadband irradiation (6.0 mW/cm^2) by exposure through TEM hexagonal and square grid masks (a). Fluorescence intensity plots for a line scan across a region (as defined by the yellow line across the image area) provides image readout in one layer (b), and changing the depth (z position) for image (signal) readout in the lower layer within a multi-layered system (c).

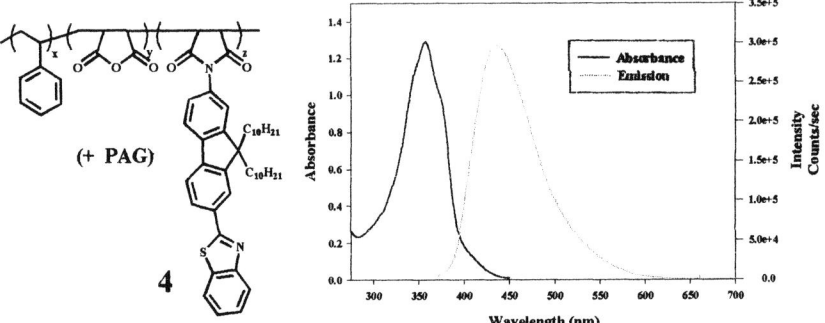

Figure 5. Two-photon absorbing fluorophore-labeled polymer (4) (left) and its corresponding linear absorption and fluorescence emission spectra (right, λex = 360 nm, DMF).

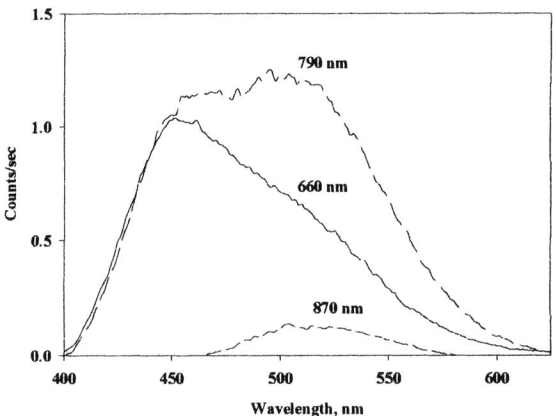

Figure 6. Two-photon upconverted fluorescence spectra of modified poly(styrene-co-maleic anhydride) in DMF. The fs excitation wavelengths are indicated.

Finally, both writing and recording were accomplished by two-photon excitation of a fluorophore/PAG photosensitive polymer film in which writing was accomplished by xy-scans at 740 nm (115 fs, 76 MHz). The written image was read by a two-photon fluorescence imaging at 800 nm (115 fs, 76 MHz), as shown in Figure 7. Thus, image writing and reading has been accomplished via near-IR two-photon excitation of polymer films containing fluorophore 1 and a photoacid generator. The behavior and relative stability of 1 makes this compound a good candidate for WORM three-dimensional memory systems with writing and reading accomplished via two-photon fluorescence imaging.

Preliminary demonstration of both two-photon writing and reading was demonstrated by writing an image on a thin film of the photosensitive mixture (fluorene 1, photoacid generator, polystyrene) under the laser scanning confocal microscope at 740 nm then reading the image at 800 nm (Figure 7).

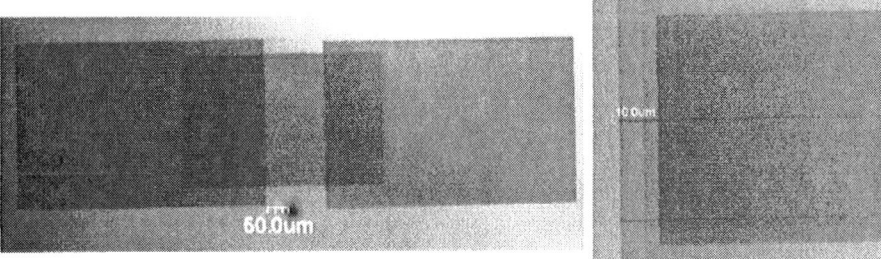

Figure 7. Image written (740 nm) and read (800 nm) in a photosensitive polymer film (1.5 μm thickness) via a two-photon excitation.

Conclusions

We have demonstrated that selective photoacid generation protonation of a two-photon absorbing fluorophore can produce "positive" and "negative" image formation in polymer films, with image contrast due to fluorescence quenching (reduction) of the neutral fluorophore using detection through a shorter wavelength broadband filter (510-550 nm, channel 1) and contrast due to fluorescence enhancement (increase) through a longer wavelength broadband filter (585-610 nm, channel 2). Both single-photon (UV) and two-photon (near-IR) imaging of the photosensitive polymer was demonstrated. Two-photon fluorescence imaging of multilayer structures provided three-dimensional signal readout. The behavior and relative stability of 1 makes this compound a good candidate for three-dimensional memory systems with writing and reading accomplished via two-photon excitation. Further, the ease of synthesis, high two-photon absorptivity, and interesting fluorescence properties of flurophore labeled polymers make them good candidates for optical power limiting and two-photon fluorescence imaging applications, aspects currently under investigation.

Acknowledgments

The National Science Foundation (ECS-9970078, DMR9975773), the National Research Council COBASE program, the Research Corporation, and the donors of The Petroleum Research Fund of the American Chemical Society are gratefully acknowledged for support of this work.

References

(1) See e.g., Cheben, P.; Calvo, M. *Appl. Phys. Lett.* **2001**, *78*, 1490.
(2) See e.g., Belfield, K. D.; Chinna, C.; Najjar, O.; Sriram, S.; Schafer, K. J. *Field Responsive Polymers, ACS Symposium Series 726*; Khan, I. M.; Harrison, J. S., Eds.; American Chemical Society: Washington, DC, 1999, Chapter 17.
(3) Belfield, K. D.; Liu, Y.; Negres, R. A.; Fan M.; Pan, G.; Hagan, D. J.; Hernandez, F. E. *Chem. Mater.* **2002**, *14*, 3663.
(4) Kim, J.-M.; Chang, T. E.; Kang, J.-H.; Park, K. H.; Han, D.-K.; Ahn, K.-D. *Angew. Chem. Int. Ed.*, **2000**, *39*, 1780.
(5) Belfield, K. D.; Schafer, K. J. *Chem. Mater.* **2002**, *14*, 3656.
(6) Belfield, K. D.; Schafer, K. J.; Liu, Y.; Liu, J.; Ren, X.; Van Stryland, E. W. *J. Phys. Org. Chem.* **2000**, *13*, 837.
(7) Denk, W.; Strickler, J. H.; Webb, W. W. *Science* **1990**, *248*, 73.
(8) Parthenopoulos, D. A.; Rentzepis, P. M. *Science* **1989**, *245*, 843.
(9) Dvornikov, S.; Rentzepis, P. M. *Opt. Commun.* **1995**, *119*, 341.
(10) Birge, R. R.; Parsons, B.; Song, Q. W.; Tallent, J. R. *Molecular Electronics*; Jortner, J.; Ratner, M. Eds.; Blackwell Science: London, 1997, Chapter 15.
(11) Belfield, K. D.; Schafer, K. J.; Mourad, W. *J. Org. Chem.* **2000**, *65*, 4475.
(12) Belfield, K. D.; Schafer, K. J.; Hagan, D. J.; Van Stryland, E. W.; Negres, R. A. *Org. Lett.* **1999**, *1*, 1575.
(13) Belfield, K. D.; Bondar, M. V.; Przhonska, O. V.; Schafer, K. J.; Mourad, W. *J. Lumin.* **2002**, *97*, 141.
(14) Pohlers, G.; Scaiano, J. C.; Sinta, R. *Chem. Mater.* **1997**, *9*, 3222.
(15) Dey, J. K.; Dogra, S. K. *Bull. Chem. Soc. Jpn.* **1991**, *64*, 3142.
(16) Arnett. E. M.; Quirk, R. P.; Burke, J. J. *J. Am. Chem. Soc.* **1970**, *92*, 1260.
(17) Photosensitive polymer film compositions typically contained 0.9 wt.% of fluorene 1 and 9 wt.% of CD1010 relative to the polymer.
(18) Belfield, K. D.; Morales, A. R.; Andrasik, S. A.; Schafer, K. J.; Yavuz, O.; Chapela, V. M.; Percino, J. *Functional Condensation Polymers*; Carraher, C. E.; Swift, G. G., Eds.; Kluwer: London, 2002, Ch. 11.

Chapter 10

Quantum Amplified Isomerization: A New Chemically Amplified Imaging System in Solid Polymers

Douglas R. Robello[1], Joseph P. Dinnocenzo[2], Samir Farid[1], Jason G. Gillmore[2], and Samuel W. Thomas III[1]

[1]Research and Development, Eastman Kodak Company, Rochester, NY 14650
[2]Department of Chemistry, University of Rochester, Rochester, NY 14627

A new imaging system based on a photoinitiated electron transfer chain reaction is reported. Specifically, irradiation of 9,10-dicyanoanthracene (sensitizer) leads to the conversion of Dewar benzene derivatives (reactants) to benzene derivatives (products) within solid polymer films. The mechanism of the reaction may involve chemical amplification with cation radicals ("holes") as the catalytic species. We present herein studies of both molecularly doped polymers and polymers containing Dewar benzene moieties attached to side chains. The refractive index of the materials could be tuned within a narrow range using this photochemical reaction, as demonstrated by the writing of persistent gratings in forced Rayleigh scattering experiments.

Introduction

Chemical amplification, the process by which a cascade of chemical reactions are triggered by absorption of a photon, represents a landmark concept in applied photochemistry. Chemical amplification leads to the efficient use of the available light and, therefore, enables imaging systems with high sensitivity. Photopolymerization and photoresists are two commercially important applications of this concept. In conventional vinyl photopolymerization, photogenerated *free radicals* are responsible for initiating the chain polymerization. In cationic photopolymerization and acid-activated photoresists, photogenerated *protons* (or the equivalent) catalyze subsequent polyaddition or deprotection reactions. We have begun to investigate an analogous process in which the *catalytic species is a cation radical ("hole")* rather than a proton, and the *reaction is an isomerization*. The cation radical is created via photoinduced electron transfer and promotes the conversion of a plurality of strained ring molecules to more stable isomers in a chain reaction. We call this process "Quantum Amplified Isomerization" (QAI).

There are a few reports in the previous literature of QAI in solution (*1,2*). In a prototypical example, the photoinitiated conversion of hexamethyl Dewar benzene (**HBDB**) to hexamethylbenzene (**HMB**) in the presence of a catalytic amount of an electron-accepting photosensitizer was found to have a quantum yield much greater than unity (*1*). The chain reaction mechanism is thought to proceed as follows:

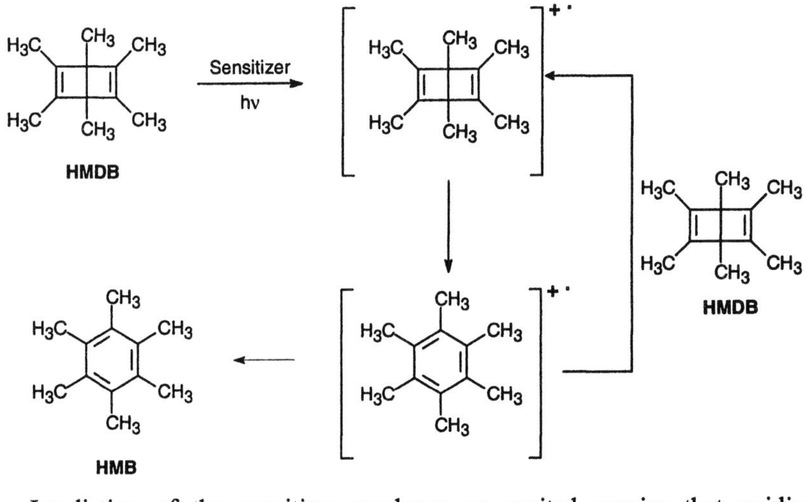

Irradiation of the sensitizer produces an excited species that oxidizes **HMDB** to its cation radical by single electron transfer. (The sensitizer is simultaneously reduced to its anion radical, not shown in the diagram for simplicity.) While neutral **HMDB** is stable, its cation radical is not and quickly isomerizes to the corresponding **HMB** cation radical. The unique feature of this QAI process is that the generated **HMB** cation radical is capable of oxidizing another molecule of the **HMDB** reactant, leading to a neutral **HMB** product and a new **HMDB** cation radical. The oxidation of **HMDB** by **HMB** cation radical has been shown to be thermoneutral or slightly exothermic in solution (*1*). Moreover, the isomerization step is highly exothermic, providing the driving force of the reaction This cyclic process constitutes a chain reaction. Termination may occur by return electron transfer from the sensitizer anion radical to the **HMDB** or **HMB** cation radical. In this respect, the QAI process shares the familiar characteristics of chain polymerization: initiation, propagation, and termination.

The highest quantum yields for QAI reactions are obtained in polar solvents that promote separation of the sensitizer anion radical and the product cation radical and, thereby, inhibit the termination step. However, in the solid state, separation of these charged species might be difficult to achieve. Therefore, at the onset of this work, we expected that conversion of reactant to product by QAI in solid media might be limited to very low levels. Herein, we report that contrary to this expectation, Dewar benzene reactants can be converted in high yield to the corresponding benzene products by QAI within a solid polymer matrix. Moreover, the QAI reaction can be carried out successfully both on Dewar benzene derivatives dissolved in host polymers and attached as side groups to polymer chains. Because Dewar benzene and isomeric benzene derivatives have differing optical properties, QAI comprises a potentially useful imaging scheme. Specifically, the refractive index of a polymer film can be tuned efficiently within a narrow range using this photoreaction.

Results and Discussion

Conversion of Dewar Benzene to Benzene Derivatives in Polymer Films

Thin films containing Dewar benzene derivatives (ca. 0.5 M) and the sensitizer 9,10-dicyanoanthracene (**DCA**, 0.01 M) in poly(methyl methacrylate) (**PMMA**) binder were prepared by casting from dichloromethane solution onto poly(ethylene terephthalate) sheets. The film thicknesses were approximately 20 µm after drying in vacuo. The films were irradiated uniformly using the 406.7 nm line of a Kr ion laser (ca. 8 mW/cm^2), a wavelength that is strongly absorbed by **DCA,** but not by the other components. The optical densities of the films were approximately 0.3 at this wavelength. The conversion of reactant to

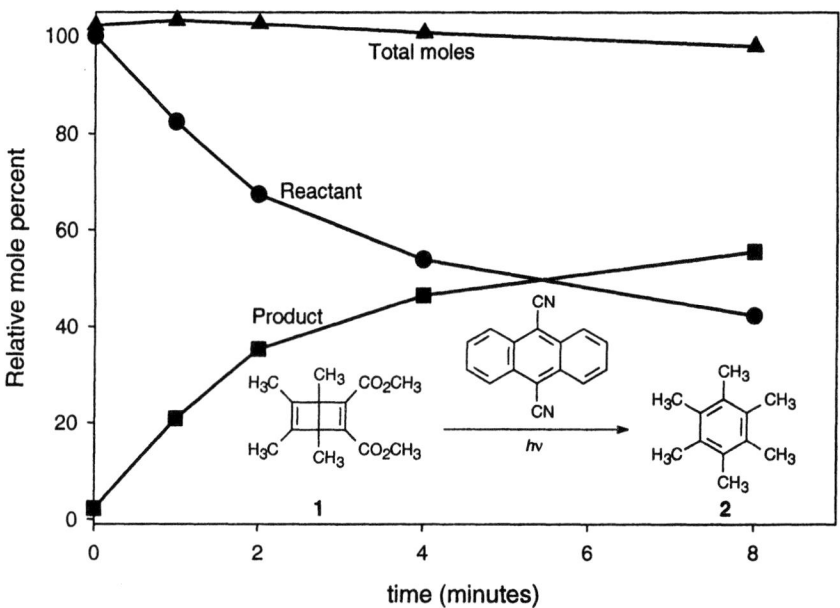

*Figure 1. Photoinitiated conversion in a **PMMA** film of a Dewar benzene reactant **1** to a benzene product **2** sensitized with 9,10-dicyanoanthracene (**DCA**). The initial concentration of reactant was 0.5 M, and sensitizer 0.01 M.. Irradiation was with a Kr ion laser at 406.7 nm, ca. 8 mW/cm².*

product was monitored by extracting films that had been irradiated for various periods of time and analyzing by HPLC. Typical results are depicted in Figure 1 for Dewar benzene derivative **1**. Remarkably, substantial conversion was achieved. No reaction occurred in the absence of **DCA** or in the dark.

A possible mechanism for the QAI reaction in solid media involves hole migration, as follows:

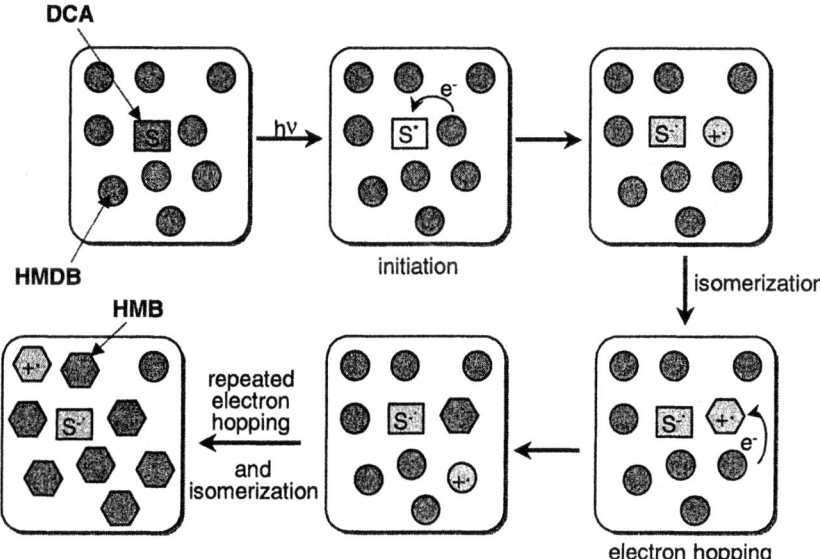

One alternative explanation for the observed high conversions is that the ingredients may have phase-separated within the film into domains of locally high concentration of reactant and **DCA**. However, we were unable to detect any evidence for phase separation of the ingredients in the film by electron microscopy. Furthermore, the T_g of the film (before irradiation) was reduced from 120 to 93 °C by the incorporation of the reactant, indicative of plasticization of the **PMMA**. These observations support the assumption that the ingredients were randomly dispersed. Another means to inhibit local phase separation of the reactant is to covalently bond the compound to the polymer, instead of dissolving the material in a polymer host.

Side Chain Polymers

For polymers randomly substituted with Dewar benzene moieties, segregation of the reactant into domains of locally high concentrations should not occur. In addition, attachment to the polymer backbone should allow for the incorporation of higher concentration of reactant without problems of phase separation (e.g., crystallization). Therefore, we undertook the synthesis of a version of this system in which **HMDB** moieties were attached covalently to the side chain of a polymer.

Because we had planned to prepare the polymer by conventional free radical polymerization using a Dewar benzene-functionalized monomer (see below), we were concerned that the presence of unsaturation and also numerous allylic hydrogen atoms in compounds such as **HMDB** might hinder polymerization. Therefore, we performed control homopolymerizations of methyl methacrylate (**MMA**) in the presence or absence of 10 mol % of **HMDB**. The molar mass distributions and glass transition temperatures of the two **PMMA** samples thus produced were indistinguishable, and the **HMDB** was recovered quantitatively (determined by GC), indicating that free radical polymerization of a methacrylate-substituted **HMDB** ought to be possible.

The requisite functionalized monomer **5** was prepared by the adaptation of the published synthesis of **HMDB** (*3-5*). First, 2-butyne was dimerized in the presence of aluminum chloride and treated in situ with ethyl tetrolate to provide ethyl pentamethyl Dewar benzoate. (**3**). Reduction of **3** with diisobutyl aluminum hydride produced the alcohol **4**, which was esterified with methacryloyl chloride to generate monomer **5**. The desired copolymers of **5** with various methacrylic esters were synthesized by solution polymerization using an azo initiator, yielding the 10/90 copolymers **6**.

Table I. Characterization Data for Dewar Benzene-Containing Copolymers

Copolymer	R	$\overline{M}_n$	$\overline{M}_w$	T_g (°C)
6a	methyl	27,100	46,900	115
6b	cyclohexyl	12,900	43,500	105
6c	*n*-hexyl	19,800	44,000	5

Copolymer molar mass determinations were made by size exclusion chromatography (SEC) using viscometric detection and universal calibration and are absolute. Glass transition temperatures were determined by differential scanning calorimetry (DSC) under nitrogen at a heating rate of 10 °C/min.

Films of copolymers **6** containing a catalytic amount of **DCA** were prepared analogously to the molecularly doped versions described above. In this case, irradiation was performed using the 405 nm line (isolated using an interference filter)of a high pressure mercury arc lamp (ca. 0.7 mW/cm^2). Irradiation of the copolymer films led to substantial conversion of the pendant Dewar benzene to the corresponding benzene moieties, as determined by ^{1}H NMR. For example,

Figure 2 illustrates the results for the copolymer **6a**. In the absence of added plasticizer, conversion was slower compared to the molecularly doped films, and leveled off near 20%. In a separate film plasticized with dibutyl phthalate, much more rapid and extensive conversion was observed.

Similar results were obtained for copolymer **6b** containing cyclohexyl methacrylate. However, copolymer **6c** containing hexyl methacrylate was found to convert at a significantly increased rate (Figure 3). These results indicate that T_g is a more important parameter affecting the rate of conversion via QAI than polarity. Copolymers **6b** and **6a** have similarly high glass transition temperatures, and exhibit comparable conversion rates. Copolymer **6c** has an equivalent polarity to copolymer **6b** (hexyl vs cyclohexyl side groups), but a significantly lower T_g, and converts much faster than **6b**, and comparably to the sample of **6a** plasticized with dibutyl phthalate. A possible explanation for this behavior is that reduced glass transition temperatures permit some limited structural reorganization and short range motion that assists charge migration or stabilizes charge separation. Diffusion of the sensitizer may also be a contributing factor.

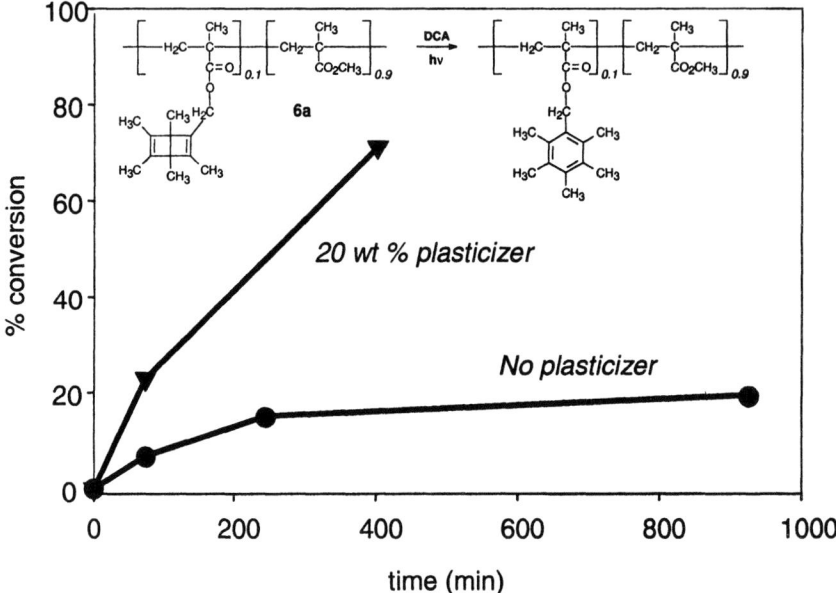

Figure 2. Photoinitiated conversion of Dewar benzene to benzene moieties in a side chain polymer film (6a), sensitized with ca. 0.01 M 9,10-dicyanoantharcene (DCA). Dibutyl phthalate was used as plasticizer. Irradiation was performed with a mercury arc lamp at 405 nm.

The power dependence of the rate of conversion appears to be in line with the applied intensity (i.e., comparing the Kr ion laser experiments to the Hg lamp experiments), however this aspect is currently under study so that definitive conclusions can be drawn.

Forced Rayleigh Scattering (FRS)

The refractive indices of **HMDB** and **HMD** are significantly different (1.47 and 1.53, respectively, for the pure materials). Therefore, QAI of such compounds might be an efficient means for tuning the refractive index of polymers. One method for detecting small changes in refractive indices of polymers is to write a grating into the material photochemically (in this case using QAI), and detect the resulting diffraction pattern. For these experiments, we adapted the technique of forced Rayleigh scattering (FRS), a common method for studying diffusion in polymers (6,7). The apparatus is illustrated in Figure 4. Irradiation was at 406.7 nm, and the power in each arm of the interfering beams was approximately 8 mW/cm^2. The period of the fringes

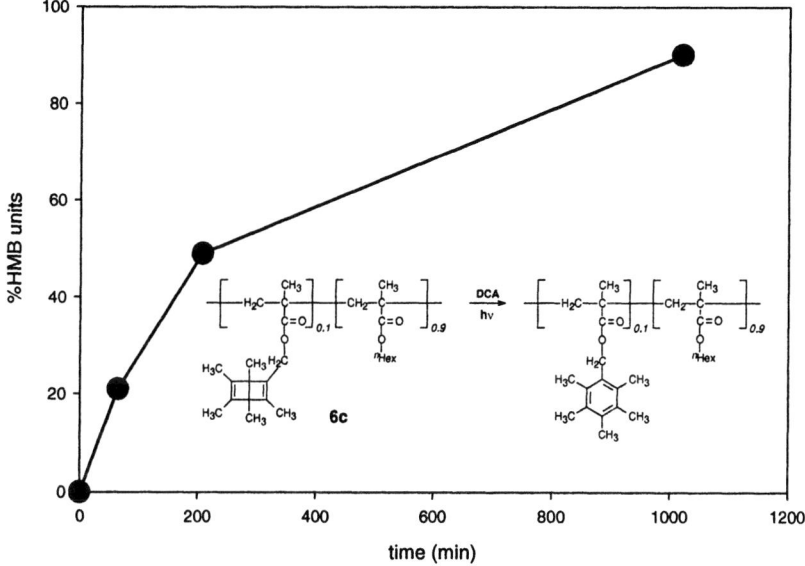

Figure 3. Photoinitiated conversion of Dewar benzene to benzene moieties in a side chain polymer film (6c), sensitized with ca. 0.01 M 9,10-dicyanoantharcene (DCA). No plasticizer was used in this case. Irradiation was performed with a mercury arc lamp at 405 nm.

produced by the interfering beams was ca. 0.7 μm. The diffraction efficiency was monitored in real time using a He-Ne laser at 632 nm, a wavelength at which the entire sample is transparent.

The QAI reaction was used to create a concentration gradient of benzene vs Dewar benzene moieties in the bright regions vs dark of the fringe pattern, respectively, thereby generating a refractive index grating in the film sample. Representative results are shown in Figure 5 for Dewar benzene derivative **1**.

The rapid increase in diffraction efficiency after exposure began is consistent with the formation of the concentration gradient. Note that the diffraction efficiency continued to increase for several minutes even after the shutter was closed, suggestive of continued conversion of Dewar benzene to benzene derivatives in the dark, in a long-lived chain process. The diffraction efficiency of the resulting grating was unchanged after several months of storage at room temperature in the dark, demonstrating that diffusion of the reactant and product is slow on the length scale of the fringes for long periods, although relatively short range diffusion during the short irradiation may occur. Notably, the grating was erased when the sample was heated to near its T_g.

Persistent gratings could also be written in films containing the copolymers **6** with pendant Dewar benzene moieties (Figure 6). Once again, the appearance of a diffraction grating is consistent with the conversion of Dewar benzene to

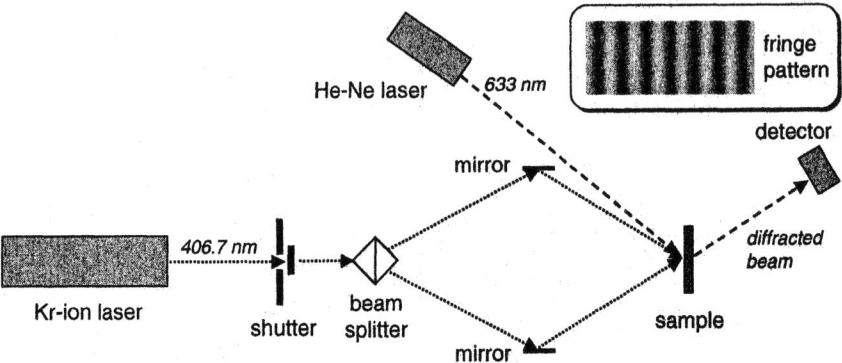

Figure 4. Forced Rayleigh scattering schematic.

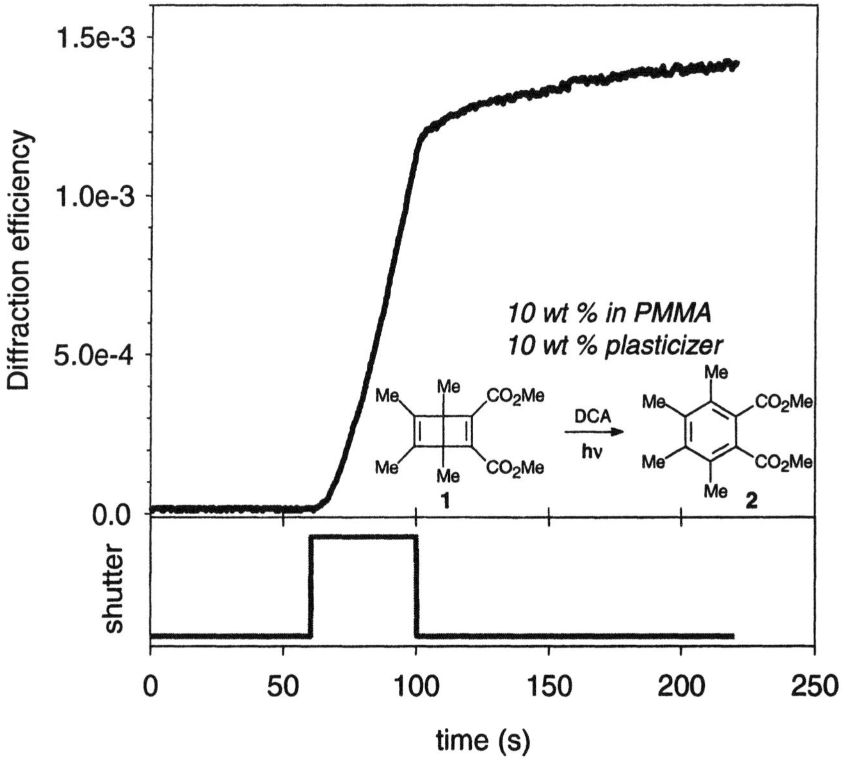

*Figure 5. Conversion of Dewar benzene **1** to benzene **2** dissolved in a solid **PMMA** film, studied by forced Rayleigh scattering.*

benzene moieties in the film by QAI. That the QAI reaction can be accomplished even when the reactant and product are tied to a polymer backbone and the permanence of the resulting gratings at ambient temperatures, provide strong evidence that long range molecular diffusion is not a necessary component of the mechanism. However, contributions to the reaction mechanism due to molecular-scale reorganization or short range diffusion of the Dewar benzene moieties or more extensive diffusion of the sensitizer molecules cannot be ruled out.

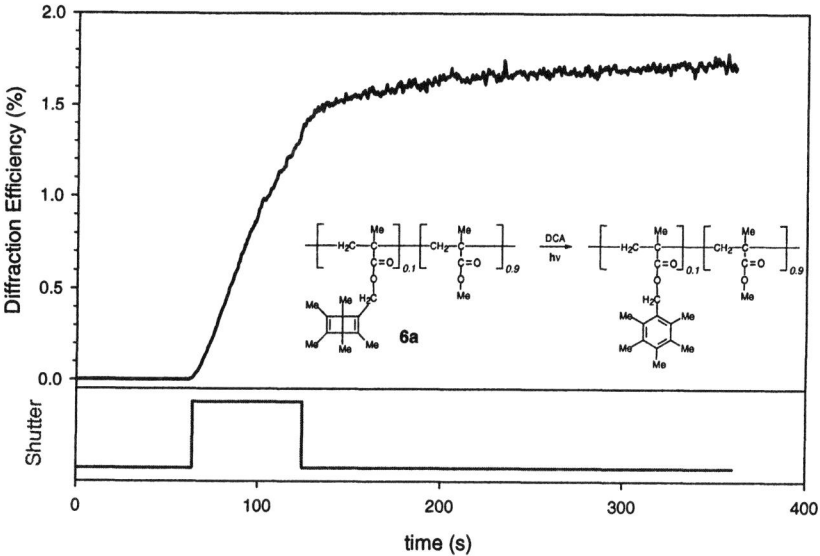

Figure 6. Conversion of Dewar benzene to benzene moieties on a side chain copolymer film (6a), studied by forced Raleigh scattering.

From the observed diffraction efficiencies, the corresponding refractive index changes can be estimated (6,7). For example, using the data in Figure 6, a diffraction efficiency of ca 2% corresponds to $\Delta n = 0.003$ at 633 nm. (The film thickness in this case was measured to be 17 μm.) Therefore, the refractive index of such films can be efficiently tuned within this narrow range by irradiation at short wavelengths.

Conclusions

The conversion of Dewar benzene to benzene moieties initiated by photoinduced-electron transfer has been achieved in a solid polymer film. The reaction proceeds to high conversion, apparently without long-range diffusion of reactant species Shorter range motions may contribute to the QAI process, particularly for systems with relatively low glass transition temperatures. Because Dewar benzene and benzene derivatives can have significantly different optical properties, the QAI reaction may be useful for applications in which sensitive tuning of these properties is important.

Acknowledgments

We thank Prof. T. Erdogan, A. Peer, and P. Adamson of the Institute of Optics, University of Rochester, for assistance with forced Rayleigh scattering experiments. The financial support of a NSF GOALI grant (DMR-0071302) is gratefully acknowledged. We thank Thomas Mourey, Trevor Bryan, and Kim Vu for SEC measurements; Roger Moody for DCS measurements; and Lorraine Ferrar and Margaret Steele for HPLC conversion studies, all members of Research and Development at Eastman Kodak Company.

References

(1) Evans, T. R.; Wake, R. W.; Sifain, M. M. *Tetrahedron Lett.* **1973**, 701.
(2) Kelley, C. K.; Kutal, C. *Organometallics* **1985**, *4*, 1351.
(3) Schaefer, W.; Hellmann, H. *Angew. Chem.* **1967**, *6*, 518.
(4) Shama, S. A.; Wamser, C. C. *Org. Synth.* **1983**, *61*, 62.
(5) Koster, J. B.; Timmermans, G. J.; Van Bekkum, H. *Synthesis* **1971**, 139.
(6) Lodge, T.; Chapman, B. *Trends Polym. Sci.* **1997**, *5*, 122.
(7) Yu, H. *Polym. Prepr.* **1991**, *32(3)*, 398.
(8) Kogelnik, H. *Bell Syst. Tech. J.* **1969**, *48*, 2909.
(9) Tomlinson, W. J.; Chandross, E. A. *Adv. Photochem.* **1980**, *12*, 201.

Chapter 11

Chromicity in Poly(aryleneethynylene)s

Uwe H. F. Bunz, James N. Wilson, and Carlito Bangcuyo

School of Chemistry and Biochemistry, Georgia Institute of Technology, Atlanta, GA 30332

> The chromicity of poly(aryleneethynylene)s (PAE) is discussed and described in this contribution. Four different types of PAEs will be scrutinized a) dialkyl-poly(*para*-phenyleneethynylene)s (dialkyl-PPEs), b) dialkoxy-PPEs, c) acceptor substituted PPEs, and d) heteroaryleneethynylenes that contain benzothiadiazole units in their main chain. We will shortly describe their syntheses and then discuss their spectral properties in absorption and emission in solution and in thin films.

Chromicity is widespread in conjugated polymers and an important topic. Prominent examples of chromic conjugated polymers are the polythiophenes, the polybithiazoles, and the polydiacetylenes (**1-3**) *(1-3)*.

Chromicity is defined as change in color or emission as a consequence of an external stimulus. Examples of such stimuli are heat, solvent change, or a specific analyte. Polymers **1-4** display Uv-vis and emission spectra that change when going from solution into the solid state, when going from a good solvent to a poor solvent, or when heating thin films. The polythiophenes *(1)* are probably the best examined examples for chromic polymers, and their chromic effects have been utilized for the sensing of a wide variety of different analytes *(4)* including DNA *(5)*. Their chromicity stems from a single chain conformational change that drives the rotationally disordered, twisted backbone (left) to a planar ordered state (right). Upon ordering/planarization λ_{max} is shifted bathochromically. The ordering process is mediated by the alignment of the side chains, i.e. any analyte that changes the order of the side chains will be detected. This property makes the chromicity of the polythiophenes useful as an analytical tool. PAEs, the subject of this review, have similar transitions as the polythiophenes and are therefore of interest as chromic materials. Concluding from the polythiophenes there should be interesting opportunities for PAEs with respect to sensory and analytical applications.

Chromic Transitions in PAEs

Synthetic Access to 4-9

There are two general methods to synthesize PAEs. One is the Pd-catalyzed coupling of a dihaloarene to a diethynylarene *(6)*. The other is alkyne metathesis, in which a dipropynylarene is treated with a mixture of Mo(CO)$_6$ and 4-chlorophenol at elevated temperatures *(7)*. The synthesis of representative examples of the polymers **4-9** is shown in Scheme 1. While the Pd-catalyzed couplings are tolerant towards functional groups and can be performed at or near room temperature, alkyne metathesis gives materials of higher degree of polymerizetion (P$_n$) and purity but at the cost of significantly higher reaction temperatures (130-150°C). However, heteroatoms are less well tolerated in this procedure. Polymers **8** and **9** can only be made by the conventional Pd-catalyzed couplings. The polymers **1-4** are obtained in high yields as film-forming powders, the color of which ranges from light yellow (**4**) to deep red (**9**).

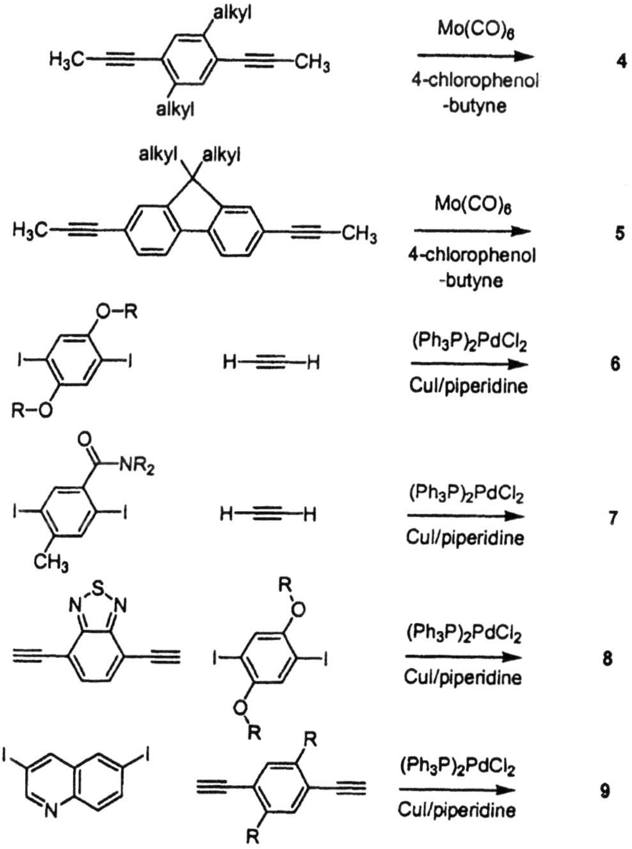

Scheme 1. Synthesis of polymers **4-9** by alkyne metathesis or Pd-catalyzed cross coupling reactions.

Thermochromism and Solvatochromism of Dialkyl-PPEs 4

Our first foray into the chromicity of PPEs was serendipitous. Inspired by Yamamoto's paper upon the aggregation of polythiophenes *(8)* we examined dialkyl-PPE's (**4a**, R = dodecyl) aggregation behavior. While PPEs had been known at that time for almost eight years and some of thir optical properties had been examined by Wrighton, Swager, and Weder *(6a,9)*, nothing was known about PPE's aggregation behavior. Aggregation in polymer solutions is induced by adding a poor solvent. In non-conjugated polymers aggregation processes do not result in a change of color, because the polymer backbone is transparent to visible light.

PPEs are almost colorless and strongly bluish-purplish fluorescent in chloroform solution. However, upon addition of methanol, a non-solvent, the PPE-solution of **4a** changed from colorless to deep yellow *(10)*. Figure 1 shows a titration experiment of **4a** with methanol. When going from the un-aggregated into the aggregated state a red shift of 40 nm is observed. Notable is the sharpness of the aggregate band and a second feature that shows a maximum in the middle and two shoulders. From a heuristic-phenomenological point of view the Uv-vis spectrum of the yellow phase of **4a** looks similar to the Uv-vis spectra of the aggregated phases of some of the poly(alkylythiophene)s **1** and to the blue phase of the poly(dicacetylene)s **3**.

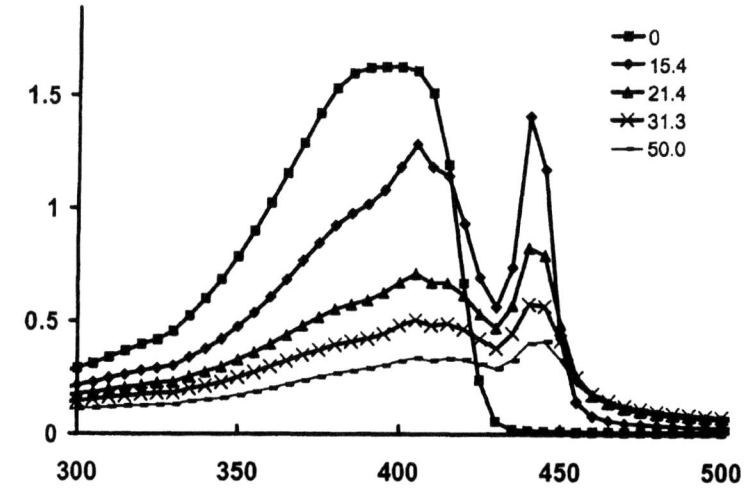

*Figure 1. Titration of **4a** with methanol. Solvatochromicity of dialkyl-PPEs. Numbers on the right side of the graph show percent amounts of methanol added to a chloroform solution of **4a**.*

The dialkyl-PPEs are not only solvatochromic (Figure 1) but they are as well thermochromic. If a thin yellow film of the PPE **4a** is heated to 140 °C it turns colorless as is displayed in Plate 1. The temperature of the chromic transition is dependent upon the degree of polymerization (P_n) and the type of the side chain. The larger P_n and the shorter the side chains, the higher the transition temperature. How do structure and chromicity of the PPEs fit together? Variable temperature powder diffraction shows that the PPEs are crystalline polymers at room temperature *(12)*. In the case of **4a** the crystalline character of the sample disappears at 110 °C. The material is liquid crystalline (smectic C) above this temperature and turns isotropic above 140 °C according to polarizing microscopy. Differential scanning calorimetry (DSC) corroborated this assignment *(13)*. The chromic transition of the PPEs is coupled to their melting behavior. We have analyzed the powder diffraction and the electron diffraction patterns of the PPEs and developed a packing model of **4** in the solid state. The dialkyl-PPEs pack in a doubly lamellar fashion that is shown in Plate 2. This model requires the conjugated backbones of the PPEs to be planar to adapt for the efficient packing of the alkyl side chains. This packing is maintained in the liquid crystalline phase and breaks down when the isotropic melt is reached. In the isotropic melt the Uv-vis resembles that of **4a** in chloroform solution.

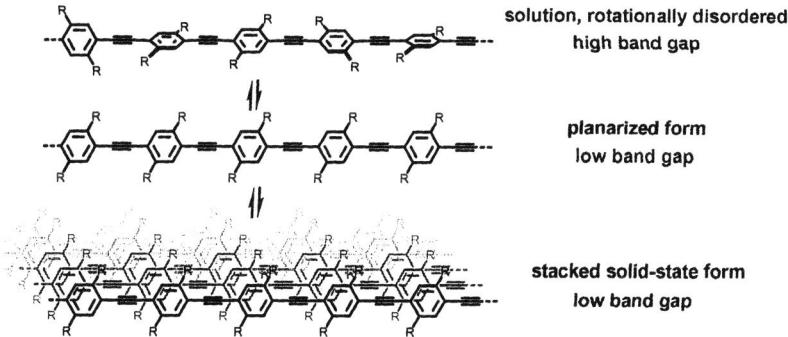

*Figure 2. Interplay between chromicity and rotational conformation in dialkyl-PPEs **4**.*

What is the conformation of the PPEs in solution? The arene units are connected by single bonds, and single bonds are not associated with large rotational barriers. Seminario and Tour *(14a)* have calculated the rotational barrier of the two benzene rings in tolane. In accordance with experimental results it is less than 1 kcal/mol in solution or in the gas phase. PPEs in solution thus do not have a preferred rotational conformation, i.e. adjacent aryl groups are

*Plate 1. Thermochromicity of diakyl-PPE **4a**. Left: Film at room temperature. Right: Film at 140° C. (See Page 4 of color insert.)*

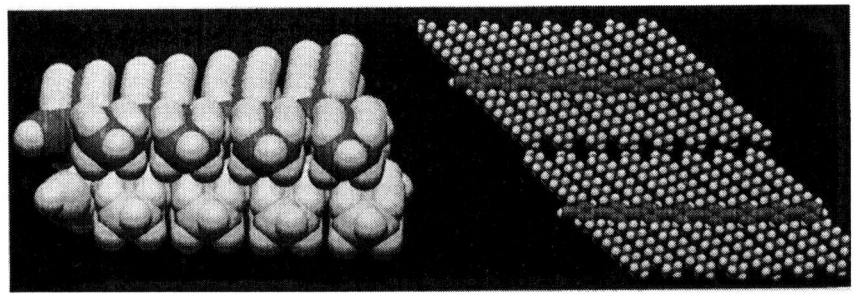

*Plate 2. Solid state structure of didodecyl-PPE **4a** at ambient temperature. (See Page 4 of color insert.)*

twisted at any arbitrary angle with respect to each other. We performed a series of bandgap calculations on an octameric phenyleneethynylene model utilizing AM 1 and PM3 semi empirical methods. The band gap of this large oligomer was significantly dependent upon the twist angle of the benzene rings with respect to each other. The larger the twist angle, the larger the band gap. Consequently we conclude that the thermochromicity and the solvatochromicity of PPEs can be explained by a simple self-consistent conformational model shown in Figure 2.

Solvatochromism of Poly(dialkylfluorenyleneethynylene)s (PFEs, 5)

The PFEs are similar to the PPEs **4**, but their Uv-vis spectra in solution are somewhat red-shifted when compared to those of **4** (λ_{max} = 392, 412 mn). The PFEs show similar aggregation behavior as the PPEs **4** do and in Figure 3 titration of bis(dimethyloctyl)-PFE with methanol, a non-solvent, is shown *(7c)*. As in the PPE case, a second red-shifted absorbance appears at 432 nm upon addition of methanol. Interestingly this behavior is not visible in spin cast thin films, and these spectra look very much like the solution spectra. The emission

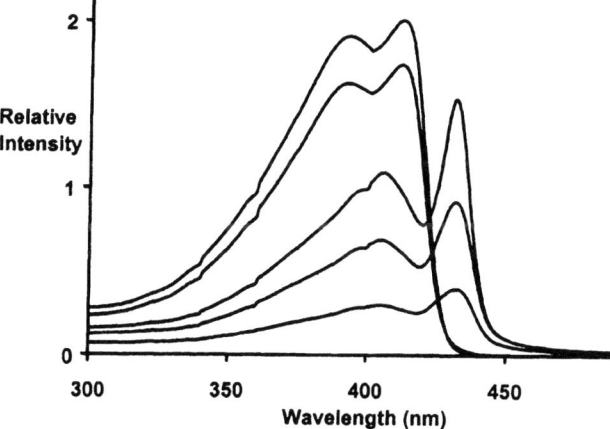

Figure 3. Absorption spectrum of 5 (CHCl₃) upon the addition of MeOH (0, 17, 38, 50, 74%, top to bottom).

of the PFEs is blue in solution and in the solid state. However, if a copolymer with 2-10% of fluorenone units is prepared, the emission in the solid state is orange as demonstrated in Figure 4 *(15)*. Efficient interchain energy transfer leads to a complete funneling of the exciton energy into the fluorenone traps. This effect is only visible in the solid state but not in solution. The manipulation of the emissive properties of the PFEs *(15)* is a doping process *(16)* that is similar to doping schemes of inorganic semiconductors *(17)*. In those a small amount of a dopant is introduced (typically by chemical vapor deposition) into the crystal lattice to form a covalently doped material, in which the minor component dominates the electronic and emissive properties. Doping of PFEs with 2 mol% of fluorenone to make a PFE-fluorenone copolymer is a life-like transfer of this process into the organic world and has approximately the same consequences on the emissive properties of the resulting material.

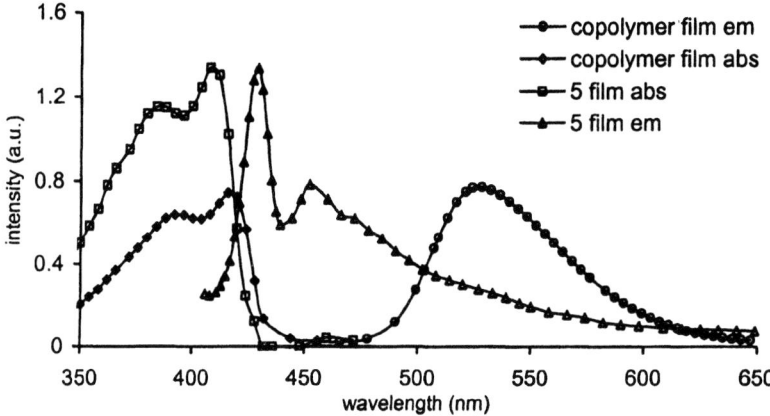

Figure 4. Absorption and emission of PFE 5 and doped PFE in thin films.

Solvatochromicity in Donor and Acceptor Substituted PPEs 6 and 7.

Swager has recently shown that Langmuir Blodgett (LB) films of PPE-types can form different aggregated phases with fascinating chromicities in absorption and emission *(18)*. He compares donor substituted PPEs with donor- acceptor substituted PPEs and finds that in LB film both, conformation and interchain interaction, play a role for the chromic properties of his PPEs. We have as well examined dialkoxy-PPEs **6** and find that they show similar solvatochromicity as the dialkyl-PPEs, but with a smaller change in λ_{max} absorption. In Figure 5, λ_{max} of **6** changes from 456 mn to 478 nm. Emission of **6** changes as well, it is likewise red shifted and greatly diminished. Thermochromicity of **6** is not known, because these materials melt under decomposition or not at all. However, annealing of thin films of **6** leads to a red shift in absorption and emission with λ_{max} values *(6a)* similar to those reported here *(19)*. It is interesting to note that exten-

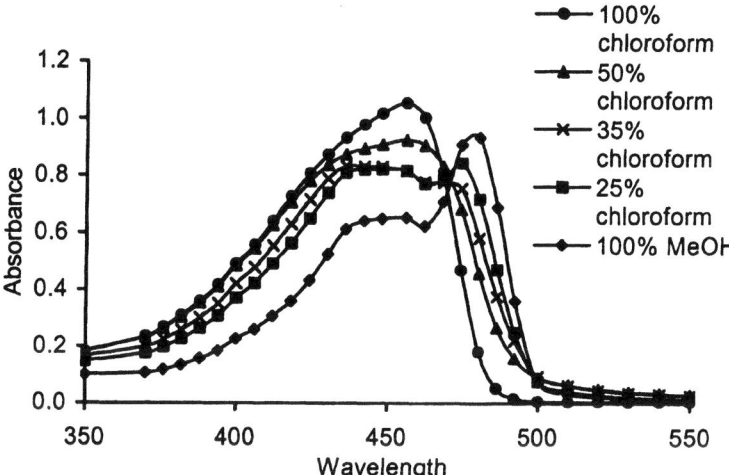

*Figure 5. Solvatochromicity of dialkoxy-PPE **6**. Upon addition of methanol λ_{max} changes from 456 nm (chloroform) to 478 nm (methanol).*

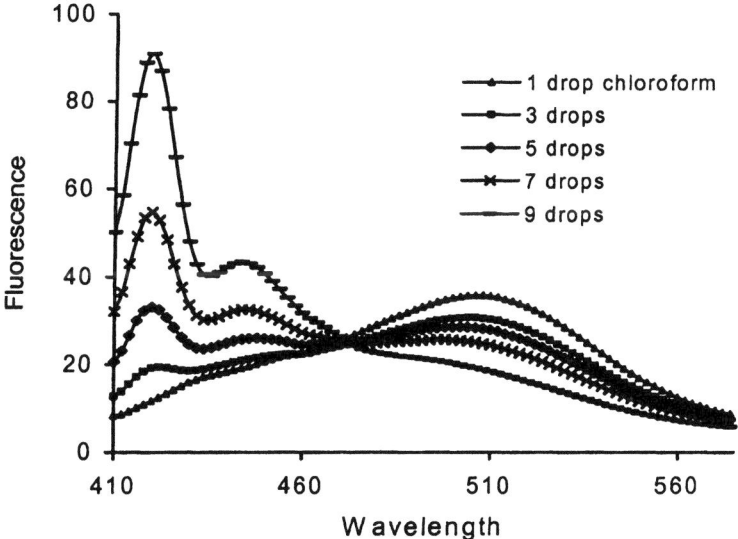

*Figure 6. Solvatochromicity (emission) of amide-substituted PPE **7**. Upon addition of chloroform λ_{max} emission changes from 508 nm (hexanes) to 420 nm (chloroform), i.e. 88 nm.*

sion of conjugation by the oxygen atoms decreases the chromic response of the PPEs significantly. The reason for this behavior is not clear.

To examine if the change in chromic behavior is due to the presence of increased conjugation in the single monomer unit or if it is influenced by the donor or acceptor quality of the substituents of the PPEs we prepared the monomer for the amide substituted PPE **7** by Swager's approach *(20)*. The formed PPE **7** is acceptor substituted throughout its conjugated backbone. Surprisingly, this material shows only small aggregative effects in absorption, but a larger solvatochromic effect in emission (Figure 6). The chromicity of the amide-substituted PPE **7** (R = 2-ethylhexyl) is dramatic and should be exploitable in sensor-type applications *(21)* in which fluorescence changes upon binding of an analyte to a suitably functionalized PPE side chain. The chromicity of **7** is due to excimer or exciplex formation. We are currently investigating the emissive lifetimes of **7**.

Chromicities in poly(heteroaryleneethynylene)s 8 and 9

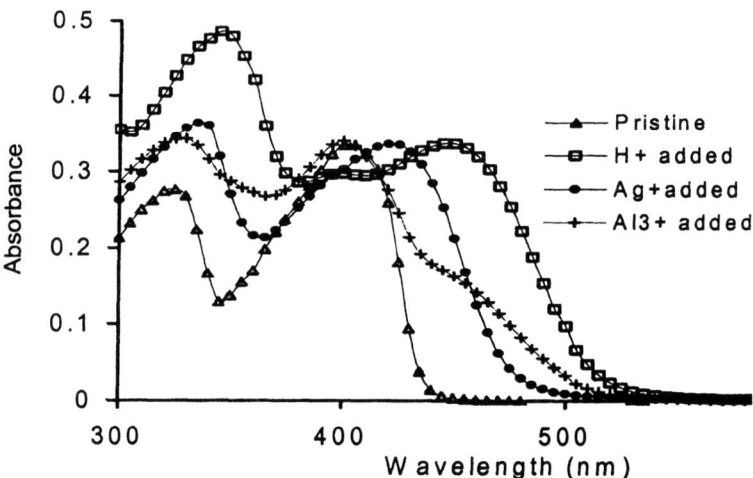

*Figure 7. Uv-vis spectra of **9a** in the pristine state and after addition of protons or silver salts. The addition of Al^{3+} has a similar effect as the addition of protons, due to partial hydrolysis.*

Once heteroatoms are introduced into the PAEs, further manipulation of their chromic behavior is possible. The two PAEs **8** and **9** contain quinoline and benzothiadiazole units in their backbone. The presence of a basic nitrogen makes these polymers susceptible to change of their optical properties by either protons or by metal salts. We have investigated the behavior of **9a** (R = 2-ethylhexyloxy) in the presence of protons and different metal cations as shown in Figure 7 *(6e)*. The addition of protons leads to a bathochromic shift of almost 50 nm and λ_{max} changes from 400 to 448 nm. The site of protonation is probably the basic nitrogen in the quinoline unit. It is known that pyridines form complexes with silver ions, and addition of a soluble silver salt to **9a** led to a significant, albeit smaller change in the spectra of **9a**. We think that the coordination of the silver takes place at the quinoline nitrogen. The polymer **8a** (R = 2-ethylhexyloxy) is the most unusual of all the PAEs that have been made by us. This PAE is reddish and slightly fluorescent in solution but thick films (see Plate 3) show a golden-metallic luster *(6d,22)*. Depending upon the side chains **8** can show chromic effects almost as dramatic as those seen in the polythiophenes according to Yamamoto *(22)*, who prepared **8b** (R=hexoxy). For **8a** the chromic effects are much pronounced, probably due to the presence of the bulky side chains.

Conclusions

In conclusion poly(aryleneethynylene)s are highly variable and fascinating chromic materials. In the case of the dialkyl-PPEs, solvato- and thermochromicity can be explained by a conformational change of the backbone. In the aggregated planarized form this PPE is yellow and almost non-fluorescent, while in the un-aggregated form the material is colorless and highly fluorescent. Conformational change of the main chains seems to explain the chromicity of **4-6**, while for **7-9** other aspects probably play a role. The distinct chromicity of the PAEs combined with their stability and processibility should be useful for their future applications in sensory schemes and devices.

Acknowledgments: We thank the National Science Foundation (NSF CHE 0138659, DMR 0138948) and the Petroleum Research Funds for generous financial support. We thank Prof. Dr. Dieter Neher, Prof. Dr. Ulli Scherf, Prof. Dr. Dvora Perahia, Dr. Günter Lieser, and Dr. Volker Enkelmann for delightful collaborations and discussions. I thank my enthusiastic coworkers Dr. L. Kloppenburg, Dr. Neil G. Pschirer, Dr. Glen Brizius, and Shelby Shuler for their experimental skill and creativity.

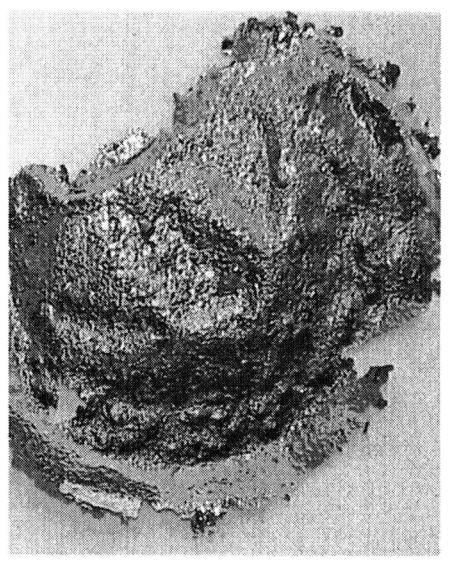

*Plate 3. Thick film of a sample of **8a**. Visible is the merocyanine luster.
(See Page 5 of color insert.)*

References

1. Hotta, S.; Rughooputh, D. D. V.; Heeger, A. J.; Wudl, F. *Macromolecules* **1987**, *20*, 212-215. Maior, R. M. S.; Hinkelmann, K.; Eckert, H.; Wudl, F. *Macromolecules* **1990**, *23*, 1268-1279.
2. Politis, J. K.; Curtis, M. D.; Gonzalez-Ronda, L.; Martin, D. C. *Chem. Mater.* **2000**, *12*, 2798-2804. Gonzalez-Ronda, L.; Martin, D. C.; Nanos, J. I.; Politis, J. K.; Curtis, M. D. *Macromolecules* **1999**, *32*, 4558-4565. Nanos, J. I.; Kampf, J. W.; Curtis, M. D.; Gonzalez-Ronda, L.; Martin, D. C. *Chem. Mater.* **1995**, *7*, 2232-2234.
3. (a) Wenz, G.; Mueller, M. A.; Schmidt, M.; Wegner, G. *Macromolecules* **1984**, *17*, 837-850. Lieser, G.; Tieke, B.; Wegner, G. *Thin Solid Films* **1980**, *68*, 77-90. (b) Sandman, D. J. *Trends Polym. Sci.* **1994**, *2*, 44-55. (c) Lee, D.-C.; Sahoo, S. K.; Cholli, A. L.; Sandman, D. J. *Macromolecules* **2002**, *35*, 4347-4355. Hankin, S. H. W.; Downey, M. J.; Sandman, D. J. *Polymer* **1992**, *33*, 5098-5101. Sandnman, D. J. *Mol. Cryst Liq. Cryst.* **1990**, *189*, 273.
4. McQuade, D. T.; Pullen, A. E.; Swager, T. M. *Chem. Rev.* **2000**, *100*, 2537-2574. Swager, T. M. *Acc. Chem. Res.* **1998**, *31*, 201-207.
5. Ho, H.-A.; Boissinot, M.; Bergeron, M. G.; Corbeil, G.; Dore, K.; Boudreau, D.; Leclerc, M. *Angew. Chem.* **2002**, *41*, 1548-1551. Hiller, M.; Kranz, C.; Huber, J.; Bäuerle, P.; Schuhmann, W. *Adv. Mater.* **1996**, *8*, 219-222.
6. a) Bunz, U. H. F. *Chem. Rev.* **2000**, *100*, 1605-1644. b) Synthesis of **6**: Wilson, J. N.; Waybright, S. M.; McAlpine, K.; Bunz, U. H. F. *Macromolecules* **2002**, *35*, 3799-3800. c) Synthesis of **7**: Wilson, J. N.; Bunz, U. H. F. *Unpublished results..* d) Synthesis of **8**: Bangcuyo, C. G.; Evans, U.; Myrick, M. L.; Bunz, U. H. F. *Macromolecules* **2001**, *34*, 7592-7594. e) Synthesis of **9**: Bangcuyo, C. G.; Rampey-Vaughn, M. E.; Quan, L. T.; Angel, S. M.; Smith, M. D.; Bunz, U. H. F. *Macromolecules* **2002**, *35*, 1563-1568.
7. a) Bunz, U. H. F. *Acc. Chem. Res.* **2001**, 34, 998-1010. b) Synthesis of **4**: Kloppenburg, L.; Song, D.; Bunz, U. H. F. *J Am. Chem. Soc.* **1998**, *120*, 7973-7974. Kloppenburg, L.; Jones, D.; Bunz, U. H. F. *Macromolecules* **1999**, *32*, 4194-4203. c) Synthesis of **5**: Pschirer, N. G.; Bunz, U. H. F. *Macromolecules* **2000**, *33*, 3961-3963.
8. Yamamoto, T.; Koniarudin, D.; Arai, M.; Lee, B. L.; Suganunia, H.; Asakawa, N.; Inoue, Y.; Kubota, K.; Sasaki, S.; Fukuda, T.; Matsuda, H. *J. Am. Chem. Soc.* **1998**, *120*, 2047.
9. Weder, C.; Wrighton, M. S. *Macromolecules* **1996**, *29*, 5157-5165. Ofer, D.; Swager, T. M.; Wrighton, M. S. *Chem. Mater.* **1995**, *7*, 418-425.

10. (a) Halkyard, C. E.; Rampey, M. E.; Kloppenburg, L.; Studer-Martinez, S. L.; Bunz, U. H. F. *Macromolecules* **1998**, *31*, 8655-8659. (b) Sluch, M. I.; Godt, A.; Bunz, U. H. F.; Berg, M. A. *J. Am. Chem. Soc.* **2001**, *123*, 6447-6448.
11. Miteva, T.; Palmer, L.; Kloppenburg, L.; Neher, D.; Bunz, U. H. F. *Macromolecules* **2000**, *33*, 652-654.
12. Bunz, U. H. F.; Enkehnann, V.; Kloppenburg, L.; Jones, D.; Shimizu, K. D.; Claridge, J. B.; zur Loye, H. C.; Lieser, G. *Chem. Mater.* **1999**, *11*, 1416-1424.
13. Kloppenburg, L.; Jones, D.; Claridge, J. B.; zur Loye, H. C.; Bunz, U. H. F. *Macromolecules* **1999**, *32*, 44604463.
14. (a) Seminario, J. M.; Zacarias, A. G.; Tour, J. M. *J. Am. Chem. Soc.* **1998**, *120*, 3970-3974. (b) Okuyama, K.; Hasegawa, T.; Ito, M.; Mikanii, N. *J. Phys. Chem.* **1984**, *88*,1711-1716.
15. Pschirer, N. G.; Byrd, K.; Bunz, U. H. F. *Macromolecules* **2001**, *34*, 8590-8592.
16. Tasch, S.; List, E. J. W.; Hochfilzer, C.; Leising, G.; Schlichting, P.; Rohr, U.; Geerts, Y.; Scherf, U.; Müllen, K. *Phys. Rev. B* **1997**, *56*, 4479-4483.
17. Ball, P. *Made to Measure, New Materials for the 21st Century*. Princeton University Press, Princeton 1997.
18. Kim J.; Swager, T. M. *Nature* **2001**, *411*, 1030-1034.
19. Wilson, J. N.; Bunz, U. H. F. *unpublished results*.
20. Zhou, Q.; Swager, T. M. *J. Am. Chem. Soc.* **1995**, *117*, 12593-12602.
21. Yang, J. S.; Swager, T. M. *J. Am. Chem. Soc.* **1998**, *120*, 11864-11873.
22. Morikita, T.; Yaimguchi, I.; Yamamoto, T. *Adv. Mater.* **2001**, *13*, 1862-1864.

Chapter 12

Novel Two-Photon Absorbing Conjugated Oligomeric Chromophores: Property Modulation by π-Center

O.-K. Kim[1], Z. Huang[1,2], E. Peterman[3], S. Kirkpatrick[3], and C. S. P. Sung[2]

[1]Chemistry Division, Naval Research Laboratory, Washington, DC 20375
[2]Institute of Materials Science, University of Connecticut, Storrs, CT 06269
[3]Air Force Research Laboratory, Wright-Patterson Air Force Base, OH 45433

A series of donor/donor (D/D), donor/acceptor (D/A) and acceptor/acceptor (A/A) pair conjugated chromophores based on a rigid conjugated linker (π-center) were synthesized (D-π-D, D-π-A and A-π-A) and two-photon absorption properties with a particular emphasis on the role of π-centers were studied. Optical and electrochemical properties of the chromophores were also investigated and correlated to two-photon absorption properties.

Introduction

Two-photon absorption (TPA) offers the advantage of high transmission at low incident intensity for fundamental optical frequencies well below the band gap frequency such that the wavelength used for two-photon excitation is roughly twice that for one-photon excitation and thus, scattering on the beam intensity is greatly reduced. Furthermore, in the TPA process, the rate of light absorption is proportional to the square root of the incident intensity while in one-photon absorption the light absorption is linearly proportional to the incident intensity. These allow a wide variety of applications such as optical power limiting (OPL) (*1-3*), two-photon upconverted lasing (*4,5*), two-photon confocal microscopy (*6*) and 3-D optical information storage (*7*).

Although there have been many dyes available, most of them exhibit insignificant two-photon absorptivity for practical applications. It is therefore a great synthetic challenge to develop new efficient TPA materials that meet the specific requirement. Significant progress has been made recently (*8-12*) but systematic studies still are lacking with respect to molecular level structure-property relations of TPA chromophores that would help develop a fine property tuning. As such, the future development of TPA-based technology relies on synthetic success in new dyes with very large TPA cross-sections at desirable wavelengths.

Molecular Concept and Synthetic Strategy

π-Conjugated organic molecules containing donor (D)/acceptor (A) moieties are electroactive, depending on the degree of charge-transfer (CT) as a function of the D/A strength and the length of π-conjugation. This situation is precisely reflected in their optical properties as well as redox properties. This concept is the basis of second-order nonlinear optical (NLO) properties of π-conjugated chromophores. Nonetheless, it has been noticed that π-conjugated linkers play a subtle but significant role in transmission of charges from D to A. Such a role of the linkers as an electron relay has been particularly noted with oligothiophenes (*13,14*). This is most likely due to their polarizability associated with sulfur d orbitals that mix well with aromatic π-orbitals, such that charge-transfer across the relay to A is facilitated. It was further noticed that among oligothiophenes, fused thiophenes such as thienothiophene, dithienothiophene (DTT), were found to be more efficient relays (*14,15*), suggesting the importance of a planar/rigid structure of the relay (*16*). This further suggests that the rigid relay, DTT, facilitates the transmission of electrons between D and A subunits. When a comparison is made with respect to molecular nonlinearities (μβs) among related chromophores (14), whose D and A subunits are the same but differing in their relays, it has been observed that DTT is the most effective relay among the oligothiophenes that were studied.

A similar CT concept has been applied to designing TPA chromophores based on D and A, which are attached to a π-conjugated linker (π-center) symmetrically (*8,11*) or asymmetrically (*9*); D-π -D and A-π -A, or D-π -A, respectively. The symmetry concept emphasizes the importance of the D (or A) strength and conjugation length of the linker bearing diphenypolyene and phenylene vinylene, and the asymmetry concept stresses, on the other hand, D/A strength and the planarity of the π-center. While in the former system, CT occurs bidirectionally, from the end to the center or from the center to the end, in the latter system, CT does occur unidirectionally (Figure 1). It is generally recognized that when a comparison is made with respect to TPAs between D/D and D/A pair chromophores bearing the same conjugated linker, D/D pair systems are more effective compared to D/A pair systems (*10,17*). This seems to suggest that a balanced charge distribution across the π-center is an important criterion of TPA. The extent of charge redistribution in the excited state is determined through the interaction of D or A with π-center. In this regard, the role of π-center is crucial for TPA of chromophores, even though it has not been well defined.

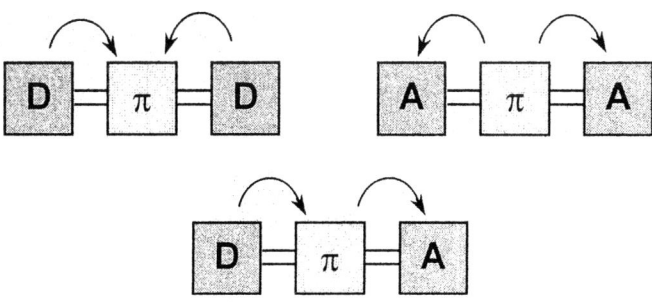

Figure 1. D and/or A pair chromophores and charge-transfer modes.

The role of π-centers in conjugated D-π-A systems has been actively investigated recently on photo-induced electron- and energy-transfers, particularly in porphyrin-based chromophores with various conjugated linkers, focusing on the role of bridging π-centers in electron-transfer and donor-acceptor electronic coupling (*18-20*). Electronic coupling of the D and A is dependent on the π-center and the largest coupling occurs when the energy-gap between the excited states of the donor and π-center is the smallest. This situation is closely related to the rigid π-center structure, which promotes overlap to a great extent with the orbitals of D and A. In other word, a rigid π-center positions two neighboring active centers at a fixed distance and with a well-defined geometry. Furthermore, the π-center provides an effective electronic communication and facilitates electron and/or energy migration between the D and A units (*19,20*).

We are particularly interested in the π-center role of DTT in TPA, due to the high polarizability and rigidity (*10,14*). It has been demonstrated that DTT-based chromophores, particularly the symmetric D/D pair system displayed one of the largest effective TPA cross-section (σ_{TPA}) (*10*). This was verified by a recent theoretical study (*17*) suggesting that the π-center plays the most crucial role for TPA, and DTT is the most efficient π-center among many related ones. However, it is still not certain which trait of DTT, polarizability or rigidity, is more influential on the molecular TPA. We were tempted to make some critical assessment of π-center role based on another planar, rigid linker, fluorene, by modifying it at the C-9 position with electron-donating bis(2-ethylhexyl) unit and with a sterically-demanding fluorenyl unit, forming the spiro-structure that gives an added rigidity to the π-center, while preserving the planarity of the π-center for the active D/D or A/A subunit. We have systematically synthesized chromophores bearing such modified-fluorene π-centers attached by a fixed D/D or A/A pair such that the rigidity effect would be more rigorously assessed by comparing measured σ_{TPA} values among the related chromophores. Here we reveal clear evidence of a pronounced rigidity effect on TPA properties of chromophores bearing fluorene and oligothiophene as well. Discussion of the TPA properties is made in conjunction with their optical properties.

Results and Discussion

Synthesis

Conjugated TPA oligomers based on DTT are prepared by attaching either a D/D or a D/A pair to DTT through conjugation, forming a D-π-D (10, 103 and 105) or a D-π-A (102 and 104) sequence. In this case, D is *N*-ethylcarbazole, *N,N,N*-triiphenylamine or *p*-(di-*n*-butylamino)benzene moiety and A is 2-phenyl-5-(4-*ter*-butyl)-1,3,4-oxadiazole. These chromophores were synthesized by Wittig reaction of dithieno[3,2-2'3'-d]thiophene-2,6-dicarboxaldehyde (DTT-2CHO) with a triphenylphosphonium functional end of D or A moiety such as *N*-ethylcarbazol-3-ylmethyl (D), *N,N*-diphenylamino-p-benzyl (D), *N,N*-dibuthylamino-*p*-benzyl or [2-(*p-ter*-butyl-phenyl-1,3,4-oxadiazol-5-yl)]benzyl (A).

Fluorene-based oligomeric chromophores were synthesized by Wittig reactions by reacting funcionalized $Ph_3P^+CH_2-$ or CHO-terminated D (or A) with a bifunctional fluorene, either bicarboxyaldehyde, CHO-π-CHO or triphenyphosphonium salt, $P^+(Ph)_3CH_2$-π-CH_2 $P^+(Ph)_3$, respectively. The introduction of bis(2-ethylhexyl) or fluorenyl moiety at the C-9 position of fluorene was carried out by modifying the literature procedure (*21*). In the case when the π-center is 9,9-bis(2-ethylhexyl) fluorene, the diformyl intermediate was synthesized by 2,7-dibromomethylation and a subsequent treatment (22) with hexamethylenetetramine, and then with acetic acid. When the π-center is 9,9-spirobifluorene, the 2,7-diformylation was carried out by treating 2,7-dibromo-9,9-spirobifluorene with BuLi in hexane and subsequently with DMF at -78 °C. The chemical structures of fluorene-based chromophores are also given in Figure 2.

$$101 = D_1 - \pi_1 - D_1$$
$$102 = D_1 - \pi_1 - A$$
$$103 = D_2 - \pi_1 - D_2$$
$$104 = D_2 - \pi_1 - A$$
$$105 = D_3 - \pi_1 - D_3$$
$$403 = D_2 - \pi_4 - D_2$$

$$201 = D_2 - \pi_2 - D_2$$
$$203 = A - \pi_2 - A$$
$$301 = D_2 - \pi_3 - D_2$$
$$302 = D_2 - \pi_3 - A$$
$$303 = A - \pi_3 - A$$

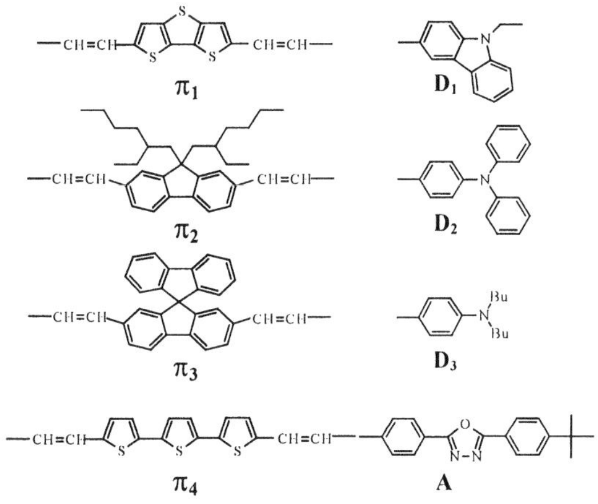

Figure 2. Oligomeric TPA chromophores

Spectroscopic Properties

As shown in Figure 3, when the chromophores comprise an asymmetric D/A pair of mixed components in the molecule (**102** and **104**), their absorption maximum, λ_{max} (of single photon absorption) is slightly red-shifted (ca. 5 nm) relative to that of their symmetric D/D pair counterpart (**101** and **103**). This is associated with a partial charge-transfer (CT) in the excited state of the asymmetric molecule. These oligomers are highly fluorescent (Φ_f of **101**: 0.47,

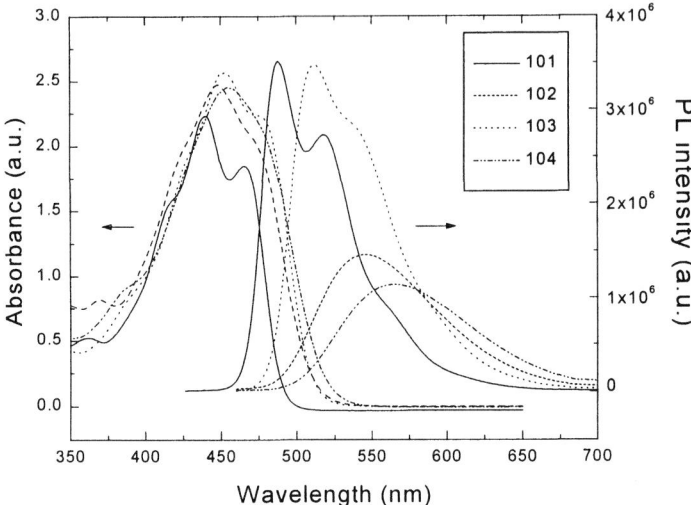

Figure 3. Absorption and emission spectra of DTT-based chromophores. (Reproduced from reference 10. Copyright 2000 American Chemical Society)

for example), particularly when DTT is linked by a symmetric D/D pair. The emission intensity (of single photon excitation) of chromophores with the asymmetric structure is significantly small relative to the symmetric counterpart. This is probably due to a partial quenching associated with intramolecular CT in the excited state (*16*), indicative of a strong coupling between the D and A due to the particular role of DTT.

Table I. Optical Properties of Oligothiophene-based Chromophores

Compound	λ_{max} (nm)	$\varepsilon \times 10^{-4}$ ($cm^{-1}\,mol^{-1}L$)	λ_{em} (nm)	E_g (eV)[a]	Φ_f[b]
101	416, 440, 465	6.62, 8.52, 6.85	487, 518	2.52	0.47
102	447	8.64	545	2.43	0.095
103	453, 479	10.4, 8.69	512, 543	2.38	0.31
104	455	9.97	565	2.36	0.055
105	456	---	585	---	0.52
403	469	1.47	535	2.22	0.070

[a]Band-gap (E_g) is estimated from the onset of absorption. [b]Fluorescence quantum yield (in methylene chloride) is determined relative to 9,10-diphenylanthracene.

Chromophores, **103** and **403** have the same D/D pair with different π-centers, DTT and terthiophene, respectively. Their optical properties are almost identical except the large difference in the quantum yield (Φ_f), which is associated with the presence/absence of planar/rigid π-center structure. Optical properties of fluorene-based (π_2 or π_3) chromophores resemble each other. As long as the D/D or A/A pair is the same, their absorption and emission maxima are very close to each other. The (Φ_f)s of fluorene-based chromophores with the symmetric structures are very high, particularly the spirofluorene (π_3)-based ones (**301** and **303**) relative to their π_2-based counterpart (**201** and **203**). This may be the consequence of the added rigidity of π_3. When compared within the same π-center, the A/A pair chromophores (**203** and **303**) have remarkably high (Φ_f)s relative to their respective D/D pair counterpart (**201** and **301**). Also, when a comparison is made between D/D and D/A pair systems (**301** and **302**), the Φ_f of the latter is much lower and is probably due to the same reason (intramolecular CT) as in the case of **101** and **102**. However, the Φ_f of **302** is relatively higher than that of **102** or **104**, suggesting that electronic coupling between D and A is stronger when the π-center is DTT instead of fluorene.

Table II. Optical Properties of Fluorene-based Chromophores

Compound	λ_{max} (nm)	$\varepsilon \times 10^{-4}$ ($cm^{-1} mol^{-1} L$)	λ_{em} (nm)	E_g (eV)^a	Φ_f^b
201	408	8.00	467	2.65	0.59
203	400, 420	8.39	444, 468	2.72	0.79
301	414	10.20	478	2.55	0.64
302	411	3.70	513	2.53	0.25
303	398, 419	13.10	439, 468	2.66	0.88

^{a}Band-gap from the onset of absorption; ^{b}Fluorescence quantum yield.

Two-Photon Absorption (TPA) Properties

TPA cross-section (σ_{TPA}) values were determined from experimentally measured two-photon absorption coefficient, β, which is obtained by measuring the nonlinear transmissivity (T_i) of chromophores in solution for a given input intensity, I_0, and a given thickness of a sample solution, l_0, from the following relationships (23,24):

$$T_i = [\ln(1+ \beta l_0 I_0)] / \beta l_0 I_0 \tag{1}$$
$$\sigma_{TPA} = h\nu\beta / N_0 = 10^3 \, h\nu\beta / N_A C \tag{2}$$

where N_0 and N_A are the density of assumed absorptive centers and Avogadro's number, respectively, and C is the molar concentration of solute. A modified adaptation of an ultrafast Z-scan and nonlinear fluorescence (NLF) experiments, detailed elsewhere (25,26), were used to measure the effective two-photon absorption coefficients of the sample. Table III summarizes σ_{TPA} values measured with femtosecond pulses for DTT-based chromophore. They all show a remarkably large TPA absorptivity although there are some significant differences in σ_{TPA} values between D/D and D/A pair systems. Such overall distinct TPA properties are assumed to be attributable to the unique electronic modulation role of DTT (17) associated with the large polarizability as well as the rigid/planar structure that contributes a great deal to the reduction of band-gap and the extended π-electron delocalization (16,27,28). Among related chromophores, which have the same D_2/D_2 pair unit but differing in the π-center, **103** exhibits a noticeably larger σ_{TPA} (8,17). The magnitude of σ_{TPA} values in DTT-based chromophores is in the following order: **105** > **103** > **104** ≥ **101** > **102**. In considering that **101**, **103** and **105** are a symmetric D/D structure with different D units, and that **102** and **104** are an asymmetric D/A structure, it can be said that the symmetric structure is more effective for TPA than corresponding asymmetric counterparts. Their effectiveness corresponds to their D strength: $D_3 \gg D_2 > D_1$.

Table III. TPA Cross-Section of various DTT-based Chromophores

	TPA Chromophore					
	101	*102*	*103*	*104*	*105*	*403*
Concentration (10^{-2} mol/L)*	0.5	0.5	0.5	0.5	0.5	0.5
Wavelength	796	796	796	796	790	790
TPA Coeff. β(cm/GW)	0.0206	0.0206	0.0523	0.0288	0.251	0.009
σ'$_{TPA}$ (10^{-20}cm^4/GW)	0.86	0.68	1.74 (1.11)$^+$	0.96	6.85	0.31
σ$_{TPA}$ (10^{-50}cm^4 S/photon)	216	171	438 (279)$^+$	241	1722	78

*in THF solution. $^+$in tetrachloroethane solution

It is interesting to note that σ_{TPA} of some asymmetric D/A pair chromophore such as **104** is larger than that of a symmetric D/D counterpart, **101**. This suggests that the important parameter of σ_{TPA} is not simply being a D/D or D/A pair system but is the electronic property of the individual D or A subunit. This seems to suggest that the energy level between π-center and each active subunit is responsible for TPA of the system. Another interesting comparison with respect to the π-center role is made between DTT (π_1) and a non-rigid version of π-center, terthiophene (π_4). With the same D_2/D_2 pair, **403** exhibits a much smaller σ_{TPA} compared to DTT counterpart (**103**), while their optical as well as redox properties are very similar (*29*) except for the quantum yield (see; Table I). One plausible reason for this is due to the rigidity and planarity that enhance the extended delocalization of π-electrons that is relatively lacking in π_4. This implies that the π-center's role in TPA can be defined not only by electronic property but also by the importance of rigidity.

For a comparative study, fluorene was selected as another rigid π-center, which was derived as two homologs with a different substitution at C_9-position: one is with di-2-ethylhexyl- and the other with spiro-fluorenyl moieties. While the former (π_2) bears electron-rich flexible dialkyl substituents the latter (π_3) features an added rigidity to the fluorene π-center, which preserves the rigidity and planarity for the chromophores. Regardless of the difference in the π-centers (either π_2 or π_3), as long as the D or A is the same, electrochemical properties of the D/D or A/A pair chromophores (**201** and **301**, or **203** and **303**) are almost identical (*30*). In other words, there is no measurable difference in the redox property modulations by π_2 and π_3.

Table IV. TPA Cross-Section of various Fluorene-based Chromophores

	TPA Chromophore				
	201	*203*	*301*	*303*	*AF-50*[+]
Concentration (10^{-2}mol/L)[*]	0.6	0.6	0.5	0.5	4.5[**]
Wavelength	790	790	790	790	796
TPA Coeff. β(cm/GW)	0.0399	0.0120	0.0692	0.130	---
σ'_{TPA} (10^{-20}cm^4/GW)	1.04	0.36	1.92	4.23	0.12
σ_{TPA} (10^{-50}cm^4 S/photon)	261	90	483	1064	30

[+]N,N-Dipheny-7[2-(4-pyridinyl)ethenyl]-9,9-di-didecylfluorene-2-amine. [*]in THF solution. [**]in toluene solution.

However, there are some interesting contrasts between the two fluorene π-centers with respect to their σ_{TPA} values. The π_3-based D/D and A/A pair chromophores, **301** and **303**, display larger σ_{TPA} values compared to the π_2-based counterparts (**201** and **203**). Furthermore, in π_2-based chromophores (**201** and **203**), D/D pair system (**201**) has a larger σ_{TPA} than the A/A pair counterpart (**203**) while in π_3-based chromophores (**301** and **303**) the A/A pair chromophore (**303**) has a larger value than the D/D counterpart (**301**). Such a contrast in TPA behavior between π_3-based and π_2-based chromophores is likely to result from the difference in the suitability of energy level between the π-center and the active subunit, regardless of being D or A. This would be associated not only with electronic energy levels of those components (π-center and D or A) but also with a long lifetime of the excited state of π-center that may be significantly influenced by rigidity. As a consequence, in fluorene-based chromophores, symmetric structures, either D/D or A/A pair system, are more favorable for enhanced TPA, as in the case of DTT-based chromophores.

Asymmetric TPA chromophores based on fluorene, AF-50, an earlier benchmark, and other homologs (*31,32*) represent charge-transfer type chromophores. When compared with a symmetric counterpart, **201** for example, σ_{TPA} of **AF-50** (Table IV) is noticeably smaller than that of **201**. This seems to suggest there is an energetic imbalance across the π-center that causes deficiency in the extended charge redistribution in the excited state. The electronic energy balance in the excited state that seems to be the determinant for a larger σ_{TPA} is contributed mainly by electronic interactions between π-centers and D or A active unit and also by π-center rigidity and their interplay.

Conclusions

We have investigated property modulation role of conjugated relays (π-centers) such as oligothiophenes and fluorenes in two-photon absorption (TPA) of oligomeric chromophores based on such a π-center, which is attached with D/D, D/A or A/A pair active components each at the opposite side. Chromophores based on the polarizable and planar/rigid dithienothiophene (DTT) display one of the largest TPA cross-section, which contrasts strongly to the one based on terthiophene, although their optical (single photon) and redox properties are very similar to each other. Two fluorenes as π-centers, which are substituted at C_9 position, one with dialkyl chains and another with an additional ☐luorine (forming a spiro-structure), have almost identical optical and redox properties. However, their TPA properties are distinctly different; regardless of D/D or A/A pair structure, spirofluorene-based chromophores have notably larger TPAs, particularly when the active unit is A/A pair, proving the effectiveness of added rigidity onto the π-center while the planarity is preserved.

Acknowledgment

The authors would like to acknowledge Office of Naval Research and Air Force Office of Scientific Research for partial funding support.

References

(1) Fleitz, P.A.; Sutherland, R.A.; Strogkendl, F.P.; Larson F.P.; Dalton, L.R. *SPIE Proc.* **1998**, *3472*, 91.
(2) He, G.S.; Bhawalkar, J.D.; Zhao, C.F. Prasad, P.N. *Appl. Phys. Lett.* **1995**, *67*, 2433.
(3) Marder, S.R.; Perry, J. *Opt. Lett.* **1997**, *22*, 1843.
(4) Bhawalkar, J.D.; He, G.S.; Prasad P.N. *Rep. Prog. Phys.* **1996**, *59*, 1041.
(5) He, G.S.; Zhao, C.F.; Bkawalkar, J.D.; Prasad, P.N. *Appl. Phys. Lett.* **1995**, *7*, 1979.
(6) Denk, W; Strickler, J.H.; Webb, W.W. *Science* **1990**, *248*, 73.
(7) Cumpston, B.H.; Ananthavel, S.P.; Barlow, S.; Dyer,D.L.; Ehrich, J.E.; Erskinne, L.L.; Heikal, A.A.; Kuebler, S.M.; Lee, S. I.-Y.; McCord-Maughon, D.; Qin, J.; Roeckel, H.; Rumi, M.; Wu, X.-L.; Marder, S.R.; Perry, J.W. *Nature*, **1999**, *398*, 51.
(8) Albota, M.; Beljonne, D.; Bredas, J.-L.; Ehrilch, J.E.; Fu, J.Y.; Heikal, A.A.; Hess, S.E.; Kogej, T.; Levin, M.D.; Marder, S.R.; McCord-Maughon, D.; Perry, J.W.; Roeckel, H.; Rumi, M.; Subramaniam, G.; Webb, W.W.; Wu, X.-L.; Xu, C. *Science* **1998**, *281*, 1653.
(9) Reinhardt, B.A.; Brott, L.L.; Clarson, S.T.; Dillard, A.G.; Bhatt, J.C.; Kannan, R.; Yuan, L.; He, G.S.; Prasad. P.N. *Chem. Mater.* **1998**, *10*, 1863.
(10) Kim, O.-K.; Lee, K.-S.; Woo, H.Y.; Kim, K.S.; He, G.S.; Swiatkiewicz, J.; Prasad, P.N. *Chem.Mater.* **2000**, *12*, 284.
(11) Rumi, M.; Ehrich, J.E.; Heikal, A.H.; Perry, J.W.; Barlow, S.; Hu, Z.; McCord-Maughon, D.; Parker, T.C.; Roeckel, H.; Thayumanavan, S.; Marder,S.R.; Beljonne,D.; Bredas, J.-L. *J. Am. Chem. Soc.* **2000**, *122*, 9500.
(12) Ventelon, L.; Charier, S.; Moreaux, L.; Mertz, J.; Blanchard-Desc, M. *Angew. Chem. Int. Ed.* **2001**, *40*, 2098.
(13) Jen, A.K.-J.; Rao, V.P; Wong, K.Y.; Drost, K.J. *J. Chem. Soc., Chem. Commun.* **1993**, 90.
(14) Kim, O.-K.; Fort, A.; Barzoukas, M.; Blanchard-Desce, M.; Lehn, J.-M. *J. Mater. Chem.* **1999**, *9*, 2227.
(15) Rao, V.P.; Wong, K.Y.; Jen, A. K.-J.; Drost, K.J. *Chem. Mater.* **1994**, *6*, 2210.
(16) Kim, O.-K.; Lehn, J.-M. *Chem. Phys. Lett.* **1996**, *255*, 147.
(17) Wang, C.-K.; Macak, P.; Luo, Y.; Agren, H. *J. Chem. Phys.* **2001**, *114*, 9813.
(18) Hayes, RT; Wasielewski, M.R.; Gosztola, D. *J. Am. Chem. Soc.* **2000**, *122*, 5563.
(19) Kilsa, K; Kajamus, J.; MacPherson, A.N.; Martensson, J.; Albinsson, B. *J. Am. Chem. Soc.* **2001**, *123*, 3069.
(20) Odobel, F.; Suresh, S.; Blart, E.; Nicholas, Y.; Quintord, J.P.; Janvier, P.;

Le Questel, J.-Y.; Illien, B.; Rondeau, D.; Richomme, P.; Haeupl, T.; Wallin, S.; Hammarstroem, L. *Chem. Eur. J.* **2002**, *8*, 3027.
(21) Wu, R.; Schum, J.S.; Pearson, D.L.; Tour, J.M. *J. Org. Chem.* **1996**, *61*, 6905.
(22) Huang, Z.; Kim. O.-K. will be published elsewhere.
(23) He, G.S.; Cheng, L.-Y.; Bhawalkar, J.D.; Prasad; Brott, L.L.; Clarson, S.J. *Opt. Soc. Am. B*, **1997**, *14*, 1079.
(24) Tutt, L.W.; Boggess, T.F. *Prog. Quantum Electron.* **1993**, *17*, 2279.
(25) Sheik-Bahae, M; Said, A.A ; Wei, T.-H; Hagan, D.J.; Van Strylan, E.W. *IEEE J. Quantum Electron.* **1990**, *26*, 760.
(26) Natarajan, L.V. ; Kirkpatrick, S. M.; Sutherland, R. L.; Sowards, L; Spangler, C.W.; Fleitz, P. A.; Cooper, P. A. *SPIE* **1998**, *3472*, 151.
(27) Roncali, J.; Thobie-Gautier, C.; *Adv. Mater.* **1994**, *6*, 846.
(28) Blanchard, P.; Brisset, H.; Riou, A.; Hierle, R.; Roncali, J. *J. Org. Chem.* **1998**, *63*, 8310.
(29) Oxidation/reduction potentials of **103** and **403** are 0.69/-1.75 V and 0.76/-1.65 V (*vs* SCE), respectively.
(30) Oxidation potentials of **201/ 301** and reduction potentials of **203/ 303**, for example, are 0.84/0.87 V and −1.71/-1.65 V (*vs* SCE), respectively.
(31) Reinhardt, B.A.; Brott, L.L.; Clarson, S.J.; Dillard, A.G.; Bhatt, J.C.; Kannan, R.; Yuan, L.; He, G.S.; Prasad, P.N. *Chem. Mater.* **1998**, *10*, 1863.
(32) Kannan, R.; He, G.S.; Yuan, L.; Xu, F.; Prasad, P.N.; Dombroskie, A.G.; Reinhardt, B.A.; Bauer, J.W.; Vaia, R.A.; Tan, L.-S. *Chem. Mater.* **2001**, *13*, 1896.

Chapter 13

Synthesis, Properties, and Applications of Photochromic Amorphous Molecular Materials and Electrochromic Polymers

Yasuhiko Shirota, Hideyuki Nakano, Ichiro Imae, Yutaka Ohsedo, Yoshiaki Yasuda, Hisayuki Utsumi, Toshiki Ujike, and Toru Takahashi

Department of Applied Chemistry, Faculty of Engineering, Osaka University, Yamadaoka, Suita, Osaka 565-0871, Japan

The synthesis, properties, and applications of photochromic amorphous molecular materials and novel organic electrochromic materials including molecular gels and polymers containing pendant oligothiophenes are described.

Introduction

Photochromic materials have recently come to receive renewed interest in view of their potential technological applications for optical data storage, visible image formation, and optical switching. Electrochromic materials have also attracted a great deal of attention because of their potential applications for display devices and smart windows that control the sun radiation. Since such practical applications usually require materials as solid films, there have been extensive studies on both photochromic and electrochromic polymers. Molecularly dispersed polymer systems, where low molecular-weight photochromic compounds are dispersed in a polymer binder, have also been studied. We report here our recent work on the creation of photochromic amorphous molecular materials, electrochromic molecular gels, and a new type of electrochromic polymers containing pendant oligothiophenes with well-defined structures.

1. Photochromic Amorphous Molecular Materials

Since the late 1980s, we have performed a series of studies on the creation of low molecular-weight organic compounds that readily form amorphous glasses above room temperature, which we refer to as amorphous molecular materials, and their structures, reactions, properties, and applications.[1-5] As a part of these studies, we have proposed a new concept, "photochromic amorphous molecular materials"[6,7] and have studied the creation of such materials.[6-11] Photochromic amorphous molecular materials are expected to constitute a new class of photochromic molecular materials that form uniform amorphous thin films by themselves. They have an advantage that there is no dilution of photochromic chromophores relative to photochromic polymers and composite polymer systems, where low molecular-weight organic photochromic compounds may crystallize at high concentrations. The synthesis, properties, and applications of two novel classes of photochromic amorphous molecular materials based on azobenzene and dithienylethene are described here.

1-1. Azobenzene-based Photochromic Amorphous Molecular Materials

(a) Synthesis and Glass-forming Properties

We have designed and synthesized a series of photochromic amorphous molecular materials based on azobenzene, 4-[bis(4-methylphenyl)amino]azobenzene (BMAB), 4-[bis(4-methylphenyl)amino]-4'-methoxyazobenzene (MeO-BMAB), 4-[di(biphenyl-4-yl)amino]azobenzene (DBAB), 4-[bis(9,9-dimethylfluoren-2-yl)amino]azobenzene (BFlAB), 4,4'-bis[bis(4-methylphenyl)-amino]azobenzene (BBMAB), and 4,4'-bis[bis(4'-tert-butylbiphenyl-4-yl)-amino]azobenzene (t-BuBBAB).[6,7,11] All these compounds readily form amorphous glasses with well-defined glass-transition temperatures (Tgs) when their melt samples are cooled on standing in air, as evidenced by differential scanning calorimetry (DSC), X-ray diffraction, and polarizing microscopy. The Tgs of BMAB, MeO-BMAB, DBAB, BFlAB, BBMAB, and t-BuBBAB are 27, 33, 68, 97, 79, and 177°C, respectively, as determined by DSC. The Tg significantly increases with increasing molecular size and weight, in particular, by the introduction of a rigid biphenyl or fluorenyl moiety. These compounds also form uniform amorphous films by vacuum deposition and spin coating.

BMAB **MeO-BMAB** **DBAB**

BFlAB **BBMAB** **t-BuBAAB**

(b) Photochromic Behavior

These novel compounds containing an azobenzene chromophore exhibit photochromism in their amorphous films as well as in solution. Figure 1 shows the electronic absorption spectral change of an amorphous film of BMAB prepared by spin coating from benzene solution. Upon irradiation with 450 nm light, the absorbance of the band around 430 nm gradually decreased due to the photoisomerization from the trans- to the cis-form. When irradiation was stopped after the reaction system had reached the photostationary state, the electronic absorption spectra of the film gradually recovered to the original one due to the backward cis-trans thermal isomerization.

The fractions of the photogenerated cis-isomers of these azobenzene derivatives at the photostationary state upon irradiation with 450 nm-light are smaller for the amorphous film (ca. 0.54 for BMAB, DBAB, and BFlAB; 0.15 - 0.17 for BBMAB and t-BuBBAB) than for solution (0.80 - 0.85), decreasing with increasing molecular size. It is suggested that the local free volume around the remaining trans-isomer in the amorphous film is not large enough to allow the isomerization from the trans-form to the cis-form to the extent as observed in solution.

It was expected that the thermal stabilization of the photogenerated cis-form can be enhanced by the incorporation of a bulky group. In fact, the photogenerated cis-isomer of t-BuBBAB was found to be fairly stable in its amorphous film; 80% of the photogenerated cis-t-BuBBAB still remained after 5 days at room temperature. This is the first example of the most stable cis-azobenzene derivative in the amorphous film.[7,11]

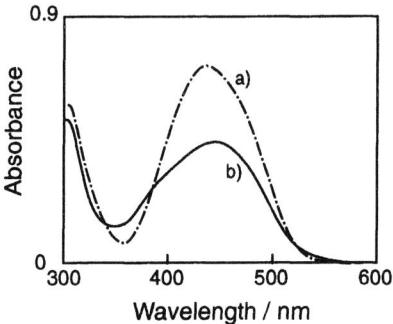

Figure 1. Electronic Absorption Spectral Change of Amorphous BMAB Film   a) before photoirradiation. b) photostationary state upon irradiation with 450 nm-light.

(c) Formation of Surface Relief Grating

Formation of surface relief grating (SRG) by irradiation of amorphous films of azobenzene-functionalized acrylate, methacrylate, and epoxy-based polymers with two coherent laser beams has been a topic of current interest.[12-20] SRG, which is formed by the mass transport induced by the photoisomerization of the azobenzene chromophore, may find potential applications for erasable and rewritable holographic memory, polarization discriminators, and waveguide couplers.

We have investigated the formation of SRG using azobenzene-based photochromic amorphous molecular materials. As compared with polymers, amorphous molecular materials are thought to provide more simple systems that are free from polymer chains and hence their entanglement for the studies of SRG formation, and to enable studies of the correlation between molecular structure and the diffraction efficiency of the resulting SRG. The SRG formation depended on the polarization direction of two writing beams; relatively high diffraction efficiency was obtained when the p-:p- and +45°:-45° (with respect to the p-polarization direction) polarized writing beams were used. The diffraction efficiencies obtained for the DBAB and BFlAB amorphous films were ca. 7 and 25%, respectively, when +45°:-45° polarized writing Ar^+ laser beams at 10mW each were used.[9,11] The modulation depths were ca. 200 and 280 nm, respectively, as measured by atomic force microscopy (AFM). These are the first examples of SRG formation with high diffraction efficiencies and large modulation depths for photochromic amorphous molecular materials.

1-2. Dithienylethene-based Photochromic Amorphous Molecular Materials

(a) Synthesis and Glass-forming Properties

We have designed and synthesized another novel class of photochromic amorphous molecular materials based on dithienylethene, 1-{5-[4-(di-*p*-tolylamino)phenyl]-2-methylthiophen-3-yl}-2-(2,5-dimethylthiophen-3-yl)-3,3,4,4,5,5-hexafluorocyclopentene (TPTTC), 1-{5-[4-(di-*p*-tolylamino)phenyl]-2-methylthiophen-3-yl}-2-(2-methylbenzo[b]thiophen-3-yl)-3,3,4,4,5,5-hexafluorocyclopentene (TPTBC), and 1,2-bis{5-[4-(di-*p*-tolylamino)phenyl]-2-methylthiophen-3-yl}-3,3,4,4,5,5-hexafluorocyclopentene (BTPTC).[8,10] These compounds readily form stable amorphous glasses with Tgs of 51, 66, and 94 °C, respectively, when the melt samples are cooled on standing in air. They also form uniform amorphous films by spin coating. Their photocyclized isomers, TPTTC-c, TPTBC-c, and BTPTC-c, which were obtained by irradiation of the corresponding open forms with 365 nm light in solution also form amorphous glasses with Tgs of 57, 73, and 104°C, respectively.

TPTTC   **TPTBC**   **BTPTC**

(b) Photochromic Behavior

TPTTC, TPTBC, and BTPTC exhibit photochromism in solution and as amorphous films, as shown in Figure 2. The quantum yields for the photocyclization reactions (Φ_{o-c}) of TPTTC, TPTBC, and BTPTC in solution were from 0.61 to 0.81, which are much higher than those reported for other dithienylethene derivatives in solution (mostly in the range from 0.3 to 0.5).[21] The values of Φ_{o-c} as amorphous films (0.28 - 0.33) were smaller than those in solution for all the compounds. On the other hand, the quantum yields for the backward photoinduced ring-opening reactions (Φ_{o-c}) of TPTTC-c, TPTBC-c, and BTPTC-c to regenerate the corresponding open forms both in solution and as amorphous films were 0.001-0.015. These Φ_{o-c} values are approximately two orders of magnitude smaller than those reported for other cyclized dithienylethene derivatives in solution, e.g. 1,2-bis(2,4-dimethylthiophen-3-yl)-3,3,4,4,5,5-hexafluorocyclopentene and 1,2-bis(2-methylbenzo[b]thiophen-3-yl)-3,3,4,4,5,5-hexafluorocyclopentene (0.13 and 0.41, respectively).[21]

The molar ratios of the photocyclized molecule to the total amount of the starting and photocyclized molecule at the photostationary state (Y_{pss}) for TPTTC, TPTBC, and BTPTC were almost 100% in solution and 0.68, 0.36, and 0.77, respectively, as amorphous films. Next, the photocyclization was studied using the amorphous films of the open-forms with only the ap-conformation, which were obtained from the 100% cyclized amorphous films of TPTTC-c, TPTBC-c, and BTPTC-c by irradiation with visible light (>580 nm). The results that the Y_{pss} values for these films (0.69, 0.38, and 0.79, respectively) are almost the same as those for the corresponding films obtained by spin coating from the solution of the unirradiated compounds of the open form indicate that almost all molecules take up the anti-parallel conformation.[10]

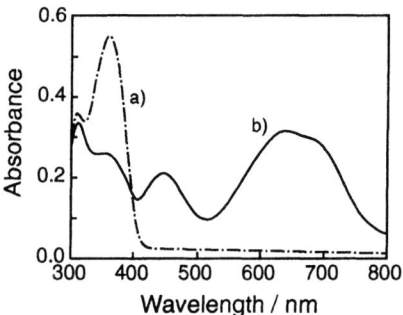

Figure 2. Electronic Absorption Spectral Change of Amorphous BTPTC Film   a) before photoirradiation. b) photostationary state upon irradiation with 365 nm-light.

(c) Application for Dual Image Formation

The amorphous films of these compounds containing the dithienyethene chlomophore exhibit clear color changes upon irradiation with UV or visible light. Dichroism was induced by irradiation of the amorphous films of TPTBC and BTPTC with linearly polarized light.[10] For example, when the colored film of TPTBC, which was obtained by irradiation of an isotropic amorphous film of TPTBC with non-polarized 365 nm-light, was irradiated with linearly polarized red light of the wavelength longer than 580 nm at ca. 20 mW cm^{-1} to induce the backward photochemical ring-opening reaction, the color of the film gradually decayed. The intensity of the electronic absorption band at 585 nm was found to depend on the angle (θ) between the polarization directions of the irradiated red light and the probe light. The absorbance of the cyclized form decreased with decreasing angle θ, being the greatest for θ = 90° and the smallest for θ = 0°.[10]

Using the phenomenon of such dichroism, dual image formation at the same location of the amorphous film was realized.[10] The blue-colored film of TPTBC, obtained by irradiation with non-polarized 365 nm-light, was irradiated with polarized red light of horizontal polarization (θ = 0°) through a mask, where the area other than the letter in the mask was exposed, followed by irradiation with another polarized red light of vertical polarization (θ = 90°) through another mask, where the area other than the letter in the mask was exposed. When the film is viewed through the polarizer with θ = 0°, the letter in the first mask is visible, while the letter in the second mask is visible through the polarizer with θ = 90°. The formation of such dual images at the same location can be applied for stereo image formation by the use of the two polarizers.

2. Electrochromic Organic Materials

Electrochromic devices have characteristics of low-voltage operation, good optical contrast, and a wide-viewing angle owing to the non-emitting nature of materials. Organic electrochromic materials are of interest because of their potential capability of multi-color display by the selection of suitable chromophores. In particular, polymeric electrochromic materials are attractive for their good cycle life owing to their ability to be cast in the film form. Our recent work on the creation of electochromic molecular gels and novel electrochromic polymers are described here.

2-1. Electrochromic Molecular Gels

There have been extensive studies on polymer gels;[22-24] however, few studies have been made of organic gels composed of small molecules. Low molecular-weight organic gels, which we refer to as "molecular gels", are expected to constitute a new class of organic functional materials that take the form of solid with solvent inclusion. We have performed studies of the creation of novel molecular gels and their structures, properties, and applications.[25-28]

We have designed and synthesized several novel classes of organic compounds for gel formation, e.g., N,N',N''-tristerayltrimesamide (TSTA), 4,4',4''-tris(stearoylamino)triphenylamine (TSATA), and 1,3,5-tris(4-stearoylaminophenyl)benzene (TSAPB), and have found that these compounds form gels with a variety of organic solvents, immobilizing solvent at low concentrations. Typically, an amount of 1.3 g TSATA incorporates 1000 ml of 1,2-dichloroethane.

TSTA **TSATA** **TSAPB**

Gels, which take the form of a solid but contain a large amount of solvent within, are thought to behave like a liquid microscopically. Therefore, it was expected that molecular gels containing an electrolyte function as a new type of ionic conductors and that molecular gels having a redox moiety function as a new type of potential electrochromic materials with dimensional stability.

The cell with the TSATA/benzonitrile gel containing tetrabutylammonium perchlorate as an electrolyte sandwitched between an indium-tin-oxide (ITO)-coated glass as a working electrode and a glass substrate with a platinum wire as a counter electrode exhibited a reversible clear color change on electrochemical oxidation and reduction, turning green and colorless when +3 V and -3 V, respectively, were applied to the ITO electrode. The electronic absorption spectrum of the anodically oxidized, green-colored gel shows a new absorption band with λ_{max} at 800 nm attributable to the TSATA radical cation, which disappears on electrochemical reduction.[25] Thus, the TSATA molecular gels function as potential electrochromic materials. Compared with a composite electrochromic system consisting of polymer gels, supporting electrolyte, and an electrochromic compound, the TSATA molecular gels have an advantage that no diffusion of the electrochromic chromophore takes place as it is immobilized by covalent bonds.

2-2. Electrochromic Polymers Containing Pendant Oligothiophenes

Electrically conducting polymers such as polypyrrole, polythiophene, and polyisothianaphthene have received attention as potential electrochromic materials, since they undergo reversible color changes on electrochemical doping and dedoping.[29-32] Recently, new electrochromic polymers have been reported, which include poly(3,4-alkylenedioxythiophene)s,[33-35] poly{2,7-bis-[2,2'-(3,4-ethylenedioxythiophenyl)]-N-alkylcarbazole},[36] and poly(3,4-propylenedioxypyrrole),[37] electrolytically polymerized acenaphthofluoranthene.[38]

As a part of our studies on the synthesis, properties, and applications of oligothiophenes with well-defined structures,[39-48] we have designed and synthesized a new class of vinyl and methacrylate polymers containing pendant oligothiophenes of varying conjugation lengths. Non-conjugated polymers containing pendant oligothiophenes are expected to have good processability and to show properties characteristics of oligothiophenes. They constitute a new class of electrochromic materials.[49-55]

(a) Vinyl Polymers Containing Pendant Oligothiophenes

A novel family of electrochemically-doped vinyl polymers containing α,α'-oligothiophenes with various conjugation lengths, poly(5-vinyl-2,2':5',2''-terthiohene) (PV3T), poly(5-vinyl-2,2':5',2'':5'',2'''-quaterthiophene) (PV4T), poly(4',3'''-dioctyl-5-vinyl-2,2':5',2'':5'',2''':5''',2''''-quinquethiohene) (PVDOc5T), and poly(4',3''''-dioctyl-5-vinyl-2,2':5',2'':5'',2''':5''',2'''':5'''',2'''''-sexiterthiohene) (PVDOc6T), have been designed and synthesized by the electrochemical polymerization of the corresponding vinyl monomers in the presence of tetra-n-butylammonium perchrolate in solution.[49,50,52] The electrochemically-doped polymers, which were obtained as deep-colored, smooth and lustrous films on the surface of the electrode, have been identified as radical-cation salts of the pendant oligothiophenes with ClO_4^- as a dopant. The doped polymers are partially cross-linked due to the coupling reaction of the pendant oligothiophene radical cations. The degree of cross-linking decreased with increasing conjugation length of the pendant oligothiophenes.[49,50,52] The occurrence of cross-linking is indispensable for making the polymer film insoluble in solvents.

The thin lustrous film of electrochemically-doped PV3T obtained by electrolytic polymerization of the corresponding vinyl monomer was found to undergo reversible clear color changes from bluish purple to pale yellow and vice versa on electrochemical dedoping and doping. The color switched from green to pale yellow and vice versa for the films of PV4T, PVDOc5T and PVDOc6T.[49,50,52]

PV3T   PV4T   PVDOc5T   PVDOc6T

(b) Methacrylate Polymers Containing Pendant Oligothiophenes and Copolymers with Methacrylate Containing an Oligo(ethyleneoxide) Moiety

In order to synthesize high molecular-weight neutral polymers that permit the formation of uniform films by either spin coating or solvent casting, we have synthesized a series of new methacrylate polymers containing oligothiophenes of varying conjugation lengths, poly[(2,2':5',2"-terthiohene-5-yl)methyl methacrylate (PMA3T), poly[3,3'''-dioctyl-2,2':5',2":5",2'''-quaterthiophen-5-yl)methyl methacrylate] (PMADOc4T), poly[(4',3'''-dioctyl-2,2':5',2":5",2''':5''',2''''-quinquethiohen-5-yl)methyl methacrylate] (PMADOc5T), and poly[(4',3''''- dioctyl-2,2':5',2":5",2''':5''',2'''':5'''',2'''''-sexiterthiohen-5-yl)-

PMA3T   PMADOc4T   PMADOc5T   PMADOc6T

methyl methacrylate] (PMADOc6T), by free radical polymerization of the corresponding new monomers. The resulting polymers with Mws from 57,000 to 67,000 are soluble in benzene, THF, and dichloromethane, forming transparent uniform amorphous films by spin coating and solvent casting.[51,55]

Reversible clear color changes from yellowish orange to purple, from orange to blue, and from orange to intense blue took place on electrochemical doping and dedoping of PMA3T, PMADOc4T, and PMADOcnT (n = 5, 6), respectively.[51,55]

It was expected that the electrochromic response time can be improved by enhancing ion transport in polymer films and that the incorporation of a polar and flexible group such as oligo(ethyleneoxide) into the polymer enhances ion transport in polymer films. Based on this concept, we have designed and synthesized novel copolymers containing pendant terthiophene and oligo(ethyleneoxide) moieties, poly{(2,2':5',2''-terthiophen-5-yl)methyl methacrylate-co-α-ethoxy-ω-methacryloyl-oligo(ethyleneoxide)] (poly(MA3T-co-MAEO)).[53]

poly(MA3T-co-MAEO)

Novel methacrylate copolymers containing pendant terthiophene and oligo(ethyleneoxide) moieties, poly(MA3T-co-MAEO) with 80 / 20 and 60 / 40 mole fraction (MA3T / MAEO), were prepared by radical copolymerization of the corresponding monomers. Poly(MA3T-co-MAEO) copolymer films exhibited reversible, clear color change from yellowish orange to purple and vice versa on electrochemical doping and dedoping. Table I lists the response times for the color change on doping and dedoping of PMA3T and poly(MA3T-co-MAEO) films. It was found that the response time decreased with increasing fraction of an oligo(ethyleneoxide) moiety. The improvement of electrochromic response time for the methacrylate copolymers relative to the homopolymer is ascribed to the improved ion-transporting properties of the copolymer films. These results show that the incorporation of a polar and flexible group such as oligo(ethyleneoxide) into the electroactive polymer by copolymerization is an effective approach for improving electrochromic response time.

Table I. Response Time for Electrochemical Doping and Dedoping of Methacrylate Homo- and Copolymers, PMA3T and Poly(MA3T-*co*-MAEO)

Polymer film	Response Time / s [a]	
	Doping	Dedoping
PMA3T	1.2	1.2
poly(MA3T-*co*-MAEO) 80 / 20	0.7	1.0
poly(MA3T-*co*-MAEO) 60 / 40	0.6	0.8

[a] The response time was defined as the time required for an absorbance change of 90% at the wavelength of the absorption maximum (550 nm) of terthiophene radical cation.

Conclusions

Two novel classes of photochromic amorphous molecular materials based on azobenzene and dithienylethene have been created. Photochromic behavior of these materials in solution and as amorphous films has been elucidated. These photochromic amorphous molecular materials based on azobenzene and dithienylethene have found applications for SRG formation and dual image formation, respectively.

Novel molecular gels and a new class of vinyl and methacrylate polymers containing pendant oligothiophenes of varying conjugation lengths have been developed. These materials undergo reversible, clear color changes on electrochemical oxidation and reduction, constituting novel classes of potential electrochromic materials.

References

(1) Shirota, Y.; Kobata, T.; Noma, N. *Chem. Lett.*, **1989**, 1145.
(2) Higuchi, A.; Inada, H.; Kobata, T.; Shirota, Y. *Adv. Mater.*, **1991**, *3*, 549.
(3) Ishikawa, W.; Inada, H.; Nakano, H.; Shirota Y., *Chem. Lett.*, **1991**, 1731.
(4) Inada, H.; Shirota, Y. *J. Mater. Chem.*, **1993**, *3*, 319.

(5) Shirota, Y. *J. Mater. Chem.*, **2000**, *10*, 1 and references cited therein.
(6) Yoshikawa, S.; Kotani, Y.; Shirota, Y. *69th Annual Meeting of the Chemical Society of Japan*, Kyoto, **1995**, prepr. No. 2, pp. 641.
(7) Shirota, Y.; Moriwaki, K.; Yoshikawa, S.; Ujike, T.; Nakano, H. *J. Mater. Chem.*, **1998**, *8*, 2579.
(8) Utsumi, H.; Nagahama, D.; Nakano, H.; Shirota, Y. *J. Mater. Chem.*, **2000**, *10*, 2436.
(9) Nakano, H.; Takahashi, T.; Kadota, T.; Shirota, Y. *Adv. Mater.*, **2002**, *14*, 1157.
(10) Utsumi, H,; Nagahama, D.; Nakano, H.; Shirota, Y. *J. Mater. Chem.*, **2002**, *12*, 2612.
(11) Shirota, Y.; Moriwaki, K.; Yoshikawa, S.; Ujike, T.; Nagahama, D.; Takahashi, T.; Nakano, H. to be submitted for publication.
(12) Rochon, P.; Batalla, E.; Natansohn, A. *Appl. Phys. Lett.*, **1995**, *66*, 136.
(13) Kim, D. Y.; Tripathy, S. K.; Li, L.; Kumar, J. *Appl. Phys. Lett.*, **1995**, *66*, 1166.
(14) Barret, C.; Natansohn, A.; Rochon, P. *J. Phys. Chem.*, **1996**, *100*, 8836.
(15) Lefin, P.; Fiorini, C.; Nunzi, J. M. *Pure Appl. Opt.*, **1998**, *7*, 71.
(16) Pedersen, T. G.; Johansen, P. M.; Holme, N. C. R.; Ramanujam, P. S.; Hvilsted, S. *Phys. Rev. Lett.*, **1998**, *80*, 89.
(17) Kumar, J.; Li, L.; Jiang, X. L.; Kim, D. Y.; Lee, T. S.; Tripathy, S. K. *Appl. Phys. Lett.*, **1998**, *72*, 2096.
(18) Lagugné Labarthet, F.; Buffeteau, T.; Sourisseau, C. *J. Phys. Chem. B*, **1998**, *102*, 2654.
(19) Viswanathan, N. K.; Kim, D. Y.; Bian, S.; Williams, J.; Liu, W.; Li, L.; Samuelson, L.; Kumar, J.; Tripathy, S. K. *J. Mater. Chem.*, **1999**, *9*, 1941.
(20) Fiorini, C.; Prudhomme, N.; de Veyrac, G.; Maurin, I.; Raimond, P.; Nunzi, J.-M. *Synth. Metals*, **2000**, *115*, 121.
(21) Irie, M.; Sakemura, K.; Okinaka, M.; Uchida, K. *J. Org. Chem.*, **1995**, *60*, 8305.
(22) Yu, X.; Tanaka, A.; Tanaka, K.; Tanaka, T. *J. Chem. Phys.*, **1992**, *97*, 7805.
(23) Osada, Y.; Okuzaki, H.; Hori, H. *Nature*, **1992**, *355*, 242.
(24) Bohidar, H. B.; Jena, S. S. *J. Chem. Phys.*, **1993**, *98*, 8970.
(25) Yasuda, Y.; Takebe, Y.; Fukumoto, M.; Inada, H.; Shirota, Y. *Adv. Mater.*, **1996**, *8*, 740.
(26) Yasuda, Y.; Iishi, E.; Inada, H.; Shirota, Y. *Chem. Lett.*, **1996**, 575.
(27) Kamiyama, T.; Yasuda, Y.; Shirota, Y. *Polym. J.*, **1999**, *31*, 1165.
(28) Yasuda, Y.; Kamiyama, T.; Shirota, Y. *Electrochimica Acta*, **2000**, *45*, 1537.
(29) Garnier, F.; Tourillon, G.; Gazard, M.; Dubois, J. C. *J. Electroanal. Chem.*, **1983**, *148*, 299.

(30) Yoshino, K.; Kaneto, K.; Inuishi, Y. *Jpn. J. Appl. Phys.*, **1983**, *22*, L157.
(31) Roncali, *J. Chem. Rev.*, **1992**, *92*, 711.
(32) Yashima, H.; Kobayashi, M.; Lee, K.-B.; Chung, D.; Heeger, A. J.; Wudl, F. *J. Electrochem. Soc.*, **1987**, *134*, 46.
(33) Sotzing, G. A.; Reddinger, J. L.; Katritzky, A. R.; Soloducho, J.;Musgrave, R.; Reynolds, J. R.; Steel, P. J. *Chem. Mater.*, **1997**, *9*, 1578.
(34) Sankaran, B.; Reynolds, J. R. *Macromolecules*, **1997**, *30*, 2582.
(35) Welsh, D. M.; Kumar, A.; Meijer, E. W.; Reynolds, J. R. *Adv. Mater.*, **1999**, *11*, 1379.
(36) Kumar, A., Welsh, D. M., Morvant, M. C.; Piroux, F.; Abboud, K. A.; Reynolds, J. R. *Chem. Mater.*, **1998**, *10*, 869.
(37) Schottland, P.; Zong, K.; Gaupp, C. L.; Thompson, B. C.; Thomas, C. A.; Giurgiu, I.; Hickman, R.; Abboud, K. A.; Reynolds, J. R.; *Macromolecules*, **2000**, *33*, 7051.
(38) Debad, J. D.; Bard, A. J. *J. Am. Chem. Soc.*, **1998**, *120*, 2476.
(39) Noma, N.; Kawaguchi, K.; Imae, I.; Nakano, H.; Shirota, Y. *J. Mater. Chem.*, **1996**, *6*, 117.
(40) Noma, N.; Kawaguchi, K.; Imae, I.; Shirota, *Synth. Met.*, **1997**, *84*, 597.
(41) Noda, T.; Imae, I.; Noma, N.; Shirota, Y. *Adv. Mater.*, **1997**, *9*, 239.
(42) Noda, T.; Ogawa, H.; Noma, N.; Shirota, Y. *Adv. Mater.*, **1997**, *9*, 720.
(43) Noda, T.; Ogawa, H.; Noma, N.; Shirota, Y. *Appl. Phys. Lett.*, **1997**, *70*, 699.
(44) Noda, T.; Shirota, Y. *J. Am. Chem. Soc.*, **1998**, *120*, 9714.
(45) Noda, T.; Ogawa, H.; Noma, N.; Shirota, Y. *J. Mater. Chem.*, **1999**, *9*, 2177.
(46) Shirota, Y.; Kinoshita, M.; Noda, T.; Okumoto, K.; Ohara, T. *J. Am. Chem. Soc.*, **2000**, *122*, 11021.
(47) Liu, P.; Nakano, H.; Shirota, Y. *Liq. Cryst.*, **2001**, *28*, 581.
(48) Liu, P.; Shirota, Y.; Osada, Y. *Polym. Adv. Technol.*, **2000**, *11*, 512.
(49) Nawa, K.; Miyawaki, K.; Imae, I.; Noma, N.; Shirota, Y. *J. Mater. Chem.*, **1993**, *3*, 113.
(50) Nawa, K.; Imae, I.; Noma, N.; Shirota, Y. *Macromolecules*, **1995**, *28*, 723.
(51) Ohsedo, Y.; Imae, I.; Noma, N.; Shirota, Y.; *Synth. Met.*, **1996**, *81*, 157.
(52) Imae, I.; Nawa, K.; Ohsedo, Y.; Noma, N.; Shirota, Y. *Macromolecules*, **1997**, *30*, 380.
(53) Ohsedo, Y.; Imae, I.; Shirota, Y.; *Synth. Met.*, **1999**, *102*, 969.
(54) Ohsedo, Y.; Imae, I.; Shirota, Y. *Electrochimica Acta*, **2000**, *45*, 1543.
(55) Ohsedo, Y.; Imae, I.; Shirota, Y., *J. Polym. Sci. Part B: Polym. Phys.*, in press.

Tunable Emission and Electroluminescence

Chapter 14

Voltage-Tunable Multicolor Electroluminescence from Single-Layer Polymer Blends and Bilayer Polymer Films

Maksudul M. Alam, Christopher J. Tonzola, Yan Zhu, and Samson A. Jenekhe*

Departments of Chemical Engineering and Chemistry, University of Washington, Box 351750, Seattle, WA 98195–1750

> Voltage-tunable orange↔yellow↔green electroluminescence was observed from light-emitting diodes made from binary blends of n-type poly(2,2'-(3,3'-dioctyl-2,2'-bithienylene)-6,6'-bis(4-phenylquinoline)) or poly(4-hexylquinoline) or poly(4-octylquinoline) with p-type poly(2-methoxy-5-(2'-ethylhexyloxy)-1,4-phenylene vinylene). Bilayer poly(p-phenylene vinylene)/poly(9,9-dioctylfluorene) light-emitting diodes had voltage-tunable green↔blue colors. A large enhancement in performance was observed in both the blend and bilayer color-tunable light-emitting diodes. The blend devices had a turn-on voltage as low as 5 V, a luminance of up to 1490 cd/m^2, and an external electroluminescence efficiency of up to 0.7%. These results demonstrate that proper choice of electroluminescent polymers with favorable electronic structures can facilitate achievement of efficient multicolor light-emitting diodes from donor/acceptor polymer blends.

Conjugated polymers have good mechanical, electrical, and luminescent properties which lend them to the development of light-emitting diodes (LEDs) (*1-14*), photovoltaic cells (*15-17*), thin film transistors (*18-20*), imaging photoreceptors (*21,22*), and other device applications. LEDs based on conjugated polymers are being developed for various applications, including flat-panel displays and lighting (*1-5*). Conjugated polymer semiconductors offer many important advantages as the emissive materials in LEDs: low cost, light weight, large-area solution-processing of thin films, flexible panel displays with wide viewing angle, and full color emissive displays (*1-6*). Electroluminescence

(EL) colors spanning blue, green, yellow, orange, and red have been obtained through synthetic manipulation of molecular and supramolecular architectures of conjugated polymers (*7,8,23*). In addition, multicomponent conjugated polymer systems such as multilayered thin films (*5,9-11*), blends (*7,13,14*), and block copolymers (*24*) offer additional capabilities and flexibilities in the design of polymer LEDs. Conceptually, in such multicomponent conjugated polymer systems, each polymer can contribute its emission spectrum by selective excitation depending on the extent of any intermolecular interactions that can lead to complications due to exciplex formation or energy transfer among the components (*21,25*).

Although color-tunable electroluminescence has previously been observed in multicomponent conjugated polymer systems (*5,7-11*), the most efficient devices reported were those comprised of multilayered conjugated polymers and other charge transport layers. Blending two or more emissive polymers seems to provide a simple and effective way to improve the EL efficiency and to obtain color-tunable EL emission. To date, polymer blend LEDs have been made from blends of the same electron donating (p-type) or electron accepting (n-type) components (*7,13,14*). In either the p-type blend or n-type blend cases, a separate charge transport layer is necessary in the devices. One avenue to improve the performance (i.e. lower turn-on voltage, higher EL efficiency and luminance) and stability of polymer LEDs is through blending of n-type (electron transport) polymers with p-type (hole transport) polymers. In the multilayered LEDs, the tunable colors were observed mainly in the red↔yellow↔green range (*5,9-11*). Proper selection of component EL polymers for other multilayered LEDs such as blue emitting poly(*p*-phenylene)s (*1,4*) or polyfluorenes (*1,4*) as one component can readily extend the accessible tunable multicolor EL throughout the CIE diagram.

In this paper, we report voltage-tunable multicolor electroluminescence from devices made from binary blends of poly(2,2'-(3,3'-dioctyl-2,2'-bithienylene)-6,6'-bis(4-phenylquinoline)) (POBTPQ) or poly(4-hexylquinoline) (P4HQ) or poly(4-octylquinoline) (P4OQ), as an n-type (electron transport) component with poly(2-methoxy-5(2'-ethyl-hexyloxy)-1,4-phenylene vinylene) (MEH-PPV) as a p-type (hole transport) component. These results represent the first achievement of efficient multicolor LEDs from electron donor/acceptor conjugated polymer blends. Bilayer LEDs made from green emitting poly(*p*-phenylene vinylene) (PPV) and blue emitting poly(9,9-dioctylfluorene) (PFO) were also investigated to obtain green↔blue color-tunable devices for the first time. The morphology of the blends was also investigated by atomic force microscopy (AFM).

Experimental

Materials

The polyquinolines, POBTPQ, P4HQ and P4OQ, used in this study were synthesized in our laboratory. The synthesis, characterization, electrochemistry, thin film processing, optical, and electroluminescent properties of POBTPQ (M_w ~ 64 900), P4HQ (M_w ~ 15 150), and P4OQ (M_w ~ 10 250) were reported by our group (26,27). Both the MEH-PPV (M_w ~ 85 000) and the PFO (M_w ~ 10 000) samples used here were purchased from American Dye Source, Inc. Poly(ethylenedioxythiophene)/poly(styrenesulfonic acid) (PEDOT; solution in water) and the sulfonium precursor of PPV in ~ 1 wt % methanol solution were purchased from Aldrich Chemical Co. and Lark Enterprises, respectively. All solvents were of spectroscopic grade and were used as received. The molecular structures of the conjugated polymers used in this study are shown in Chart 1.

Chart 1

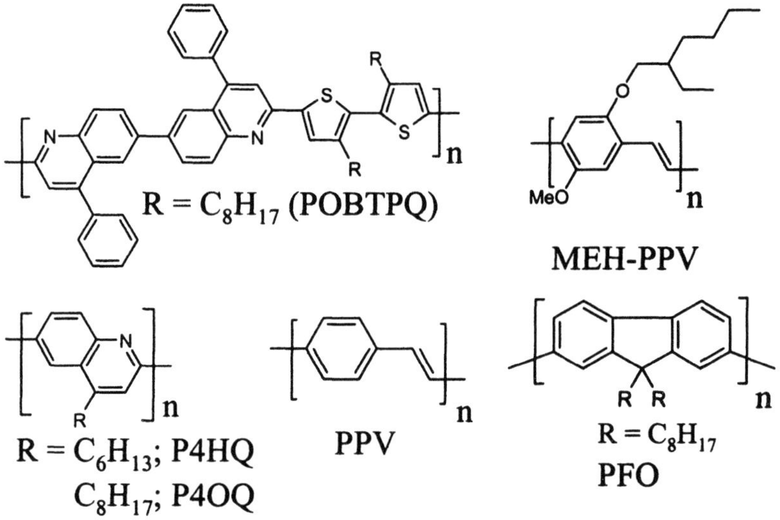

R = C_8H_{17} (POBTPQ)

MEH-PPV

R = C_6H_{13}; P4HQ
C_8H_{17}; P4OQ

PPV

R = C_8H_{17}
PFO

Preparation of Blends and Thin Films

Binary blends of POBTPQ and MEH-PPV were prepared by dissolving them with appropriate weight percent ratio in chloroform in which both polymers are very soluble. The resulting solutions were homogeneous. Compositions of blends in this paper refer to weight percentage of MEH-PPV. Thin films of the homopolymers or binary blends were obtained by spin-coating from their $CHCl_3$ solutions (0.5 wt%). Similarly, the thin films of binary P4HQ:MEH-PPV and P4OQ:MEH-PPV blends were obtained from their blend solutions (0.5 wt%) in $CHCl_3$. The thin films used for optical absorption and photoluminescence (PL) measurements were spin-coated onto silica substrates. All the films were dried in vacuum at 60 °C for 8 h to remove any residual solvent. The thin films of binary blends were homogeneous and showed excellent optical transparency. No visible phase separation was observed. For morphological investigations, the polymer blends were spin coated from their solutions in $CHCl_3$ onto polished silicon substrates and the resulting thin films were dried at 60 °C in vacuum for 12 h.

Photophysics and Surface Morphology

Optical absorption spectra were obtained by using a Lambda-900 UV/vis/near-IR spectrophotometer (Perkin-Elmer). Photoluminescence (PL) studies were carried out on a Spex Fluolog-2 spectrofluorimeter. The thin films were positioned such that the emitted light was detected at 22.5° from the incident beam (5,8). For surface morphology of the binary polymer blends, tapping-mode Atomic Force Microscopy (AFM) images were obtained using a NanoScope III instrument (Digital Instruments Inc., Santa Barbara, CA).

Fabrication and Characterization of LEDs

Electroluminescent (EL) devices were fabricated and investigated as sandwich structures between two electrodes where aluminum (Al) was used as the cathode and indium-tin oxide (ITO) was used as the anode. Schematic structures of the polymer EL devices are shown in Figure 1. The thin films of homopolymers or their binary blends were spin coated from their solution in $CHCl_3$ onto pre-cleaned ITO-coated glass substrates (Delta Technologies, Ltd., Stillwater, MN; R_s = 8 –12 Ω/□) and dried at 60 °C in vacuum for 8 h. In some devices, a thin layer of poly(ethylenedioxythiophene):polystyrenesulfonate (PEDOT) (< 40 nm) was first spin coated from its solution in water onto ITO and dried at 80 °C in vacuum for 10 h. Then, the homopolymers or their blend layers were spin coated onto the PEDOT layer and dried in vacuum at 60 °C for 8 h. For bilayer LEDs, PPV thin films were spin coated onto ITO from a sulfonium precursor solution in methanol followed by thermal conversion in vacuum at 250 °C for 1.5 h (6). Thin films of PFO were spin coated from its 1.0 wt % solution in $CHCl_3$ onto the PPV layer and then dried in vacuum at 60 °C for 10 h. The film thicknesses were measured by an Alpha-step profilometer (Model 500, KLA Tencor, San Jose, CA) with an accuracy of ± 1 nm and confirmed by an optical absorption coefficient technique. Finally, 100 – 120 nm thick aluminum layer was thermally deposited under high vacuum (~ 3×10^{-6} torr) onto the resulting polymer layer to form an active diode area of 0.2 cm² (5-mm diameter).

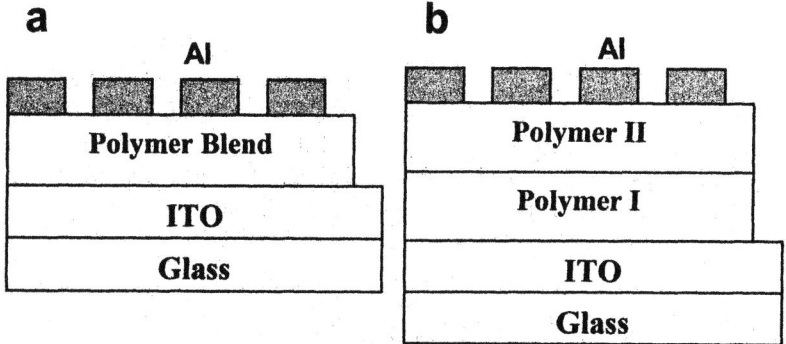

Figure 1. Schematic structure of (a) single-layer and (b) bilayer polymer LEDs.

Electroluminescence (EL) spectra were measured on a SPEX Fluorolog-2 spectrofluorimeter. Electroluminescence microscopy of the LEDs was done by using a Leica fluorescence microscope. The true color images of light emitted from the LEDs under applied bias voltages was captured by a Hamamatsu Orca II CCD camera (C4742-98) using an Openlab 2.2.5 imaging software in a PC. Current-voltage (I–V) and luminance-voltage (L–V) curves were recorded simultaneously by hooking up an HP4155A semiconductor parameter analyzer (Yokogawa Hewlett-Packard, Tokyo) together with a Grasby S370 optometer (Grasby Optronics, Orlando, FL) equipped with a calibrated luminance sensor head (Model 211) (5,6,8). The EL quantum efficiencies of the diodes were estimated by using procedures similar to that previously reported (5,6,8). All the fabrication and measurements were done under ambient laboratory conditions.

Results and Discussion

Morphology and Photophysics of Binary Blends of Conjugated Polymers

The 5 μm × 5 μm AFM phase images of the surface morphology of a POBTPQ:MEH-PPV blend (60:40 wt% ratio) along with the homopolymers are shown in Plate 1. The AFM images of both homopolymers were featureless. Two distinct phases were observed in the morphologies of POBTPQ:MEH-PPV blend as exemplified by the AFM image of the 60:40 wt% blend shown in Plate 1. The phase separation length-scale was in the range of 90 to 120 nm. The nanophase separation observed for this blend is due to spinodal decomposition. Similar phase separation behavior was also observed in the surface morphologies of P4HQ:MEH-PPV and P4OQ:MEH-PPV blend systems. The observed length-scale of the phase-separation is in the range for which voltage-tunable EL color can be expected (5,7,9,13).

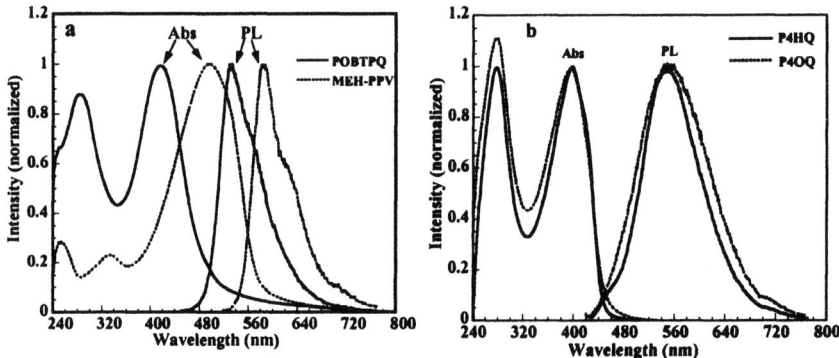

Figure 2. Optical absorption and photoluminescence spectra of thin films: (a) POBTPQ and MEH-PPV; (b) P4HQ and P4OQ.

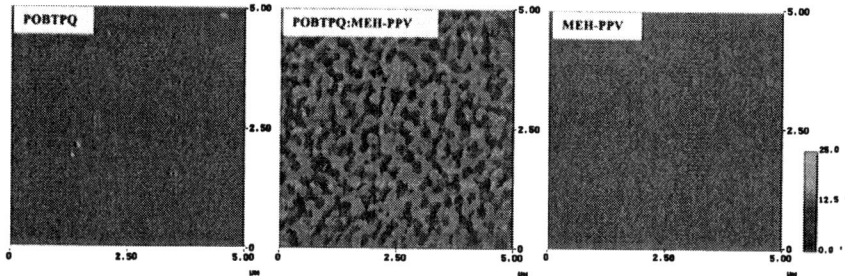

Plate 1. Tapping-mode AFM phase images of POBTPQ, POBTPQ:MEH-PPV blend (60:40 wt%) and MEH-PPV on polished silicon substrate.

Optical absorption and PL spectra of thin films of POBTPQ and MEH-PPV homopolymers are shown in Figure 2a. The lowest energy π -π* transition bands of POBTPQ and MEH-PPV are at 414 and 500 nm, respectively. POBTPQ has green PL emission with peak at 530 nm whereas MEH-PPV has orange-red PL emission with peak at 585 nm. The PL spectrum of POBTPQ and the absorption spectrum of MEH-PPV overlap to a reasonable extent in the 450 – 600 nm region (Figure 2a) from which energy transfer from POBTPQ to MEH-PPV can be expected (*14*). Figure 2b shows the absorption and PL spectra of thin films of P4HQ and P4OQ. These two polymers have identical lowest energy π -π* transitions with peak maxima at 398 nm. They emit yellow light in thin films with emission peak at 550 nm. Similar energy transfer from P4HQ (or P4OQ) to MEH-PPV can be expected due to their PL and absorption spectral overlap in the 460 – 660 nm region.

Figure 3a shows the absorption spectrum of a binary blend thin film of POBTPQ:MEH-PPV (60:40 wt%). The absorption spectrum of the binary blend is a simple superposition of those of POBTPQ and MEH-PPV. No new absorption features were observed in the wavelength range of 200 – 2000 nm, suggesting that the two blend components have no observable interactions in their electronic ground states. Similar results were observed in P4HQ:MEH-PPV (Figure 3b) and P4OQ:MEH-PPV blends.

The PL emission spectrum of a thin film of POBTPQ:MEH-PPV blend (414 nm excitation) is shown in Figure 4. Only an orange PL emission band at 585 nm, characteristic of MEH-PPV with enhanced PL intensity compared to that of pure MEH-PPV, was observed. The enhancement in PL intensity of MEH-PPV is due to energy transfer from POBTPQ to MEH-PPV. Similar energy transfer from P4HQ (or P4OQ) to MEH-PPV was observed in the binary blend systems. These results are contrary to previously reported bipolar (n-type/p-type) conjugated polymer blends in which photoinduced charge transfer and luminescence quenching were significant (*28*).

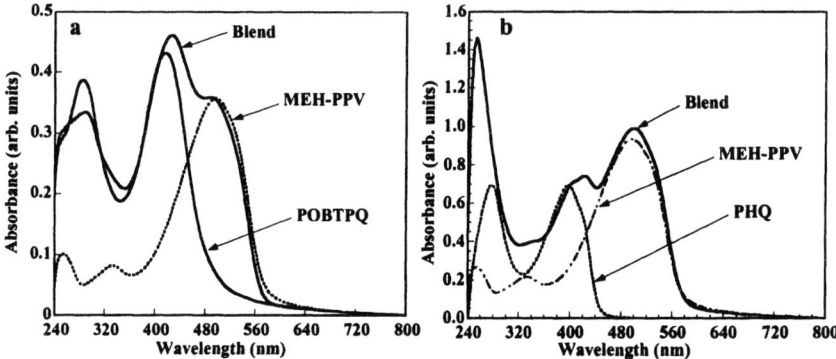

Figure 3. Optical absorption spectra of thin films of (a) POBTPQ:MEH-PPV and (b) P4HQ: MEH-PPV blends.

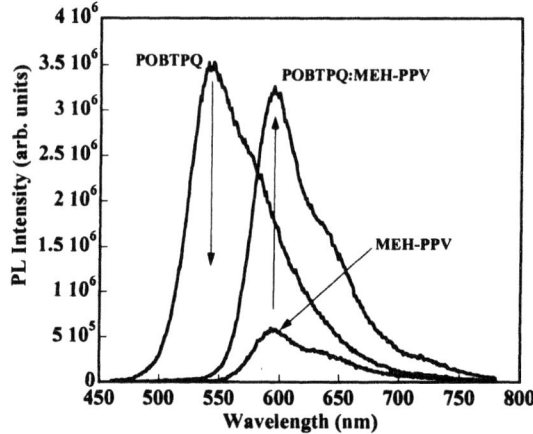

Figure 4. PL spectra of thin films of POBTPQ, MEH-PPV and a POBTPQ:MEH-PPV blend (60:40 wt%).

Voltage-Tunable Multicolor Polymer LEDs

Green electroluminescence with a peak at 530 nm was observed for POBTPQ in LEDs of the type ITO/PEDOT/POBTPQ/Al at all forward bias voltages. The observed EL emission spectrum of POBTPQ is identical to the PL emission spectrum discussed above. Yellow EL emission with a peak at 552 nm was observed for both P4HQ and P4OQ diodes of the type ITO/PEDOT/P4HQ (or P4OQ)/Al. Single-layer EL devices made from POBTPQ:MEH-PPV blends showed a continuous color change from green to yellow to orange under a varying bias voltage. Such blend LEDs, ITO/PEDOT/blend/Al, had a bright orange color at low bias voltages (10-12 V) and yellow to green colors at higher bias voltages. EL micrographs of a color-tunable POBTPQ:MEH-PPV blend

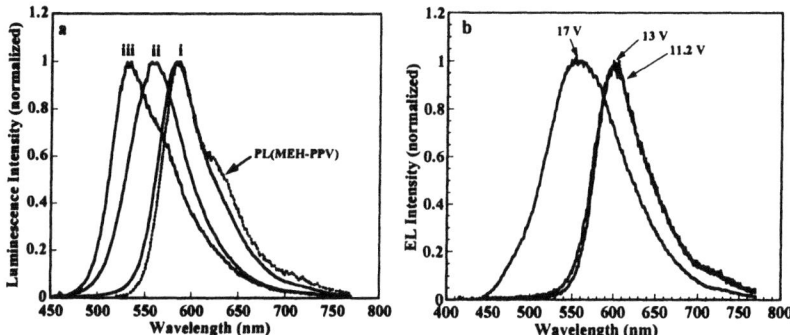

Figure 5. (a) EL Spectra of a POBTPQ:MEH-PPV blend (60:40 wt%) LED at various bias voltages: i) 11, ii) 14 and iii) 16 V. The PL spectrum of MEH-PPV thin film is also shown. (b) EL spectra of a P4HQ:MEH-PPV blend (70:30 wt%) LED at various forward bias voltages.

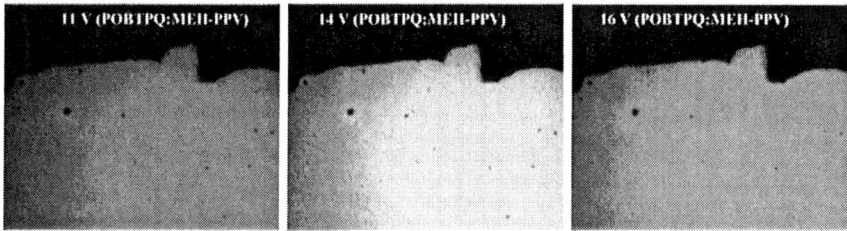

Plate 2. EL micrographs (x10) of a POBTPQ:MEH-PPV blend (60:40 wt%) diode at 11, 14, and 16 V. (See Page 5 of color insert.)

(60:40 wt%) diode under different bias voltages are shown in Plate 2. The voltage-tunable EL spectra corresponding to the EL micrographs of Plate 2 are shown in Figure 5a. The orange EL emission ($\lambda_{max} \approx$ 585 nm) observed at low bias-voltages is identical with that of the single-layer MEH-PPV diodes which suggests that this EL emission was contributed from the MEH-PPV component in the blend. The green EL emission was similarly contributed from the POBTPQ component in the blend at the highest bias voltages (> 15 V). The yellow EL emission observed at an intermediate bias voltage in the blend LEDs is a result of simultaneous emission of green and orange lights from both blend components, leading to the observed yellow color.

Voltage-tunable multicolor EL emission (orange to yellow) was similarly observed from P4HQ:MEH-PPV (Figure 5b) and P4OQ:MEH-PPV blend LEDs.

The orange EL emission was contributed from MEH-PPV component and the yellow EL emission was contributed from the P4HQ (or P4OQ) component in these blend LEDs.

Voltage-tunable EL emission was also observed from a bilayer ITO/PPV(30 nm)/PFO(50 nm)/Al diode which switches colors reversibly from green at 10 - 14 V to blue at 17 V. The multicolor switching by the applied bias voltage was observed visually and also by EL microscopy as shown in Plate 3. The voltage-tunable EL spectra corresponding to the EL micrographs of Plate 3 are shown in Figure 6. The EL spectrum at 12 V has peaks at 508 and 540 nm, showing the characteristic green PPV emission (5,6). The EL spectrum at 17 V, with dominant peaks at 436 and 465 nm, is due to the blue PFO emission (4). The EL spectra at intermediate bias-voltages have contribution from both PPV and PFO emission.

The single-layer LEDs made from POBTPQ as an emissive material had a turn-on voltage of 9 V and a luminance of 53 cd/m^2 at 17.5 V. The external EL quantum efficiency was 0.06 %. The P4HQ and P4OQ LEDs had a turn-on voltage of 13–14 V and a luminance of 15–21 cd/m^2 at 20 V. The external quantum efficiencies were 0.007% for P4HQ and 0.005% for P4OQ. The poor electroluminescence efficiency of these polyquinoline LEDs is due to their low solid state PL quantum yield (27).

Figure 7 shows the current-voltage and luminance-voltage curves of the POBTPQ:MEH-PPV blend LEDs of the type ITO/PEDOT/Blend/Al. A large enhancement in the performance of the blend LEDs was observed compared to the single-layer MEH-PPV and POBTPQ devices. The single-layer ITO/MEH-PPV/Al diode had a turn-on voltage of 12 V and a maximum luminance of 186 cd/m^2 at 16 V and current density of 500 mA/cm^2. The POBTPQ:MEH-PPV blend LEDs showed bright orange EL emission at low bias voltage (10 – 12 V) and yellow (14 V) to green (16 V) at higher bias voltages. The turn-on voltage of these blend LEDs was 5 V. The luminance of the orange color at 11 V was about 400 cd/m^2, the luminance of the yellow color at 14 V was about 1400 cd/m^2, and that of the green color at 16 V was about 1200 cd/m^2. The blend (60:40 wt%) diode had an external quantum efficiency of 0.7%, a factor of 35 times greater than the single-layer MEH-PPV diodes. Similar enhancement in LED performance was observed in devices made from P4HQ:MEH-PPV (70:30 wt%) and P4OQ:MEH-PPV (70:30 wt%) blends. The P4HQ:MEH-PPV blend (70:30 wt%) diode had a turn-on voltage of 7.0 V and a luminance of 700 cd/m^2 at 17 V with a current density of 198 mA/cm^2. A turn-on voltage of 7.5 V and a luminance of 635 cd/m^2 at 16 V were observed for the P4OQ:MEH-PPV blend (70:30 wt%) diode. The external quantum efficiencies of these two blend diodes were 0.4 - 0.55%. The external quantum efficiency is thus enhanced by a factor of 20 – 27 times compared to that of the single-layer MEH-PPV diode. The enhanced performance (lower turn-on voltage, higher luminance, and higher EL

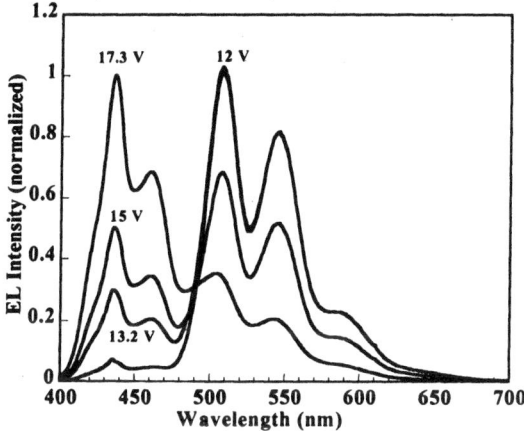

Figure 6. EL spectra of a bilayer ITO/PPV(30 nm)/PFO(50 nm)/Al diode.

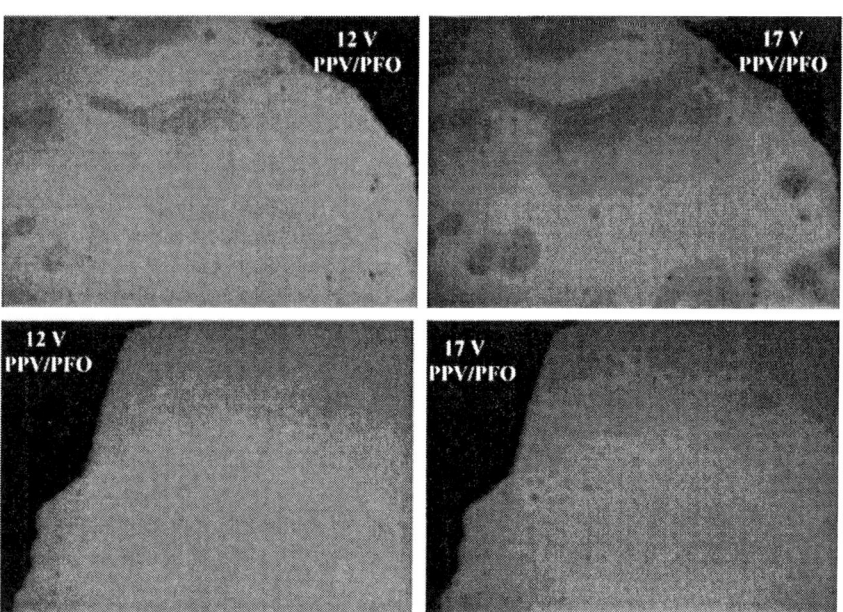

Plate 3. EL micrographs (x 10) of a bilayer ITO/PPV(30 nm)/PFO(50 nm)/Al diode. (See Page 6 of color insert.)

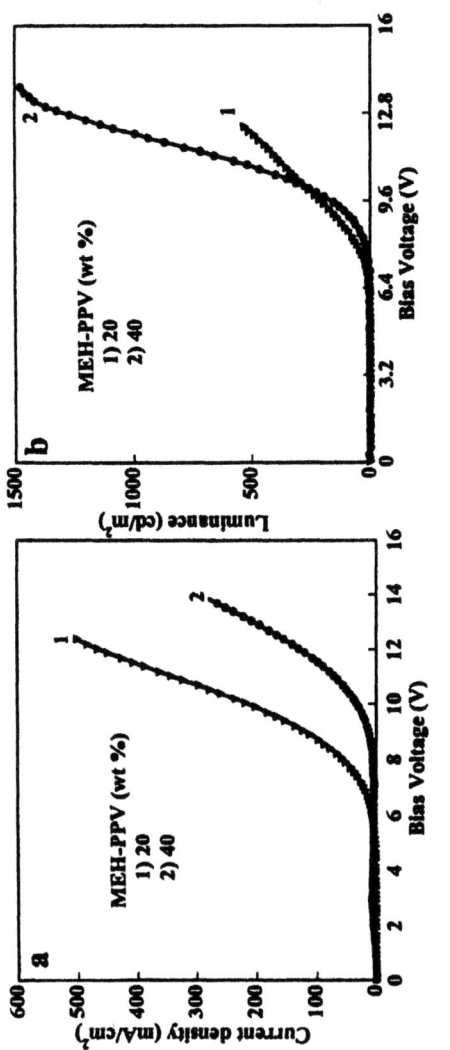

Figure 7. (a) Current-voltage and (b) luminance-voltage curves of ITO/PEDOT/Blend(POBTPQ:MEH-PPV; 60:40 wt%)/Al diodes.

efficiency) of these n-type/p-type binary polymer blend LEDs implies improved charge injection and recombination compared to that of the single-layer component polymer diodes.

The performance of the color-tunable bilayer ITO/PPV(30 nm)/PFO(50 nm)/Al diode was also substantially enhanced compared to the component single-layer devices. The turn-on voltage of the bilayer LED was 8 V. The maximum luminance was 420 cd/m^2 at a current density of 236 mA/cm^2 (17 V), and the external EL efficiency was 0.22%. The luminance of the PPV/PFO LED in the green state at 12 V was about 120 cd/m^2 and the luminance in the blue state at 17 V was about 420 cd/m^2. In contrast, the single-layer PPV diode had a turn-on voltage of 9.5 V and a luminance of 11 cd/m^2 at 11 V with a current density of 500 mA/cm^2. The external quantum efficiencies of single-layer PPV and PFO diodes were 0.002% and 0.01%, respectively. The external quantum efficiency of the bilayer PPV/PFO diode is thus enhanced by a factor of 28 – 110 times compared to the single-layer diodes. These results show that voltage-tunable EL emission from the bilayer diode is both efficient and bright.

Summary

The recent availability of organic-solvent soluble n-type conjugated polyquinolines such as POBTPQ, P4HQ and P4OQ has allowed us to demonstrate nanophase-separated bipolar blends with p-type MEH-PPV, leading to efficient and bright voltage-tunable multicolor light emitting diodes. Voltage-tunable green↔blue EL emission was also observed in bilayer LEDs containing PPV and PFO layers. The tunable EL colors obtained from a bilayer LED by varying the applied voltage originate from the two different emissive layers. The observed performance (turn-on voltage, luminance level, and EL quantum efficiency) of both blend and bilayer LEDs was significantly enhanced compared to the single-layer homopolymer LEDs.

Acknowledgement

This research was supported by the U.S. Army Research Laboratory and the U.S. Army Research Office under Grant DAAD 19-01-1-0676 and in part by the Office of Naval Research.

References

(1) Kraft, A.; Grimsdale, A. C.; Holmes, A. B. *Angew. Chem. Int. Ed.* **1998**, *37*, 402.

(2) Friend, R. H.; Gymer, R. W.; Holmes, A. B.; Burroughes, J. H.; Marks, R. N.; Taliani, C.; Bradley, D. D. C.; Dos Santos, D. A.; Bredas, J. L.; Logdlund, M.; Salaneck, W. R. *Nature* **1999**, *397*, 121.
(3) Cao, Y.; Parker, I. D.; Yu, G.; Zhang, C.; Heeger, A. J. *Nature* **1999**, *397*, 414.
(4) Bernius, M. T.; Inbasekaran, M.; O'Brien, J.; Wu, W. *Adv. Mater.* **2000**, *12*, 1737.
(5) Zhang, X.; Jenekhe, S. A. *Macromolecules* **2000**, *33*, 2069.
(6) Alam, M. M.; Jenekhe, S. A. *Chem. Mater.* **2002**, *14*, 4775.
(7) Berggren, M.; Inganas, O.; Gustafsson, G.; Rasmusson, J.; Andersson, M. R.; Hjertberg, T.; Wennerstrom, O. *Nature* **1994**, *372*, 444.
(8) Zhang, X.; Shetty, A. S.; Jenekhe, S. A. *Macromolecules* **1999**, *32*, 7422.
(9) Jenekhe, S. A.; Zhang, X.; Chen, X. L.; Choong, V.-E.; Gao, Y.; Hsieh, B. R. *Chem. Mater.* **1997**, *9*, 409.
(10) Wang, Y. Z.; Gebler, D. D.; Fu, D. K.; Swager, T. M.; Epstein, A. J. *Appl. Phys. Lett.* **1997**, *70*, 3215.
(11) Hamaguchi, M.; Yoshino, K.; *Appl. Phys. Lett.* **1996**, *69*, 143.
(12) Tarkka, R. M.; Zhang, X.; Jenekhe, S. A. *J. Am. Chem. Soc.* **1996**, *118*, 9438.
(13) Zhang, X.; Kale, D. M.; Jenekhe, S. A. *Macromolecules* **2002**, *35*, 382.
(14) Yu. G.; Nishino, H.; Heeger, A. J.; Chen, T.-A.; Rieke, R. D. *Synth. Met.* **1995**, *72*, 249.
(15) Brabec, C. J.; Sariciftci, N. S.; Hummelen, J. C. *Adv. Funct. Mater.* **2001**, *11*, 15.
(16) Antoniadis, H.; Hsieh, B. R.; Abkowitz, M. A.; Jenekhe, S. A.; Stolka, M. *Synth. Met.* **1994**, *62*, 265.
(17) Jenekhe, S. A.; Yi, S. *Appl. Phys. Lett.* **2000**, *77*, 2635.
(18) Bao, Z.; Dodabalapur, A.; Lovinger, A. J. *Appl. Phys. Lett.* **1996**, *69*, 4108.
(19) Dimitrakopoulos, C. D.; Malenfant, P. R. L. *Adv. Mater.* **2002**, *14*, 99.
(20) Babel, A.; Jenekhe, S. A. *Adv. Mater.* **2002**, *14*, 371.
(21) Jenekhe, S. A.; Osaheni, J. A. *Science* **1994**, *265*, 765.
(22) Zhang, X.; Jenekhe, S. A.; Perlstein, J. *Chem. Mater.* **1996**, *8*, 1571.
(23) Agrawal, A. K.; Jenekhe, S. A. *Macromolecules* **1993**, *26*, 895.
(24) Chen, X. L.; Jenekhe, S. A. *Appl. Phys. Lett.* **1997**, *70*, 487.
(25) Osaheni, J. A.; Jenekhe, S. A. *Macromolecules* **1994**, *27*, 739.
(26) Tonzola, C. J.; Alam, M. M.; Jenekhe, S. A. *Adv. Mater.* **2002**, *14*, 1086.
(27) Zhu, Y.; Alam, M. M.; Jenekhe, S. A. *Macromolecules* **2002**, *35*, 9844.
(28) Halls, J. J. M.; Walsh, C. A.; Greenham, N. C.; Marseglia, E. A.; Friend, R. H.; Moratti, S. C.; Holmes, A. B. *Nature* **1995**, *376*, 498.

Chapter 15

Blue Light Emitting Polymers and Devices

Qibing Pei[1], S. Pyo[2], Shun-Chi Chang[2], and Yang Yang[2]

[1]SRI International, 333 Ravenswood Avenue, Menlo Park, CA 94025–3493
[2]Department of Materials Science and Engineering, University of California, Los Angeles, CA 90095

Blue light-emitting polymers are critically important for the development of full-color polymer displays. A number of such polymers have been synthesized, with high photoluminescence and electroluminescence efficiencies. Both polymer light-emitting diodes (LEDs) and electrochemical cells have been fabricated, but various materials issues remain to be solved. Device lifetimes have been limited, due to factors such as (1) the instability of the polymers which usually have low glass transition temperature, and (2) accelerated failure at the polymer/electrode interfaces, due to high charge injection barriers and the blue polymers' low charge carrier mobility. Using dual functional triarylamine moieties as the side groups, we have prepared new blue light-emitting poly(paraphenylenes) exhibiting high luminescent efficiency, high glass transition temperature, good environmental stability, and enhanced carrier mobility. Blue LEDs have been demonstrated with 4.2 cd/A efficiency and 360 cd/m^2 brightness at 8 V.

Introduction

One of the most prominent conjugated polymer for blue electroluminescence is poly(paraphenylene) (*1*). PPP is insoluble in any solvent. It has been rendered soluble in organic solvents by the attachment of flexible side groups, such as in poly(2-octyloxy-1,4-phenylene) (*2*). Spin-coating was used to prepare high-quality thin films of alkoxy-PPP. Blue LEDs were demonstrated with high EL quantum efficiency. However, the LEDs' operating voltages were high. The alkoxy side groups, which are electronically passive, considerably reduce the conductivity of the polymer. They separate the PPP backbones farther from each other and further twist the phenylene rings from being coplanar. Carrier mobility is consequently hindered. The device lifetime was short. The alkoxy-PPP has a low glass transition temperature. The PPP chains tend to aggregate, due to interaction between delocalized π-electrons, causing red shift of the emission spectrum and reducing luminescent efficiency.

We explored poly(fluorenes) (PFs), derivatives of PPP wherein every two neighboring phenyl rings are locked in a plane by the C-9. PFs have better semiconductivity than PPPs (*3*). PFs with long-chain alkyl or polyether side groups attached to the C-9 were soluble. Blue LEDs based on PFs, with high quantum efficiency and lower operating voltages, were fabricated. Recently, significant progress has been made in improving the performance of blue LEDs based on PF (*4*). However, device operating lifetime is still unsatisfactory. The planar aromatic rings in PFs tend to aggregate. The emission color readily shifts toward white or red, due to eximer emission (*3,5*). Molecular motion is the main driving force for the formation of excited-state aggregates. Several approaches have been taken to overcome this problem. An example approach was to introduce co-monomers into the PF main chain (*6*). Unfortunately, the resulting copolymers required high voltages to operate and still showed some eximer emission in the electroluminescent spectrum. Ladder-PPPs were also investigated which exhibited similar problems (*7*).

Based on available results, we think that the general strategy of using flexible alkyl or alkoxy side groups or chain segments to solubilize light-emitting polymers is impractical. The preferred approach should not employ solubilizing groups that would either reduce the conductivity of the conjugated main chain, or reduce the polymer's thermal stability, or both.

On the other hand, triarylamine compounds have been among the best organic/polymer materials in terms of charge injection and carrier mobility. These compounds have been widely used in organic LEDs as the hole transport material. We have demonstrated that triarylamine side groups are effective in rendering poly(paraphenylene vinylene) (PPV) soluble. The resulting PPV polymers have good solubility and high photoluminescent efficiency (8). In this article we report that the triarylamine-type side groups are also effective in rendering PPP soluble. Blue LEDs have been fabricated with high luminescent efficiencies and low operating voltages.

Results and discussion

A representative PPP with triarylamine side groups is TA-PPP (9) with the following repeating units:

$$\left(\!\!\left\langle\bigcirc\right\rangle\!\!\right)_x\!\!\left(\!\!\left\langle\bigcirc\right\rangle^{TA}\!\!\right)_{1-x}$$

wherein TA stands for a triarylamine or diarylamino side group, and x varies between 0 and 1. TA-PPP was synthesize by Yamamoto condensation polymerization using dichloro- or dibromo- monomers (10), and by Suzuki coupling polymerization using dibromo- and diborate mixed monomers (11). Gel permission chromatography (GPC) analysis (polystyrene standard) showed a moderate weight-average molecular weight of 30,000 and polydispersity of 4.

TA-PPP is readily soluble in certain organic solvents. Spin-cast thin films are optically clear, with intense blue fluorescence. Figure 1 displays the absorption and PL spectra of the polymer in both solution and thin film. The spectra of the solution and thin films are almost identical, indicating little aggregation or excimer formation in the solid state, contrary to most blue light emitting polymers including poly(9,9-dioctylfluorene) (DO-PF). Figure 1 also shows the PL spectra of DO-PF in a solution and thin film. The sub-band emission, due to excited-state aggregation, becomes dominant in the soid films.

Certain conjugated polymers with alkyl side groups, such as poly(3-alkylthiophene) exhibit solvatochromism, that is, the absorption spectra of the polymers in solution is significant shifted to shorter wavelength compared those in solid thin films (12). The driving force is molecular motion of the flexible side groups that twists the conjugated polymer chain from being coplanar. The twisting is enhanced in solution. Similar effect was also observed when the solid thin film was heated. Polymers with higher glass transition temperature exhibit less solvatochromism. The negligible solvatochromism observed in TA-PPP is consistent with its rigid polymer chain and side groups.

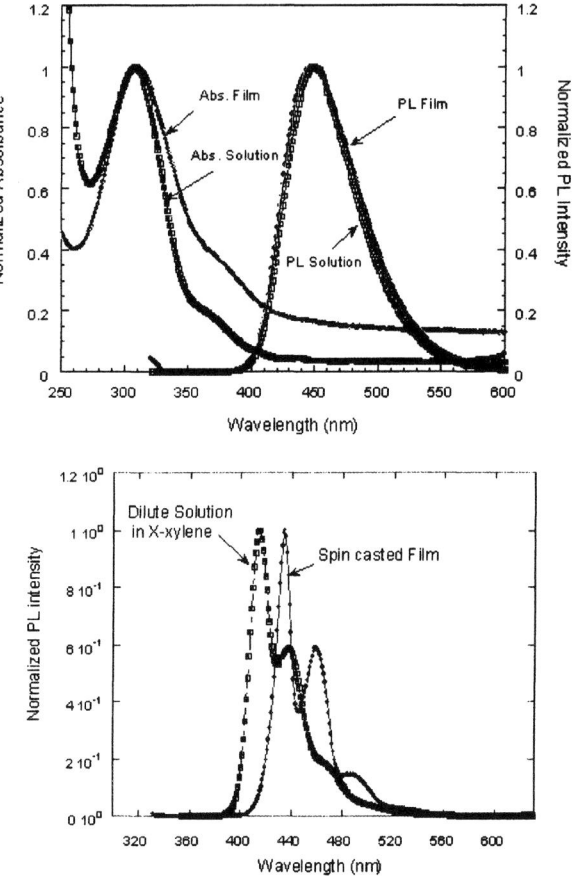

Figure 1. Top: UV-Vis absorption and PL spectra for TA-PPP in 1,4-dioxane and in the solid state film. Bottom: PL spectra of DO-PF in p-xylene and in the solid state film. The thin films were spin-coated and baked at 70 °C for 30 min.

From the absorption onset of TA-PPP, the polymer's band gap is estimated to be 3.4 eV. Through electrochemical cyclovoltametry, the oxidation potential or HOMO was determined at 5.2 eV. This puts the LUMO at 1.8 eV. Therefore, TA-PPP is hole-predominant.

The absorption and PL spectra of TA-PPP do not overlap, indicative of a good luminescent material. The PL quantum efficiency, though not quantitatively measured, is comparable to that of pristine DO-PF, which is in the range of 70–100%. A thin film of TA-PPP cast on glass was heated on a hot plate at 100 °C in laboratory air and normal room light for 1 month. No PL degradation (quantum efficiency and emission color) was observed. Even after treatment at 150 °C for 2 hr, no spectral change occurred, as shown in Figure 2. Slight color shift was observed at 200 °C. In comparison, similar treatment with DO-PF showed remarkable color shift at 100 °C. The PL of DO-PF was almost completely quenched at 200 °C, as shown in Figure 3.

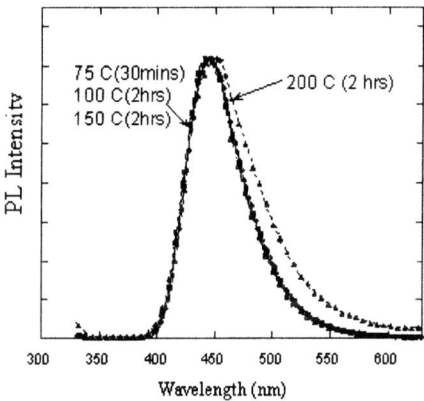

Figure 2. PL spectrum of TA-PPP thin film after thermal treatment in air

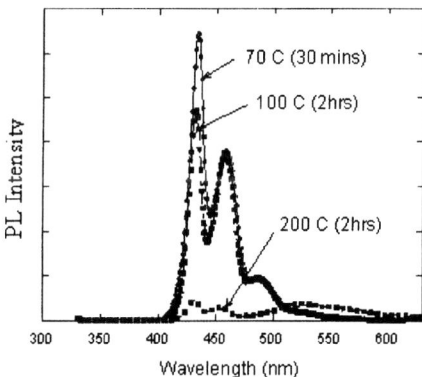

Figure 3. PL spectrum of DO-PF thin film after thermal treatment in air

To verify whether the high stability of PL is from TA-PPP's rigid structure, thermogravimetric analysis (TGA) and differential scanning calorimetry (DSC) were carried out. The TGA diagram shown in Figure 4 showed little weight loss until about 560 °C. Under similar condition, DO-PF decomposes at less than 400 °C. The decomposition of DO-PF likely starts with the alkyl side groups. The mass spectrum of 2,7-dichloro-9,9-dioctylfluorene, a monomer for DO-PF, are abundant of fragments attributable to broken alkyl chains.

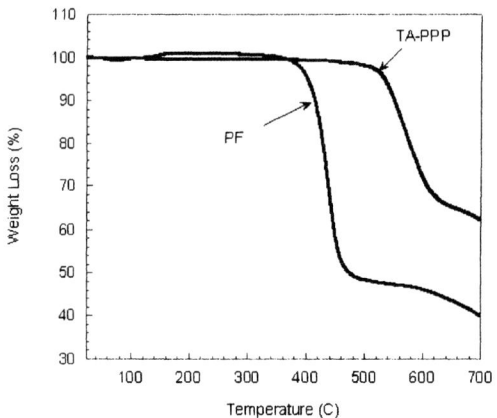

Figure 4. TGA diagram of TA-PPP and DO-PF with a ramping rate of 10 °C/min under dry N_2 gas flow.

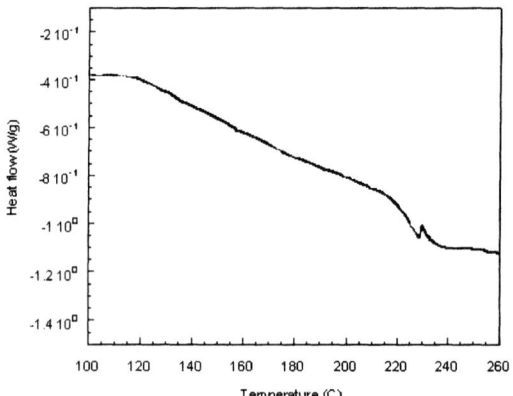

Figure 5. DSC thermogram of TA-PPP at a heating and cooling rate of 10 °C/min.

DSC thermogram of TA-PPP (Figure 5) shows a glass transition at about 225 °C, confirming the high thermal stability of TA-PPP.

Blue light emitting polymers and oligomers based on spiro-PPP has also been reported by Salbeck et al. with high glass transition temperature (13). The polyquinolines also exhibit high efficiency blue luminescence and high glass transition temperature (14). However, these polymers have lower conductivity than PPP with flexible alkoxy side groups. The resulting LEDs require high voltages to operate. The TA-PPP, with its dual functional side groups, is expected to overcome the problems caused by high operating voltages.

LEDs were fabricated with TA-PPP as the emissive layer. Single-layer devices of ITO/PEDOT/TA-PPP/Ca/Al were fabricated. PEDOT, poly(3,4-ethylenedioxythiophene), was used to enhance hole injection from the anode. Charge injections of the single layer LEDs were clearly hole dominant. The barrier for electron injection, around 1.0 eV, is too high. Electron dominant materials such as DO-PF and 2-(4-t-butylphenyl)-5-biphenyloxadiazole (t-PBD) were used to enhance electron injection. The thin film of a TA-PPP and PF blend (95:5 weight ratio) was phase separated. Atomic force microscopy (AFM) showed PF spheres, close to 1 μm in diameter, dispersed in the TA-PPP matrix (Figure 6). This type of phase separation is common in blends of stiff and soft polymers. The PL emission of the blend film was characteristic of TA-PPP. However, once thermally treated, the spectrum shifted bathochromically much like PF. The EL spectrum from LEDs based on the blend thin film contained much emission from PF in the 500–700 nm regime. The device efficiency was about 0.43 cd/A. TA-PPP/PF double layer LEDs were also fabricated. But the efficiency was not improved because when PF was spin coated onto TA-PPP, the PF solution washed out most of the TA-PPP layer.

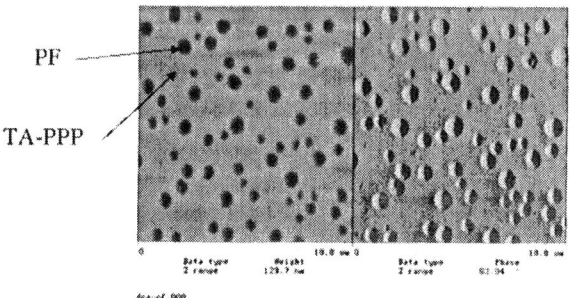

Figure 6. AFM image of a thin film of TA-PPP:DO-PF blend (90:10, by weight).

Thin films of the blend of TA-PPP with t-PBD (80:20 by weight) were also phase separated, even to the naked eyes, albeit to a lesser degree than the TA-PPP and DO-PF blend. Remarkable device performance was achieved with the TA-PPP and t-PBD blend. Figure 7 shows a typical current-light-voltage response of an ITO/PEDOT/(TA-PPP + *t*-PBD)/Ca/Al LED. Light emission turns on at 4.5 V, and reaches 100 cd/m^2 at 7.2 V and 10^4 cd/m^2 at 12 V. The highest efficiency is 4.2 cd/A at 1 mA and 360 cd/m^2. The emission color is nearly identical to the PL of TA-PPP spectrum, which corresponds to (0.167, 0.157) on the CIE chart.

The high efficiency is probably due to charge confinement at the TA-PPP/*t*-PBD interfaces. Figure 8 illustrates the band diagrams of TA-PPP, *t*-PBD, and the electrode materials. Apparently, electrons and holes injected from the electrodes are confined near the TA-PPP/*t*-PBD interfaces formed between the TA-PPP and *t*-PBD microdomains. The confined charges recombine and produce photons. The charge confinement may also be responsible for the higher-than-expected driving voltages (*14*). By reduction of the size of the microdomains and the thickness of the blend film, the voltages may be lowered.

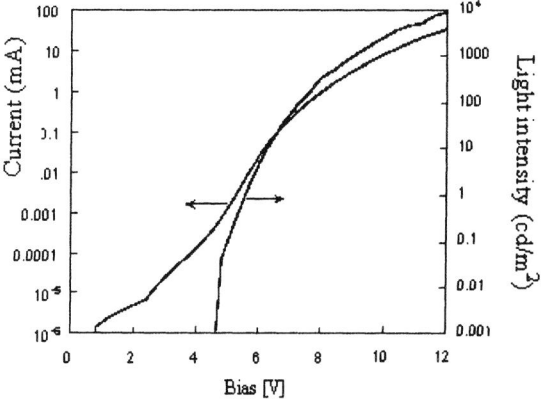

Figure 7. Current-light intensity-voltage response of an ITO/PEDOT/(TA-PPP + t-PBD)/Ca/Al PLED

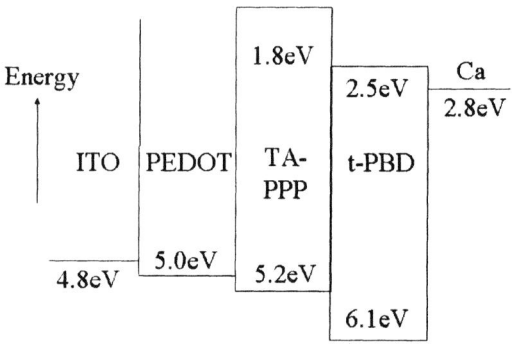

Figure 8. Band diagrams of varioua materials used in the efficient blue LED

Conclusion

Using the rigid, dual-functional triarylamine moieties as the side groups, we have obtained a PPP derivative with good solubility, good film-forming properties, and high PL and EL efficiencies. The rigid TA-PPP exhibit high thermal stability while retaining good semiconductivity, essential for high-performing polymer LEDs. The efficiency and color quality of the blue LEDs based on TA-PPP compare favorably with most other blue polymer LEDs. There is much room for further improvement. We have also extended this strategy for preparing processable light-emitting polymers to other conjugated polymers including poly(paraphenylene vinylene) with lower band gaps and bathochromically-shifted emission colors (8).

Acknowledgement

The work reported here was in part supported financially by the Office of Naval Research, ONR Contract N00014-99-C-0274.

References

(1) Grem, G.; Leditzky, B.; Ullrich, B.; Leising, G. *Adv. Mater.* **1992**, *4*, 36.
(2) Yang, Y.; Pei, Q.; Heeger, A. J. *J. Appl. Phys.* **1996**, *79*, 934.
(3) Pei, Q.; Yang, Y. *J. Am. Chem. Soc.* **1996**, *118*, 7416.
(4) (a) Inbasekaran, M., 3rd International Conference on Electroluminescence, September 5-8, **2001**, Los Angeles, California. (b) Grell. M.; Bradly, D. D. C.; Inbasekaran, M.; Woo, E. P. *Adv. Mater.* **1997**, *31*, 2465. (c) Fukuda, M.; Sawada, K.; Yoshino, K.; *J. Polym. Sci. Polym. Chem.* **1993**, *31*, 2465
(5) Bliznyuk, V. N.; Carter, S.A.; Scott, J. C.; Klärner, G.; Miller, R. D.; Miller, D. C. *Macromolecules* **1999**, *32*, 361.
(6) Kreyenschmidt, M., Klaerner, G.; Fuhrer, T.; Ashenhurst, J.; Karg, S. W.; Chen, D.; Lee, V. Y.; Scott, J. C.; Miller, R. D. *Macromolecules* **1998**, *31*, 1099.
(7) Kreyenschmidt, M.; Uckert, F.; Mullen, K. *Mocromolecules* **1995**, *28*, 4577
(8) Pei, Q. US Patent 6,414,104. **2002**.
(9) Pei, Q. US Patent Application Serial No. 09/864,704, **2001**.
(10) Yamamoto, T. *Prog. Polym. Sci.* **1992**, *17*, 1153. (b) Yamamoto, T.; Morita, A.; Miyazaki, Y.; Maruyama, T.; Wakayama, H.; Zhou, Z.-H.; Nakamura, Y.; Kanbara, T.; Sasaki, S.; Kubota, K. *Macromolecules*, **1992**, *25*, 1214.

(*11*) (a) Rehahn, M.; Schlüter, A.-D.; Wegner, G.; Feast, W. J. *Polymer* **1989**, *30*, 1060. (b) Rehahn, M.; Schlüter, A.-D.; Wegner, G. *Makromol. Chem.* **1990**, *191*, 1991.

(*12*) (a) Hotta, S.; Rughooputh, S.D.D.; Heeger, A. J.; Wudl, F. *Macromolecules* **1987**, *20*,212. (b) Inganäs, O.; Salaneck, W.R.; Österholm, J.-E.; Laakso, J. *Synth. Met.* **1989**, *28*, 377.

(*13*) (a) Salbeck, J.; Yu, N.; Bauer, J.; Weissörtel, F.; Bestgen, H. *Synth. Met.* **1997**, *91*, 209. (b) Salbeck, J.; Weissörtel, F.; Bauer, J. *Macromol. Symp.* **1997**,*125*, 121.

(*14*) Parker, I. D.; Pei, Q.; Marrocco, M. *Appl. Phys. Lett.* **1994**, *65*, 1272.

Chapter 16

Polarized Electroluminescence from Double-Layer LEDs with Active Film Formed by Two Perpendicularly Oriented Polymers

A. Bolognesi[1], C. Botta[1], D. Facchinetti[1], C. Mercogliano[1], M. Jandke[2], P. Strohriegl[2], K. Kreger[2], A. Relini[3], and R. Rolandi[3]

[1]Istituto per lo Studio delle Macromolecole, CNR, Via E. Bassini 15, 20133 Milano, Italy
[2]Bayreuther Institüt fur Makromoleküforschung, Universität Bayreuth, D–95440 Bayreuth, Germany
[3]Istituto Nazionale di Fisica della Materia and Dip. Fisica Università di Genova, Via Dodecaneso 33, 16146 Genova, Italy

The aim of this work is to study the feasibility of a double layer device, where the two active materials are formed, respectively, by two oriented polymers, whose orientation direction is orthogonal, and which can emit simultaneously in different regions of the visible spectrum. We demonstrate that the anisotropy of the polymers is not lost when they are perpendicularly oriented, obtaining polarized light in a large spectral region, extending from the green to the red. The emitted light, observed through a polarizer, can be varied from green to red by simply rotating the polarization axis, obtaining polarized light of variable colour. This peculiar device design is particularly appealing as it can increase the versatility of organic LEDs providing polarized light with easily variable colour emission.

In recent years, organic light-emitting diodes (OLEds) are gathering a lot of interest. Polymers used as active materials in OLEDs have conjugated chains which can be aligned by low-cost techniques and so that it's easy to prepare well oriented thin films giving polarized electroluminescence (EL) *(1)*. Polarized EL is useful for application such as backlights in liquid crystal displays *(2)*. The most used methods for aligning polymeric films have been recently described in a review by Grell and Bradley *(3)*. Depending on the optical properties of the polymers and on the orienting technique employed, different emission colours and EL polarization ratios (R_{EL}) were obtained.

For emission in the blue, disubstituted polyfluorene aligned by LC self-organization on pre-oriented substrates gives a polarization ratio of about 15 in EL *(4)*.

A further improvment in the orientation of polyfluorene has been obtained by orienting monodomains of poly(9,9-dioctylfluorene) on an alignment layer of segmented poly(p-phenylenevinylene) (PPV), reaching the highest polarization ratio of 25 in EL *(5)*.

For green emission, unsubstituted PPV oriented by the rubbing technique gives a value of $R_{EL} \approx 12$ *(6)*. Red polarized emission with a dichroic ratio $R_{EL} \approx 8$ has been obtained with a poly(3-alkylthiophene) derivative oriented by a combination of rubbing and thermal annealing *(7)*.

In this paper we report the preparation and characterization of a device in which the active material is composed of two layers of two different polymers emetting in the yellow-green and in the red, oriented perpendiculary to each other.

Results

The yellow-green emitting polymer is a segmented PPV (Fig. 1), obtained as reported in ref. 8 and 9.

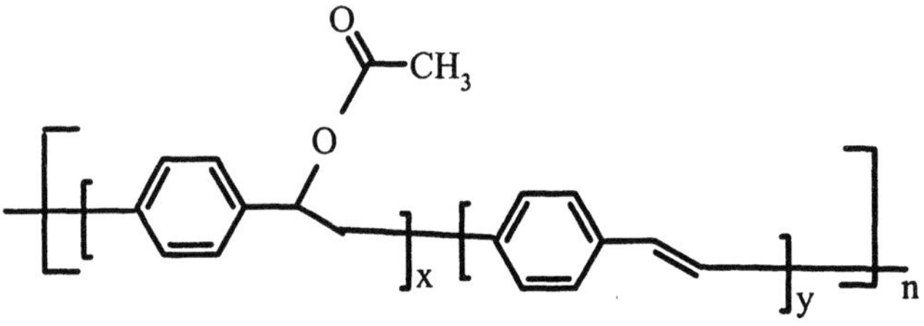

Figure 1: Structure of segmented PPV used in this work.

The red emitting polymer is a regioregular poly[3-(6-methoxyhexyl)thiophene] (P6OMe, Fig. 2).

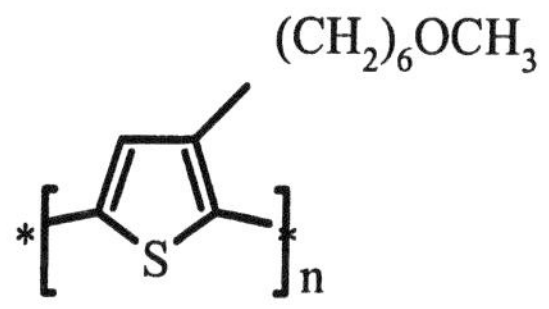

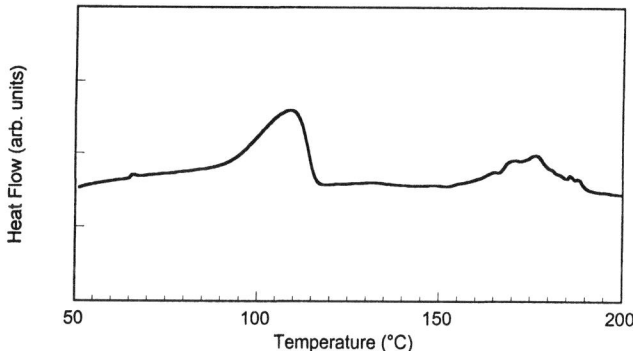

Figure 2: Structure of P6OMe

The differential scanning calorimetry (DSC) trace of P6OMe powders, reported in Fig. 3, consists of two broad endotherms whose maxima are at 110°C and 170°C. The higher temperature transition has been attribuited to the melting of the polymer. At 110°C a thermal phase transition takes place from a three dimensional ordered semicrystalline phase to a liquid crystalline phase *(7)*.

Figure 3: DSC scan of P6OMe

P6OMe can be aligned by means of the rubbing technique. Higher dichroic ratios were obtained by heating the oriented samples at 105°C in vacuum, for 1 h, followed by a slow cooling *(7)*.
The procedure to prepare the heterostructures was:

1   deposition of PPV in its precursor form onto ITO (indium-tin-oxide) coated glass substrate, orientation and conversion, as described in ref. 6;

2. spin coating of P6OMe from chloroform solution on the top of the converted and oriented PPV film;
3. rubbing of the surface of the P6OMe film, with a cloth mounted on a rotating cylinder *(7)* in the orthogonal direction with respect to the orientation of the PPV chains;
4. thermal annealing of the films, as previously described *(7)* for P6OMe single layer devices.

The final thickness of the P6OMe layer was 10-15 nm. In Fig. 4 we report a sketch of the PPV/P6OMe interface.

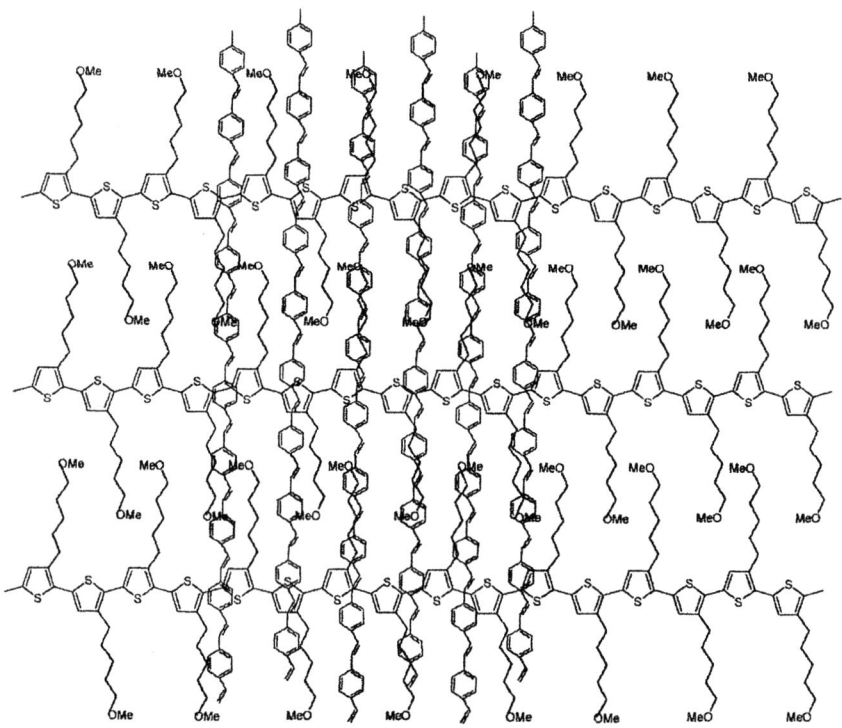

Figure 4: Sketch of the PPV/P6OMe interface

The heterostructure of the active double layer LED was been characterized by atomic force microscopy (AFM) *(10)*: domains oriented orthogonally are clearly recognized.

The polarized UV-Vis absorption spectra at room temperature of the heterostructures, prepared following the above reported steps, are shown in Fig.5.

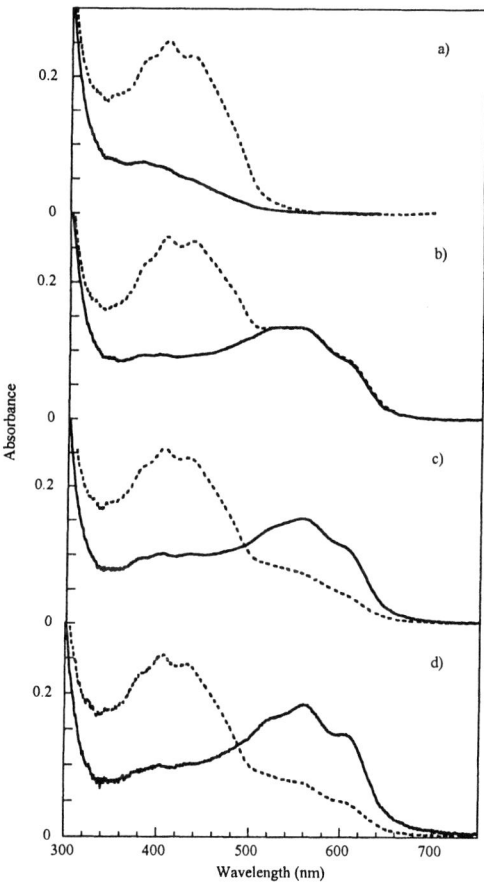

Figure 5: Polarized absorption spectra of a double-layer film of oriented PPV/P6OMe for polarization parallel (dotted line) and orthogonal (solid line) to the PPV orientation during the preparation procedure: a) step1, b)step 2; c)step3, d)step 4.

We have verified that the anisotropy of the PPV layer is not reduced after the deposition (see Fig. 5b) and rubbing (see Fig. 5c), orthogonally to the PPV orientation, of the P6OMe layer, provided that the rubbing is soft enough. The thermal annealing of the films increases the anisotropy of the P6OMe layer without affecting the PPV layer (see Fig. 5d). The increase in the PAT orientation, due to the thermal treatment, is independent of the direction of the PPV orientation, as we observed the same increase in polarization ratio after rubbing of the PAT layer in the same direction of PPV. The alignment procedure used for P6OMe permits a two-layer film to be prepared with any tipe of relative orientation of the layers, because alignment of spin-coated P6OMe on pre-oriented polymeric substrates by thermal annealing does not occur.

Even though the degree of the anisotropy reached for this double layer system is lower than that found for the single polymeric layers, it shows photoluminescence and electroluminescence emission with sensitive polarization depencences.

In Fig. 6 we report the low temperature (100 K) polarized photoluminescence measurements for a double layer of PPV and PAT oriented perpendicularly.

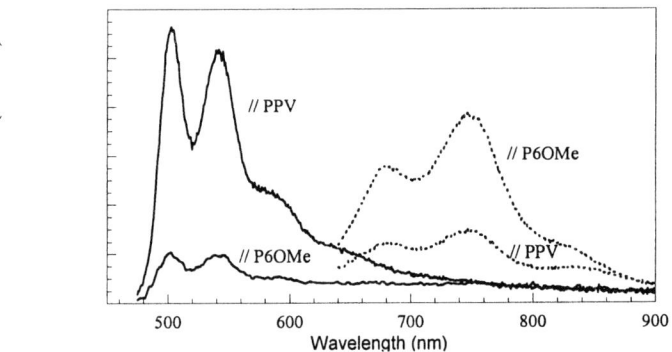

Figure 6: Polarized photoluminescence of a double layer film of oriented PPV/P6OMe. The films are excited at 363 nm (solid line) and 514 nm (dotted line).

The PL polarization ratio is 5 and 3 for PPV and PAT, respectively. The PL of P6OMe was obtained by exciting the film at 514.5 nm, where PPV has a negligible absorption, while by exciting at 363 nm the PL spectrum corresponds to the PPV emission, with a weak contribution from PAT only for the light analysed for polarization orthogonal to the PPV orientation. This is consistent with the presence of negligible energy transfer from PPV to PAT.

EL measurements were performed on a simple device structure ITO/PPV/P6OMe/Al where the PPV layer consists of two isotropic layers of PPV and a third thin layer of oriented PPV, that represents the interface with the perpendicularly oriented thin layer of P6OMe. Figure 7 shows polarized EL from the two-layer device, at different voltages.

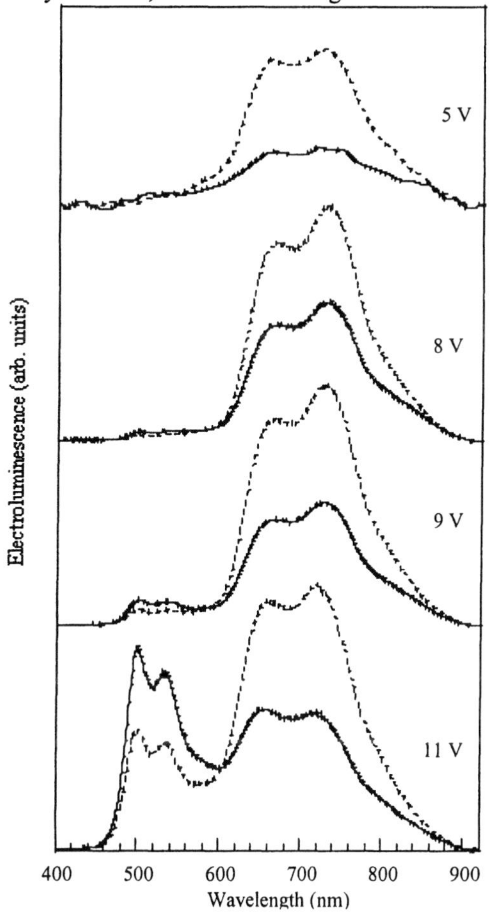

Figure 7: Polarized electroluminescence of a double layer device, at different voltages. The light is analized with a polarizer with axis parallel to the PPV (solid line) and P6OMe (dotted line) orientations.

At low voltages (< 7V) only the red emission of P6OMe is detected, showing an anisotropy of about 3. By increasing the voltage the emission from the PPV layer is also observed, with an onset voltage of about 8V. The EL emission from PPV increases, with respect to that from P6OMe, by increasing the bias voltage, reaching nearly the same intensity at 11-12V.

The polarization properties of the EL from the double structure allows to clearly observe a variation in the colour from the red to the green, at high voltages, by simply rotating the axis of a polarizer.

In conclusion, we have realized a simple heterostructure device that can emit light changing reversibly from red to orange-green by increasing the voltage from 4 to 12V. Moreover, the polarization properties of the emitted light allow to tune the colour from red to green by simply rotating the axis of a polarizer, while the device is operating at 10-12V.

Experimental

PPV was prepared from a precursor polymer containing acetate side groups, which partially remain in polymer during thermal elimination (6). Orientation of the PPV layer was obtained by rubbing during thermal elimination, and subsequently converting to PPV by annealing at 180°C for 2 h.

Details on the preparation of P6OMe have been reported previously (11). The weight average molecular weight, Mw, as detected from GPC, and referred to a calibration curve on polystyrene standards was 24.000 with a Mw/Mn of 1.6. All the data reported in the text are referred to the polymer obtained as residue to hot acetone extraction. The regioregularity of the polymer, as determined by ^{1}H-NMR investigation, was 98% (11).

After deposition by spin coating of the PAT layer, the films were repeatedly rubbed by a velvet cloth on a rotating cylinder. After this treatment the samples were heated at 105°C in vacuum (10^{-3} mmHg) for 1 h. The temperature was then slowly decreased down to room temperature in 1 h

Polarized absorption spectra were measured with a Cary 2400 spectrometer using a couple of sheet polarizers on both the sample and the reference beam, whose baseline was previously recorded for both the polarizations. Polarized PL and EL were obtained with a SPEX 270M polychromator equipped with a liquid N_2 cooled CCD detector. The emission was analysed with a sheet polarizer and a polarization scrambler was used to avoid polarization dependence of the gratings and detection system. The sample was kept under nitrogen atmosphere during the PL measurements, performed in the backscattering geometry by exciting with the 514.5 and 363.8 nm lines of an Ar^+ ion laser. The spectra were corrected for the spectral response of the instrument, measured with a calibrated lamp.

The devices were obtained by spin coating the chloroform solution of P6OMe (13mg/ml) onto ITO ($50\Omega/cm^2$) coated glass which was previously covered with PPV. The film on ITO was then rubbed and annealed as described above. The second electrode was formed by aluminium evaporated (10^{-5} mmHg) on the top of the rubbed-annealed film (aluminium thickness 100 nm). The onset voltages of the LEDs are in the range of 3 - 4.5 V and the external efficiencies are about 10^{-4} %.

References

(1) Bradley, D. D. C.; Friend, R. H.; Lindenberger, H.; Roth, S. *Polymer* **1986**, *27*, 1709.
(2) Friend, R. H.; Gymer, R. W.; Holmes, A. B.; Burroughes, J. H.; Matks, R. N.; Taliani, C.; Bradley, D. D. C.; Dos Santos, D. A.; Brédas, J. L.; Lögdlun, M.; Salaneck, W. R. *Nature* **1999**, *397*, 121
(3) Grell, M.; Bradley, D. D. C. *Adv. Mater.* **1999**, *11*, 895
(4) Grell, M.; Knoll, W.; Lupo, D.; Meisel, A.; Miteva, T.; Neher, D.; Nothofer, H. G.; Scherf, U.; Yasuda, A. *Adv. Mater.* **1999**, *11*, 671.
(5) Whitehead, K. S.; Grell, M.; Bradley, D. D. C.; Jandke, M.; Strohriegl, P. *Appl. Phys. Lett.* **2000**, *20*, 2946.
(6) Jandke, M.; Strohriegl, P.; Gmeiner, J.; Brütting, W.; Schwoerer, M. *Adv. Mater.* **1999**, *11*, 1518.
(7) Bolognesi, A.; Botta, C.; Martinelli, M.; Porzio, W. *Organic Electronics* **2000**, *1*, 27.
(8) Herold, M.; Gmeiner, I.; Schwoere, M. *Acta Polymerica* **1994**, *45*, 392.
(9) Loerner, E.; Meier, M.; Gmeiner, I.; Herold, M.; Brütting, W.; Schowere, M. *Opt. Mater.* **1998**, *9*, 109.
(10) Bolognesi, A.; Botta, C.; Facchinetti, D.; Jandke, M.; Kreger, K.; Strohriegl, P.; Relini, A.; Rolandi, R.; Blumstengel, S. *Adv. Mater.* **2001**, *13*, 1072.
(11) Bolognesi, A.; Porzio, W.; Bajo, G.; Zannoni, G.; Fanning, L. *Acta Polymerica* **1999**, *50*, 151.

Chapter 17

Tuning Optical and Electroluminescent Properties of Poly(thiophene)s via Post-Functionalization

Steven Holdcroft[1,*], Yuning Li[2], George Vamvounis[1], Hany Aziz[3], and Zoran D. Popovic[3]

[1]Department of Chemistry, Simon Fraser University, 8888 University Drive, Burnaby, British Columbia V5A 1S6, Canada
[2]Institute for Chemical Process and Engineering Technology (ICPET), National Research Council of Canada (NRC), 1200 Montreal Road, Ottawa, Ontario K1A 0R6, Canada
[3]Xerox Research Center of Canada, 2660 Speakman Drive, Mississauga, Ontario L5K 2L1, Canada

Electrophilic substitution of **P3HT** was conducted using N-bromosuccinimide (NBS), N-chlorosuccinimide (NCS), and fuming nitric acid under mild conditions to afford quantitatively brominated, chlorinated, and nitro-substituted polymers, **Br-PHT**, **Cl-PHT**, and **NO$_2$-PHT**, respectively. **Br-PHT** was further transformed to various aryl, vinyl, and alkylnyl-substituted poly(thiophene)s *via* Suzuki, Stille, and Heck coupling reactions. The optical properties of polymers could be tuned by the steric and electronic effects of incorporated substituents. Polymers possessing *ortho*-alkylphenyl or 2-(3-alkyl)thienyl groups exhibit significantly enhanced fluorescence yield (Φ_{fl}) in the solid state (~22%), which is one order of magnitude greater than that of **P3HT** (1.6%). Electroluminescence study of *o*-tolyl substituted polymers (50 – 100% of substitution) revealed that the EL efficiency of these polymers was much enhanced (internal Φ_{fl}, 0.7 – 1.8%), compared with **P3HT**.

Introduction

Polymer light-emitting diodes (PLEDs) have many attractive features over other LEDs, e.g., low cost, tunable structures and properties, and possibility of fabricating large-area displays(*1*). Previous studies on poly(thiophene)s indicated that they are generally poor emitters for PLEDs(*1b,2*). Interchain interactions (e.g., π-π stacking, excimer formation) are considered largely responsible for the drastic decrease of quantum yield of the polymer films compared with polymer solution(*1b*). Efforts have been made to suppress these interchain interactions, including the use of polymer blends(*3*), incorporation of sterically encumbered side-chain groups(*4*), etc. However, re-aggregation of poly(thiophene)s would occur in polymer blends over time and the attachment of desired side-chain groups is often restricted by the synthetic limitations either in the monomer synthesis stage or in the polymerization. In this study we applied an alternate route, post-fuctionalization, to approach poly(thiophene)s with favored structures and improved luminescence properties. This method simplifies the procedures for acquiring complex structures and broadens the spectrum of available functional groups on the polymer side chains. From a single precursor poly(3-hexylthiophene) (**P3HT**), a variety of novel disubstituted poly(thiophene)s with diverse structures were derived. The steric and electronic effects of incorporated functional groups on the electrooptical properties of the polymers were discussed.

Results and Discussion

Electrophilic Substitution of P3HT

Aromatic hydrogens on the polymer backbone of poly(thiophene)s can be activated electrochemically to the overoxidized state and substituted by nucleophiles Cl^-, Br^- or CH_3O^-(*5*). However, poly(3-alkylthiophene)s in the neutral state are electron rich due to the extended π-system and the electron-donating effect of the 3-alkyl side chain, and thus the 4-hydrogen of the thienyl ring should be more susceptible to electrophilic substitution.

Electrophilic substitution was attempted on regioregular **P3HT** using *N*-bromosuccinimide (NBS). After reaction at room temperature for 15 h and at 50 °C for 2h in chloroform, a yellow solid was obtained in 99% yield. A ^{1}H NMR spectrum of the product (Figure 2) shows the disappearance of 4-hydrogens of

P3HT (at 6.99 ppm) and the chemical shift of the α-methylene hydrogen of the hexyl group decreased from 2.80 ppm to 2.71 ppm. This result, along with ^{13}C NMR and IR spectroscopy, and elemental analysis, indicate clearly the product to be poly(3-bromo-4-hexylthiophene) (**Br-PHT**). Similarly, *N*-chlorosuccinimide (NCS) produced poly(3-chloro-4-hexylthiophene) (**Cl-PHT**) in 98% yield based on 100% substitution after reaction for 4 h at room temperature (Figure 1). No aromatic hydrogens corresponding to **P3HT** (at 6.99 ppm) remained in the product in the ^{1}H NMR spectra (Figure 2). Two minor peaks at 5.02 and 2.10 ppm were assigned to α-methylene and β-methylene hydrogens of the side chain, respectively, using DEPT and ^{1}H–^{13}C COSY. A small peak at 7.31 ppm was considered due to the 4-hydrogen on the thiophene ring on which the α-methylene is chlorinated (Figure 2). It is estimated from ^{1}H NMR that 85 % of the aromatic hydrogens of **P3HT** are substituted with chlorine, while 15 % of α-methylene groups are chlorinated. Nitration of **P3HT** was conducted in the presence of fuming nitric acid (93% yield). Complete substitution of the 4-hydrogens was confirmed by ^{1}H NMR (Figure 2), ^{13}C NMR, and IR spectroscopies. Elemental analysis also confirmed the product to be poly(3-hexyl-4-nitrothiophene) (**NO$_2$-PHT**).

Transformation of Br-PHT *via* Pd-Catalyzed Coupling Chemistry

Suzuki coupling(*6*) of phenylboronic acid with **Br-PHT** was conducted in THF at 80 °C for 48 h in the presence of Na$_2$CO$_3$ and 2 % equiv. of Pd(PPh$_3$)$_4$ (Figure 1). ^{1}H NMR spectroscopy of the product (**Ph-PHT**) clearly showed the emergence of phenyl protons at 7.27 and 7.15 ppm (Figure 3). From the ratio integrals of aromatic protons to –C*H$_3$* protons on the hexyl groups, the substitution of Br by phenyl was estimated to be > 99 %. This was also confirmed by elemental analysis (Br < 0.3 %). Other *para*-, *meta*-, and *ortho*-substituted phenylboronic acids, 1-naphthylboronic acid, and 2-thiopheneboronic acid also reacted quantitatively with **Br-PHT** to produce the corresponding disubstituted poly(thiophene)s (Figure 1).

Several tributyltin compounds were used for Stille-type coupling(*7*) with **Br-PHT** (Figure 1). Reactions of 2-thiophene- and 2-furantributyltin with **Br-PHT** at 100 °C in toluene gave products **2-Th-PHT** and **2-Fu-PHT** with complete substitution of Br (> 99 %) within 24 h. Reaction of vinyl- and phenylethynyltributyltin with **Br-PHT** afforded highly substituted polymers **Vi-PHT** (91 %) and **PhE-PHT** (> 99%). Phenyltributyltin was found reluctant to react with **Br-PHT**. Only 42 % of the Br groups in **Br-PHT** were replaced by phenyl even after 120 h (**Ph-PHT**). This result is in agreement with the known poor reactivity of stannylbenzenes under Stille coupling conditions(*8*).

223

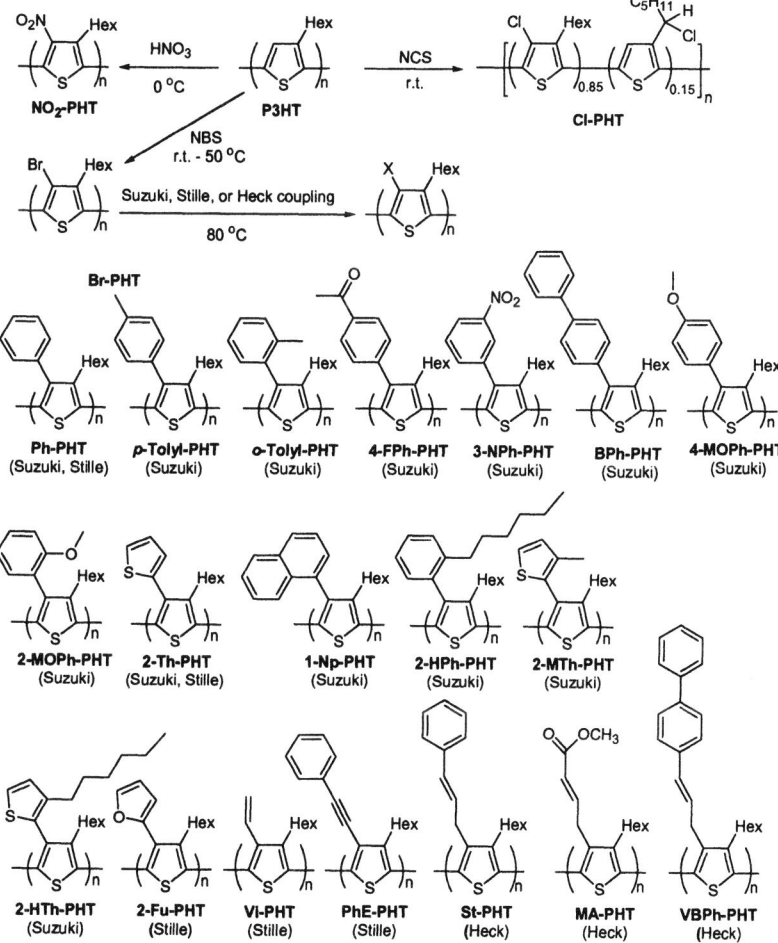

*Figure 1. Schematic diagram of post-fuctionalization of **P3HT**.*

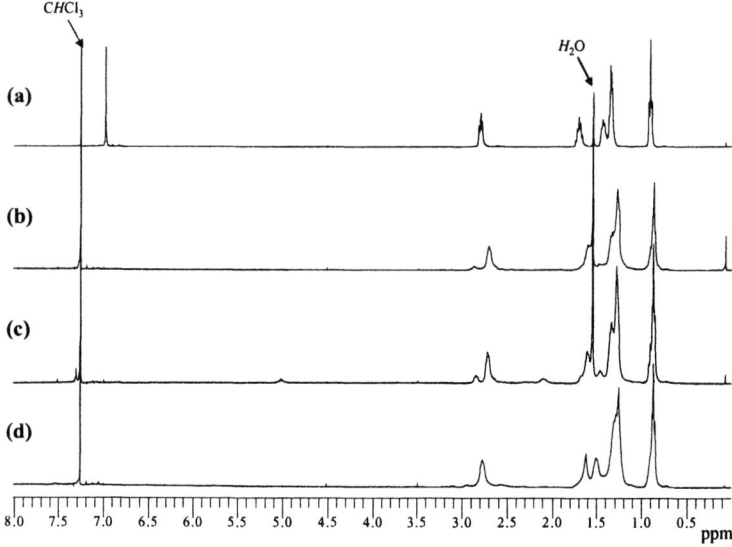

*Figure 2. 400 MHz ^{1}H NMR Spectra of (a) **P3HT**, (b) **Br-PHT**, (c) **Cl-PHT**, and (d) **NO$_2$-PHT**.*

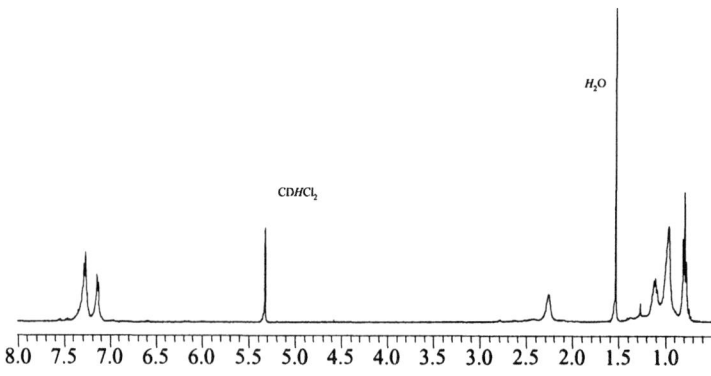

*Figure 3. 400 MHz ^{1}H NMR spectrum of **Ph-PHT**.*

Heck reactions(9) of **Br-PHT** using methyl acrylate, styrene, and 4-vinylbiphenyl could also occur, but ^{1}H NMR analysis showed that the extent of substitution was relatively low (29 – 51 % for **MA-PHT**, **St-PHT**, and **VBPh-PHT**).

Optical Properties

As shown in Table 1, the absorption maximum of **Br-PHT** (λ_{max}, 339 nm) is blue shifted, with respect to **P3HT** (λ_{max}, 442 nm), indicating increased twisting of the polymer backbone. A decrease in percentage Φ_{fl} to 4% in solution (40.1% for **P3HT**) is observed due to the shortened conjugation length and the heavy atom effect of bromine(*10*). The solid-state Φ_{fl} of **Br-PHT** is also not appreciably improved (Φ_{fl}, 1.8%) compared to **P3HT** (Φ_{fl}, 1.6%). On the other hand, the chlorinated product **Cl-PHT** shows a smaller blue shift in absorption λ_{max} and a higher Φ_{fl} (solution, 12%; solid sate, 5.1%), compared to **Br-PHT**. The weaker heavy-atom effect of chlorine and less twisting of backbone (due to the smaller size of chloro group) might both contribute to above result. Introduction of a nitro group at the 4-position of the thiophene ring (**NO₂-PHT**) results in the virtual absence of fluorescence emission, suggesting that the nitro group is a strong quencher of excitation.

Table I. Optical Properties of 3,4-Disubstituted Poly(thiophene)s.

Polymers	Solution			Films		
	λ_{max} Abs, nm	λ_{max} Em, nm	Φ_{fl} %	λ_{max} Abs, nm	λ_{max} Em, nm	Φ_{fl} %
P3HT	442	571	40.1	550	660,730	1.6
Br-PHT	339	504	4.0	344	520	1.8
Cl-PHT	357	516	12.0	365	527	5.1
NO₂-PHT	338	556	0.3			
Ph-PHT	363	528	7.9	369	553	3.2
p-Tolyl-PHT	364	529	8.0	368	550	3.3
o-Tolyl-PHT	400	545	20.8	410	556	19.4
4-MOPh-PHT	367	531	8.2	375	538	3.0
2-MOPh-PHT	367	535	9.7	376	545	4.4
4-FPh-PHT	350	540	3.4	360	540	2.5
3-NPh-PHT	354	n.d.	~0	356	n.d.	~0
BPh-PHT	275	360	1.1			
	360	531	5.8	367	538	3.3
1-Np-PHT	370	540	8.8	378	542	2.6
2-Th-PHT	370	540	8.8	378	542	5.0
PhE-PHT	403	567	7.0	415	602	4.0
2-HPh-PHT	413	555	25.1	422	583	19.6
2-MTh-PHT	401	544	12.8	405	558	9.0
2-HTh-PHT	404	544	14.2	410	555	13.9

Table II. Optical Properties of Partially Substituted P3HT.

Polymers	Solution			Films		
	λ_{max} Abs, nm	λ_{max} Em, nm	Φ_{fl} %	λ_{max} Abs, nm	λ_{max} Em, nm	Φ_{fl} %
Br-PHT0[a]	442	571	40.1	550	660, 730	1.6
10	439	570	36.7	535	710	1.8
20	425	570	32.3	515	718	3.2
38	408	561	27.6	433	656	1.8
50	388	556	20.9	405	620	5.0
67	367	548	11.9	378	590	3.6
75	359	535	9.4	370	571	4.0
89	352	520	6.3	359	559	4.4
100[b]	339	504	4.0	344	520	1.8
o-Tolyl-PHT0[a]	442	571	40.1	550	660, 730	1.6
10	444	570	35.7	537	710	2.5
20	442	570	34.2	528	708	3.1
38	437	569	28.2	492	700	4.4
50	432	566	24.2	460	640	12.8
67	430	566	31.0	462	638	13.2
75	420	563	30.2	437	610	20.5
89	418	561	27.8	432	593	22.3
100[c]	400	545	20.8	410	556	19.4

Note: a) P3HT, b) Br-PHT, and c) o-Tolyl-PHT in Table 1.

Substitution of Br in **Br-PHT** with phenyl causes a red shift in absorption λ_{max} of the polymer **Ph-PHT** in solution from 339 nm to 363 nm. The percentage Φ_{fl} increases from 4 % to 7.9 % in solution and from 1.8% to 3.2% in the solid state (Table 1). The attachment of electron-donating 4-methyl and 4-methoxy groups to the phenyl ring has little influence on the absorption and emission $\lambda_{max's}$, and on Φ_{fl} of the resultant polymers (***p*-Tolyl-PHT** and **4-MOPh-PHT**) compared to **Ph-PHT**, indicating that the phenyl group is electronically isolated from the π-conjugated polythiophene backbone. **BPh-PHT**, possessing biphenyl substituents, exhibits two absorption peaks at 275 nm and 360 nm, which represent the biphenyl group and the backbone π-system, respectively. The absorption and emission profiles originating from excitation of the polymer backbone are similar to **Ph-PHT**. Like phenyl, *p*-tolyl, and 4-methoxyphenyl substituents, biphenyl groups in **BPh-PHT** are isolated chromophores. This can only be possible if the chromophores lie perpendicular to the thiophene ring. Introduction of electron-withdrawing 4-formyl group on the phenyl ring (**4-FPh-PHT**) causes a blue shift in absorption λ_{max} (350 nm) and a decrease in Φ_{fl} in solution (3.4%) relative to **Ph-PHT**. Substitution of Br with 3-nitrophenyl group (**3-NPh-PHT**) causes a slight blue shift in absorption λ_{max} (354 nm), but it also results in the absence of fluorescence, even though the nitro group at the *m*-position is virtually electronically isolated from the polymer backbone. The 2-thienyl-substituted polymer (**2-Th-PHT**) possesses slightly red-shifted absorption and emission maxima compared with **Ph-PHT**. Another derivative, the phenylethylenyl substituted polymer (**PhE-PHT**), displays large red shifts in absorption (403 nm, solution; 410 nm, solid state) and emission λ_{max} (567 nm, solution; 602 nm, solid state) due to the linear ethylenyl group alleviating steric repulsion. The phenylethylenyl group facilitates planarity of the polymer backbone, which, together with the electron-donating effect of the phenylethylenyl groups, narrows the band-gap. The observed electronic effect of the ethylenyl group, more pronounced than other substituents, is due to efficient overlapping of linear sp orbital with the main chain π-system. Φ_{fl} of this polymer (7.0%, solution; 4.0%, solid state) is not varied significantly compared with **Ph-PHT**.

A significant deviation from the above trends is observed for the *o*-tolyl-substituted polymer (***o*-Tolyl-PHT**) that exhibits not only a large red shift in λ_{max}, but also an unusually high Φ_{fl} in solution (20.8%) and in the solid state (19.4%). The solid state Φ_{fl} is an order of magnitude higher than **P3HT** (1.6%) and **Br-PHT** (1.8%). We attribute this to the steric effect of the *o*-tolyl substituent. As illustrated in the space-filling model (Figure 4), the steric repulsion of the *o*-tolyl group with the neighboring hexyl substituent and sulfur atom forces the phenyl ring perpendicular with respect to the thienyl ring. The *o*-methyl group in ***o*-Tolyl-PHT** resides by the sulfur atom and hence to lock the rotation or twisting of the neighboring thienyl ring. This conformation planarizes the polymer backbone, promotes a red shift in λ_{max}. In addition, the methyl groups, together with the phenyl rings, function as spacers that enlarge the distance between stacks of polymer chains, hence result in the increase in the solid state Φ_{fl}. Apparently, the phenyl group in **Ph-PHT** is not capable of preventing the main chain from rotating (Figure 4) that leads to a shorter absorption wavelength.

2-MOPh-PHT and **1-Np-PHT** possessing *o*-methoxyphenyl and 1-naphthyl, respectively, display only a slight red shift in λ_{max} (367 nm for **2-MOPh-PHT** and 370 nm for **1-Np-PHT** in solution) and a less notable increase in solution Φ_{fl} (9.7% for **2-MOPh-PHT** and 8.8% for **1-Np-PHT**) than **Ph-PHT**. Although the 2-methoxy group in **2-MOPh-PHT** is structurally similar to the *o*-tolyl group in ***o*-Tolyl-PHT**, the methoxy group seems not effectively interact with the neighboring hexyl group and sulfur atom to prevent the rotation of the backbone. The planar 1-naphthyl group in **1-Np-PHT** also seems unlikely to limit the rotational freedom of the neighboring unit. The above observations indicate that a slight variation in the size or shape of the substituents greatly impacts the backbone conformation and ultimately its optical properties.

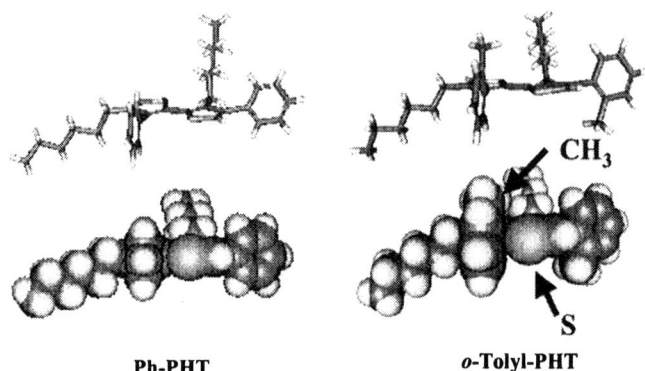

Ph-PHT   o-Tolyl-PHT

Figure 4. Space-filling models of **Ph-PHT** *and* **o-Tolyl-PHT**.

Several structural analogs of *o*-**Tolyl-PHT** (**2-HPh-PHT**, **3-MTh-PHT**, and **3-HTh-PHT**) were also designed and synthesized using **Br-PHT** as the starting polymer (Scheme 1). A common structural feature of these polymers is that the phenyl, or thienyl, side group possesses an alkyl substituent juxtapositioned to the point of attachment to the polymer chain. **2-HPh-PHT** solutions are even red shifted (λ_{max}, 413 nm) compared to *o*-**Tolyl-PHT** and Φ_{fl} further increases to 25.1%, indicating that the longer alkyl side chain on the phenyl group further increases the planarity and rigidity of the backbone. The solid state Φ_{fl} (19.6%), however, remains similar to *o*-**Tolyl-PHT**. A similar steric effect was observed for 2-(3-methyl)thienyl and 2-(3-hexyl)thienyl groups, and is manifest by red shifts in absorption λ_{max} (401 nm, **3-MTh-PHT**; 404 nm, **3-HTh-PHT** in solution) compared to the 2-thienyl substituted polymer, **2-Th-PHT** (370 nm). Φ_{fl} of **3-MTh-PHT** and **3HTh-PHT** were 12.8% and 14.2% in solution and 9.0% and 13.9% in the solid state, respectively. These values are higher than **2-Th-PHT** (8.8%, solution; 5.0%, solid state). This enhancement in Φ_{fl} can be also interpreted by the increased main chain coplanarity and intermolecular distance, which are, as in the case of *o*-**Tolyl-PHT**, originated from the steric repulsion of the 3-alkylthienyl group with the neighboring hexyl group and sulfur atom.

Although *o*-alkylphenyl and 2-(3-alkyl)thienyl substituted polymers exhibit a larger red shift in absorption λ_{max} relative to **Ph-PHT** or **2-Th-PHT**, the overall effective conjugation length of these polymers is reduced compared to **P3HT**, as judged by the absorption and emission wavelengths. The conjugation length influences not only the emission wavelength but also the quantum efficiency (Φ_{fl} of oligomers increases asymptotically with increasing degree of oligomerization up to six contiguous thiophene units)(*11*). In order to increase

Φ_{fl}, and at the same time obtain longer emission wavelengths, π-stacking must be suppressed while backbone planarity and rigidity are maintained. Since bromination of **P3HT** and subsequent Suzuki coupling are near quantitative, the conjugation length of polymers may be controlled by partial substitution of **P3HT** in order to optimize Φ_{fl} and achieve longer wavelength emission. As shown in Table 2, a series of polymers with degrees of substitution with bromo groups ranging from 0% to 100% were synthesized. The absorption λ_{max} of the brominated polymer films is controlled between 550 nm (for 0% bromination, **P3HT**) and 344 nm (for 100% bromination, **Br-PHT**). Φ_{fl} in solution decreases from 40.1% to 4% as the substitution increases from 0 to 100%, although the solid state Φ_{fl} shows no clear trends except that substituted polymers generally exhibit higher values.

Substitution of the brominated polymers with o-tolyl groups yields the corresponding partially substituted **o-Tolyl-PHT** polymers. λ_{max} decreases with degree of substitution (442 – 400 nm, solution; 550 – 410 nm, solid state) but the decrease is less pronounced than the corresponding brominated derivatives. Φ_{fl} of the polymer solution decreases with degree of substitution up to 37.5%, but does not decrease further when the degree of substitution exceeds 50%. In contrast, the solid state Φ_{fl} increases with degree of substitution, and increases more substantially when substitution levels exceed 50% to a maximum value of 22.3% (for **o-Tolyl-PHT89**). The emission wavelengths of polymers in the solid state are substantially red shifted, compared with the corresponding Br-substituted polymers. When the degree of substitution lies between 50% and 100%, the polymers are highly luminescent (Φ_{fl} = 12.8 – 22.3%) and the emission wavelengths range from 556 nm to 640 nm.

Electroluminescence Results

PLEDs using polymers **P3HT**, **o-Tolyl-PHT50**, **o-Tolyl-PHT75**, **o-Tolyl-PHT88**, and **o-Tolyl-PHT100** as emitting layers were fabricated. The devices were configured with indium-tin-oxide (ITO) anode/ polymer emitting layer/ triphenyltriazine (TPT)(*12*) electron transport layer(*2*) (300Å)/ magnesium-silver alloy (9:1) cathode (1200 Å). Figure 4 shows the EL spectra measured from the devices. It was found that the electroluminescence maxima of these polymers (EL λ_{max}: 660 and 730 nm for **P3HT**, 642 nm for **o-Tolyl-PHT50**, 594 nm for **o-Tolyl-PHT88**, and 558 nm for **o-Tolyl-PHT100**) agree well to their photoluminescence maxima (PL λ_{max}) (See Table 2), except of **o-Tolyl-PHT75** which EL λ_{max} (626 nm) was longer than its PL λ_{max} (610 nm). The device made with **P3HT** showed very poor electroluminescence (internal Φ_{EL}: 0.01%). On the other hand, devices made with o-tolyl-substituted poly(thiophene)s exhibited much enhanced electroluminescence efficiency (internal Φ_{EL}: 1.8% for **o-Tolyl-**

PHT50, 0.7% for *o*-Tolyl-PHT75, 1.7% for *o*-Tolyl-PHT88, and 1.3% for *o*-Tolyl-PHT100).

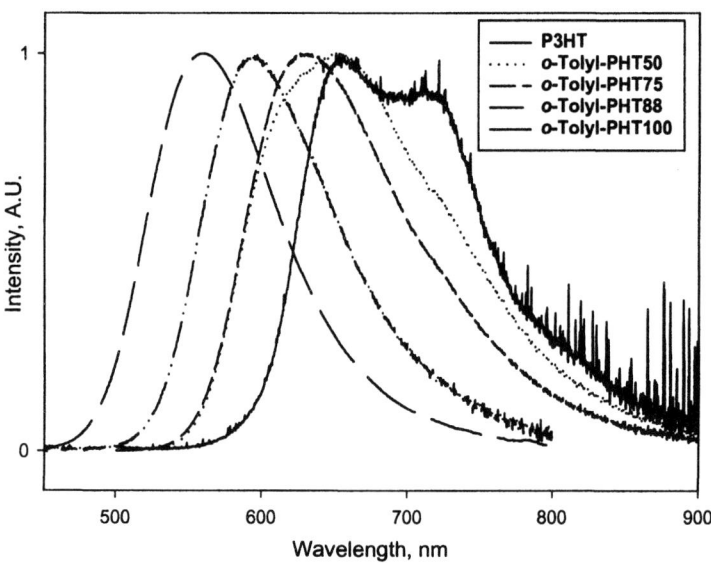

Figure 4. Eletroluminescence spectra of o-tolyl-substituted P3HT.

Conclusion

Post-functionalization of **P3HT** by using electrophilic substitution and Pd-catalyzed coupling chemistries provides an opportunity to systematically study the influence of various substituents on the photophysical properties of 3,4-disubstituted poly(thiophene)s. Poly(thiophene)s possessing *o*-tolyl, *o*-hexylphenyl, 2-(3-methyl)thienyl, and 2-(3-hexyl)thienyl groups exhibited a significant increase in the solid state Φ_{fl} (9 – 20%), compared to **P3HT** (1.6%). Partial substitution of **P3HT** with *o*-tolyl groups at the 4-postion (ranging from 50% to 100%) showed significantly enhanced solid state Φ_{fl} (13 – 22%) and tunable emission wavelengths (640 – 556 nm). PLED devices made with these *o*-tolyl-substituted poly(thiophene)s (50 – 100% of substitution) showed much improved electroluminescence performance (internal Φ_{EL}= 0.7 – 1.8%).

Experimental

Measurements. 400 MHz ^{1}H NMR spectra were obtained in CDCl$_3$ on a 400 MHz Bruker AMX400 spectrometer. IR spectra were recorded on a Bomen Michelson MB series spectrophotometer. UV-vis absorption spectra were obtained on a Cary 3E (Varian) spectrophotometer. Fluorescence measurements were carried out on a PTI QuantumMaster model QM-1 spectrometer. Polymer solutions in THF with OD = 0.05 –0.1 were deoxygenated prior to fluorescence measurement and the quantum yield of polymers ($\leq$ 10% error) was determined against quinine bisulfate standard (Φ_{fl} = 0.546 in 1.0 N H$_2$SO$_4$). Spin-coated polymer films with OD = 0.1 – 0.2 were protected under an argon flow during fluorescence measurement and the quantum yield ($\leq$ 30% error) was reported against 9,10-diphenylanthracene in PMMA (< 10^{-3} M) (Φ_{fl} = 0.83). The fabrication of PLED devices was conducted as follows. Polymer (3 mg) was dissolved in chloroform (0.5 ml) and filtered through a 0.2 μm filter and spin coated (2000 RPM) on a UV ozone pre-cleaned patterned ITO substrate. Triphenyltriazine (TPT) and cathode were deposited subsequently using vacuum evaporation at 6 × 10^{-6} Torr.

Synthesis. Regioregular poly(3-hexylthiophene) (**P3HT**) was prepared according to the method reported by McCullough *et al.*(*13*) Details on electrophilic substitution of **P3HT** and Pd-catalyzed coupling reaction of **Br-PHT** were published elsewhere(*14-16*).

Acknowledgment

The authors are thankful for the financial support of this study by NSERC. G. V. is grateful to NSERC and Xerox Research Center of Canada for the generous scholarship.

References

(1) (a) Burroughes, J. H.; Bradley, D. D. C.; Brown, A. R.; Marks, H. N.; Mackay, K.; Friend, R. H.; Burns, P. L.; Holmes, A. B. *Nature* **1990**, *347*, 539. (b) Samuel, I. D. W.; Rumble, G.; Friend, R. H. In *Primary Photoexcitations in Conjugated Polymers: Molecular Excitation versus Semiconductor Band Model*, Sariciftci, N. S. Ed.; World Scientific Publishing Co.: Singapore, 1997. (c) Yu, G.; Wang, J.; McElvain, J.; Heeger, A. J. *Adv. Mater.* **1998**, *10*, 1431.

(2) (a) Xu, B.; Holdcroft, S. *Macromolecules* **1993**, *26*, 4457. (b) Greenham, N. C.; Samuel, I. D. W.; Hayes, G. R.; Phillips, R. T.;

Kessener, Y. A. R. R.; Moratti, S. C.; Holmes, A. B.; Friend, R. H. *Chem. Phys. Lett.* **1995**, *241*, 89. (c) Chen, F.; Mehta, B.; Takiff, L.; McCullough, R. D. *J. Mater. Chem.* **1996**, *6*, 1763.

(3) (a) Lemmer, U.; Mahrt, R. F.; Wada, Y.; Greiner, A.; Bässler, H.; Göbler, E. O. *Appl. Phys. Lett.* **1993**, *62*, 2827. (b) Nishino, H.; Yu, G.; Heeger, A. J.; Chen, T. -A.; Rieke, R. D. *Synth. Met.* **1995**, *68*, 243. (c) Berggren, M.; Bergman, P.; Fagerström, J.; Inganäs, O.; Anderson, M. R.; Weman, H.; Granström, M.; Stafström, S.; Wennerström, O.; Hjertberg, T. *Chem. Phys. Lett.* **1999**, *304*, 84.

(4) Anderson, M. R.; Berggren, M.; Olinga, T.; Hjertberg, T.; Inganäs, O.; Wennerström, O. *Synth. Met.* **1997**, *85*, 1383.

(5) (a) Harada, H.; Fuchigami, T.; Nonaka, T. *J. Electroanal. Chem.* **1991**, *303*, 139. (b) Qi, Z.; Rees, N. G.; Pickup, P. *Chem. Mater.* **1996**, *8*, 701.

(6) Miyaura, N.; Suzuki, A. *Chem. Rev.* **1995**, *95*, 2457 and references cited therein.

(7) Stille, J. K. *Angew. Chem. Int. Ed. Engl.* **1986**, *25*, 508 and references cited therein.

(8) Bao, Z.; Chan, W. K.; Yu, L. *J. Am. Chem. Soc.* **1995**, *117*, 12426.

(9) (a) Heck, R. F. *Org. React.* **1981**, *27*, 345. (b) Beletskaya, I. P.; Cheprakov, A. V. *Chem. Rev.* **2000**, *100*, 3009.

(10) Saadeh, H.; Goodson, T. III; Yu, L. *Macromolecules* **1997**, *30*, 4608.

(11) (a) Chosrovian, H.; Rentsch, S.; Grebner, D.; Dahm, D. U.; Brickner, E. *Synth. Met.* **1993**, *60*, 23. (b) Herrema, J. K.; van Hutten, P. F.; Gill, R. E.; Wildeman, J.; Wieringa, R. H.; Hadziioannou, G. *Macromolecules* **1995**, *28*, 8102.

(12) Popovic, Z.P.; Aziz, H.; Hu, N.X.; Ioannidis, A.; dos Anjos, P.N.M. *J. Appl. Phys.* **2001**, *89*, 4673.

(13) McCullough, R. D.; Low, R. D.; Jayaraman, M.; Anderson, D. L. *J. Org. Chem.*, **1993**, *58*, 904.

(14) Li, Y.; Vamvounis, G.; Holdcroft, S. *Macromolecules* **2001**; *34*, 141.

(15) Li, Y.; Vamvounis, G.; Yu, J.; Holdcroft, S. *Macromolecules* **2001**, *34*, 3130.

(16) Li, Y.; Vamvounis, G.; Holdcroft, S. *Macromolecules* **2002**, *35*, 6900.

Chapter 18

Site-Isolated Luminescent Lanthanide Complexes with Polymeric Ligands

Jessica L. Bender and Cassandra L. Fraser

Department of Chemistry, University of Virginia, Charlottesville, VA 22904-4319

Luminescent lanthanide-centered star-shaped polymers, both homopolymers and block copolymer analogues, were readily accessed by a modular metal template synthesis. Dibenzoylmethane (dbm) macroligands with pendant poly(lactic acid) (PLA) and poly(ε-caprolactone) (PCL) chains were synthesized by ring opening reactions, whereas poly(methyl methacrylate) (PMMA) derivatives were accessed via atom transfer radical polymerization. Reaction of these polymeric ligands with metal salts in the presence of base, yielded M(dbm-polymer)$_n$ complexes in a single step. Examples include Ln(dbmA)$_3$X (Ln = Eu, Tb; A = PLA, PCL, PMMA; X = additional donor such as H$_2$O, solvent, α-diimines) and heteroarm stars of the type, Ln(dbmA)$_3$(bpyB$_2$), (B = PCL, PLA) with both dbm and bipyridine (bpy) macroligands. Emission spectra and luminescence lifetimes provide valuable information about sample homogeneity and other features of the Ln coordination environment. Block copolymers with labile metal crosslinks form nanostructured films that change their shape upon thermal treatment.

Introduction

Due to their unique physical properties, lanthanide metals find wide use in many technological applications (*1*). Paramagnetic gadolinium complexes are commonly found as magnetic resonance imaging (MRI) contrast agents in medicine (*2*), whereas many paramagnetic chiral shift reagents are based on europium (*3*). Europium and terbium luminescence is of great utility in sensors and molecular probes(*4, 5*), solid state phosphors (*6*), and emitting layers for OLEDs (*7*) and photonic materials (*8, 9, 10*). Compared to transition metals, which are affected in significant ways by their ligand field, lanthanides exhibit weak covalent bonding interactions with ligands due to buried f orbitals, and their coordination chemistry is governed largely by electrostatics. This influences their spectroscopic properties, and gives rise to distinctive emission colors for certain lanthanide ions, that vary little with their particular ligand set (*11*). For example, europium(III) emission is typically red, and terbium emits green, while erbium radiates in the near IR and has been exploited for signal amplification in fiber optics (*12*). In photonic crystals, lanthanide dyes (*13*) are sometimes added to modulate the propagation of light or to probe the photonic band gap structure.

Many applications benefit from the introduction of metals into polymers (*14-16*), which lend processibility and, in some cases, lead to enhanced physical properties. Conjugated polymers serve as electron and hole transporting layers in OLEDs for next-generation display technologies, for which lanthanide chromophores may be tuned to provide optimal energy transfer (*17*). Luminescent materials benefit from site isolation of chromophores in polymeric shells, which can inhibit self quenching (*18*). In photonic materials, polymers are often used as structural (*19*) and templating elements (*20, 21*), to generate materials of the correct refractive index and nanoscale periodicity for interaction with light of specific wavelengths of interest (including visible light). Currently, there is much interest in the assembly of block copolymers into ordered assemblies which can function as photonic crystals (*22, 23*). Locating chromophores at regions of high field intensity within a photonic device will provide dramatically higher coupling efficiencies between the light in the photonic material and the chromophore. In order to modulate materials properties, the ability to selectively position metal functionalities within macromolecular architectures and hierarchically assemblied composite matrices is advantageous.

Lanthanide complexes with polymeric ligands combine the favorable optical properties of lanthanide metals with the processing advantages of polymer films. Polymeric metal complexes (PMCs), comprised of macroligands coordinated to metal ions, provide a unique platform offering a wide range of possibilities for

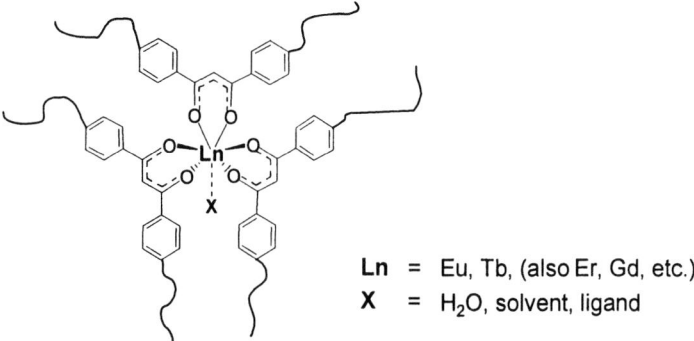

Figure 1. General Eu dbm polymeric metal complex structure.

architectural and compositional diversity (Figure 1). The metal ion core serves as an interchangeable template for synthesis, and the polymeric ligands coordinated to it may be mixed and matched to generate an array of PMCs. The properties of the complexes may be modulated through the use of a variety of β-diketonate (bdk) ligands of different steric and electronic demand, and different degrees of functionalization. Utilizing living polymerization methods, the composition, size, and architecture of the polymeric ligands can be adjusted. These macroligands can be combined with various lathanide metal salts to form tris and tetrakis complexes. Through the addition of a second type of ligand in addition to three bdks, adducts can be formed. In particular, bipyridine (bpy) ligands can be used as the adduct-forming component, allowing utilization of our expertise in bpy macroligand synthesis (24). Either the bdk, the bpy, or both ligands can be polymeric. In the latter case, with different bdk and bpy macroligands, a block copolymer is formed, with the lanthanide metal at the block junction. These heteroarm stars, prepared by a convenient one-pot synthesis, can self-assemble to form discrete nanoscale structures as shown in Figure 2. What is more, because lanthanide metals are labile, thermally induced bpy ligand dissociation corresponds to a change in film morphology.

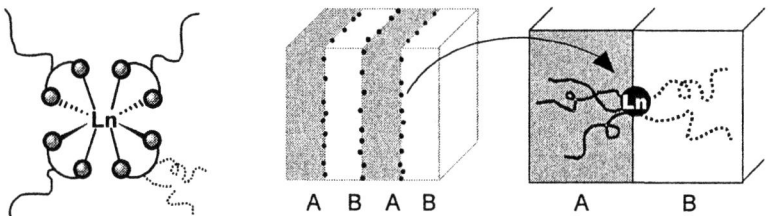

Figure 2. Block copolymer microphase separation with lanthanide complex at the block junction (lamellar morphology shown).

Dibenzoylmethane Initiators

The dibenzoylmethane (dbm) ligand is a versatile platform for polymerization initiation, due to the multifunctionality that is possible for the aromatic rings, allowing functionalization at 1-6 sites on the rings. These ligands and related bdk ligands can be prepared by Claisen condensation reactions in which ester and ketone halves are combined. In this sense, the synthesis provides a modular approach to a vast library of diketonate ligands through further variation of the ester and ketone components.

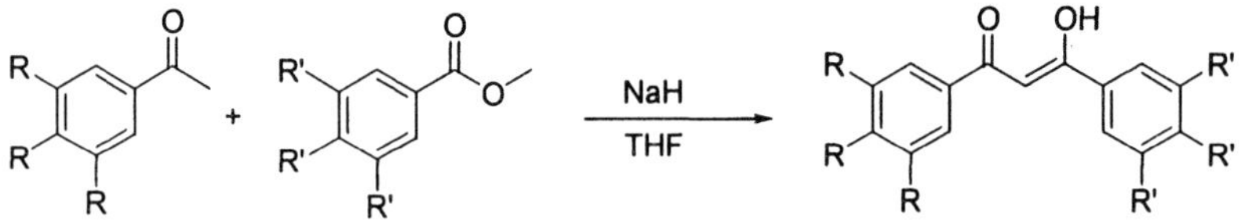

Figure 3. Claisen condensation of ester and ketone to afford dbm.

In order to achieve controlled polymerization from the bdk ligand, an ethanol moiety is introduced to the phenol site to provide a primary alcohol site, and a less sterically crowded environment for initiation (Figure 4). The alcohol requires protection in order to prevent deprotonation under the basic reaction conditions required for the ester and ketone condensation. A number of protecting groups were screened for this purpose, and one promising strategy involves methoxymethyl (MOM) protection. The formation of the MOM ether proceeds in nearly quantitative yield. After the condensation, the MOM group can be removed with aqueous HCl in THF.

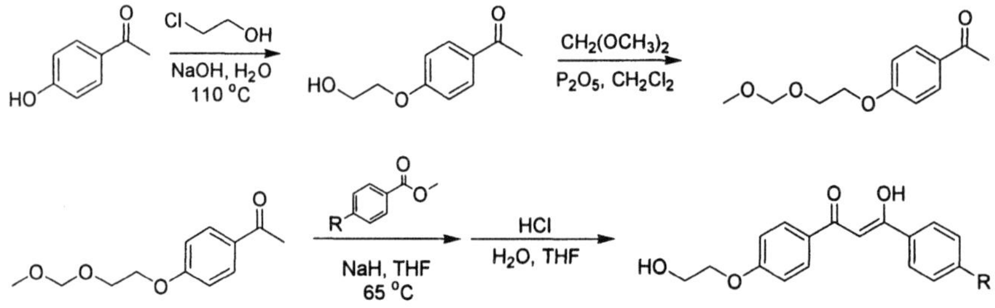

Figure 4. Synthesis of alcohol initiators: monofunctional, dbmOH (R=H), and difunctional dbm(OH)$_2$ (R=OCH$_2$CH$_2$OH).

Alcohols serve as initiators for the ring-opening polymerization (ROP) of lactide and ε-caprolactone (25), whereas a bromo ester functionality is required for the atom transfer radical (ATRP) polymerization of methyl methacrylate (26). To date, monofunctional (dbmOH, dbmBr), difunctional (dbm(OH)$_2$, dbm(Br)$_2$), and heterodifunctional initiators (dbm(OH)(Br)) have been prepared by combining and elaborating appropriately functionalized ester and ketone halves.

Dibenzoylmethane Macroligands

We have demonstrated controlled bulk lactide (27) and ε-caprolactone polymerizations from the monfunctional dbm initiator using tin octoate, Sn(oct)$_2$, as the catalyst. Polyesters with low polydispersity indices (PDIs), dbmPLA, and dbmPCL, with dbm at the chain end were produced. The effects of the dbm enol functionality on the polymerizations is also under exploration. In theory, this site too, could serve as an initiator, but in practice, this side reaction seems to be negligible. In control reactions with dbm (i.e. enol but no pendant primary OH groups), Sn(oct)$_2$, and ε-caprolactone, no polymer was observed for eight hours, though after this time, some was produced. The kinetics plot in Figure 5 outlines the details of these findings. The initiator and monomer loadings were kept constant at 1/500 for dbm, dbmOH, and dbm(OH)$_2$ with a catalyst loading of 1/40 for dbm and dbmOH and for dbm(OH)$_2$, 2/40 to accommodate both alcohol initiating sites. The data for each of the reactions

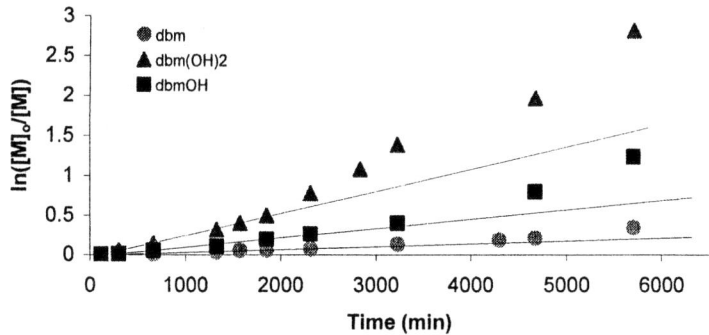

Figure 5. First order kinetics plot of dbmOH, dbm(OH)$_2$ and dbm ε-caprolactone polymerizations.

fit to a straight line up to ~30-40% conversion, after which point the polymerization becomes less controlled. The rate of monomer consumption for ε-caprolactone polymerization from the difunctional ligand is roughly two times that of the monofunctional ligand, as expected. Studies are currently underway to determine if shoulders in the GPC traces of samples collected at later time points are attributable to polymerization initiated from the dbm enol site or from an unknown impurity. If this side reaction arises from the enol functionality, it could be prevented with a protecting group.

Although tin-catalyzed lactide polymerizations work well, newer metal-free methods using 4,4'-dimethylaminopyridine (DMAP) (28) are also of interest. Trial runs with DMAP and dbmOH as the initiator produced PLA macroligands in very short reaction times (15-45 minutes). DbmPLA may be separated from DMAP by dissolving the product mixture in THF and running the solution through a short plug of neutral alumina.

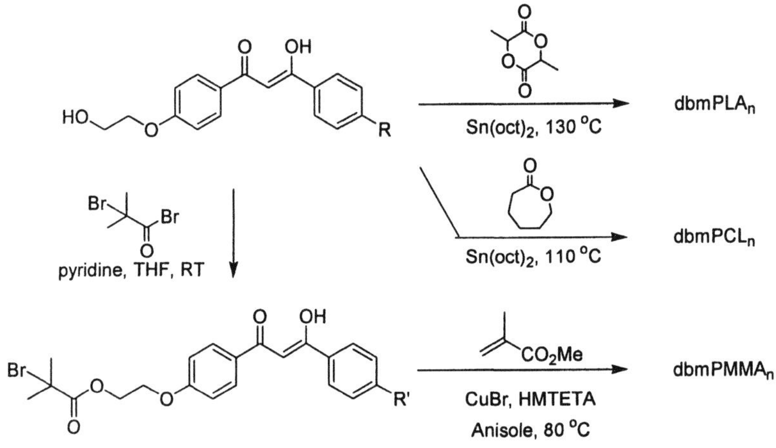

Figure 6. Routes to three different dbm macroligands (initiators: dbmOH,R = H, n=1; dbm(OH)$_2$, R=OCH$_2$CH$_2$OH, n=2; dbmBr, R'=H, n=1; dbm(Br)$_2$, R'=OCH$_2$CH$_2$OC(O)C(CH$_3$)$_2$Br, n=2).

PMMA macroligands are also being explored for chelation to lanthanide ions. In order for dbm to act as a suitable initiator for MMA, the alcohol site must first be functionalized with an α-bromo ester moiety as shown in Figure 6. DbmPMMA macroligands of various molecular weights have been obtained using the method shown in Figure 6, and future work will entail optimization of the synthesis by screening different ATRP catalysts and reaction conditions.

Polymeric Metal Complexes

Homopolymers

Eu complexes were chosen as a starting point for the exploration of lanthanide PMCs for a number of reasons. The local coordination environment of Eu can be easily probed by luminescence spectroscopy due to the sensitivity of Eu emission to the structural details of its surroundings (29). In addition, many related Eu complexes have already been studied in detail (30, 31), providing comparisons to the polymeric complexes.

Initially, Ln PMCs were synthesized in a mixed solvent system consisting of acetone, CH_2Cl_2, and MeOH to accommodate both the metal salt and macroligands. More recently, it has been found that THF can be used as a solvent for the chelation reaction, simplifying the synthesis. By varying the equivalents of macroligand in the reaction mixture, both tris or tetrakis Eu dbm complexes can be targeted.

The Eu and Tb complexes formed in this manner are studied by luminescence spectroscopy, generally as 1 mM solutions or as thin films. Although lanthanide complexes generally have absorption bands in the visible region of the spectrum, their extinction coefficients are very weak, and thus, they are best characterized by their emission properties. Metal centers may be excited directly, or sensitized by energy transfer from excited ligand states (i.e. the "antenna effect") (18). Luminescence lifetimes provide valuable information about sample homogeneity. If the luminescence decay curve can be fit to a single exponential equation, it is likely that one luminescent component is present, whereas a double exponential implies two or more species present in solution.

For tris Eu dbm complexes in CH_2Cl_2, the relative weighting of the longer lifetime component of the luminescence decay is significantly larger for the polymeric $Eu(dbmPLA)_3$ complex as opposed to its small molecule analogue, $Eu(dbm)_3$, as shown in Table I. This may be due to the shielding effect of the polymer, which prevents Ln-Ln self quenching, and helps to protect the Ln coordination sphere from water and solvent molecules which can dissipate the energy non-radiatively through stretching modes, diminishing the luminescence of the Eu species. A $Eu(dbmPLA)_3$ thin film formed by evaporation of 1 mM CH_2Cl_2 solution shows an even higher relative weighting of the long lifetime component, possibly due to the lower concentration of water and solvent molecules present to release the energy non-radiatively.

Table I. Luminescence Lifetimes for Tris Complexes
(Adapted from reference 27. Copyright 2002 American Chemical Society.)

Compound	τ_1 (ms)	Relative Weight of τ_1	τ_2 (ms)	Relative Weight of τ_2
Eu(dbm)$_3$	0.021	86%	0.302	14%
Eu(dbmPLA)$_3$	0.108	52%	0.349	48%
Eu(dbmPLA)$_3$ film	0.119	13%	0.371	87%

The solvent used for solution spectroscopy can have a large impact on the equilibrium of the system. Because Lns are labile metals, solvents may stabilize the complex, or promote ligand dissociation, shifting the equilibrium toward or away from the desired complex. Complexes prepared by identical synthetic methods can appear as multiple species in one solvent, and homogenous in another. Luminescence spectroscopy has been performed for these samples in both THF and CH_2Cl_2. When tested in CH_2Cl_2, tris Eu dbm complexes appear as multiple species according to luminescence decay, while tests in THF indicate a single component. The presence of differing amounts of bound solvent or trace H_2O in these solvents offers an alternate explanation for these observations.

Block Copolymers and Other Adducts

Due to the high coordination numbers (7-9) possible for lanthanide metals, the addition of a second ligand type to Ln tris bdks allows for the synthesis of adducts. In particular, bipyridine ligands are of interest to us since much is known about various bpy PMCs and a wide array of bpy macroligands have already been synthesized in our laboratory. Adducts, including those with polymeric ligands, may be formed in one of two ways. The first method is to form the tris Eu bdk complex, and mix this with the adduct-forming ligand. Alternatively, adducts may be prepared through a one-pot synthesis in which stoichiometric amounts of Ln salt, base, bdk macroligand/ligand, and bpy macroligand/ligand are stirred at room temperature. Even though lanthanides are labile, it is still possible to form discrete heteroleptic complexes; for electrostatic reasons, ligands do not exchange to form statistical mixtures of all possible ligand combinations. This serves as the basis for a modular metal template approach to polymer synthesis, as outlined below in Figure 7.

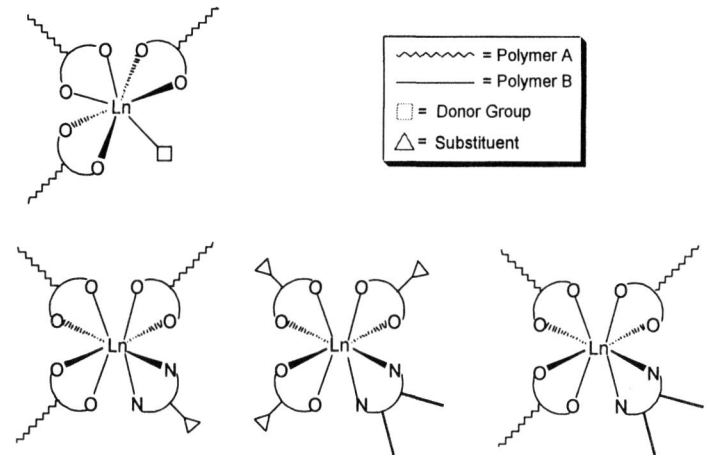

Figure 7. The modular approach to Ln tris and adduct complexes.

To date, bdk and bpy macroligands combined with Ln systems have been based mainly on PLA, PCL, and PMMA but it is apparent from the diversity of the bpy macroligand toolkit (Figure 8) that there are many possibilities available just by varying the bpy macroligand. Ultimately, a library such as this will be created for the bdk ligand system so that an abundance of materials can be formed by mixing and matching subunits from the two polymeric ligand sets.

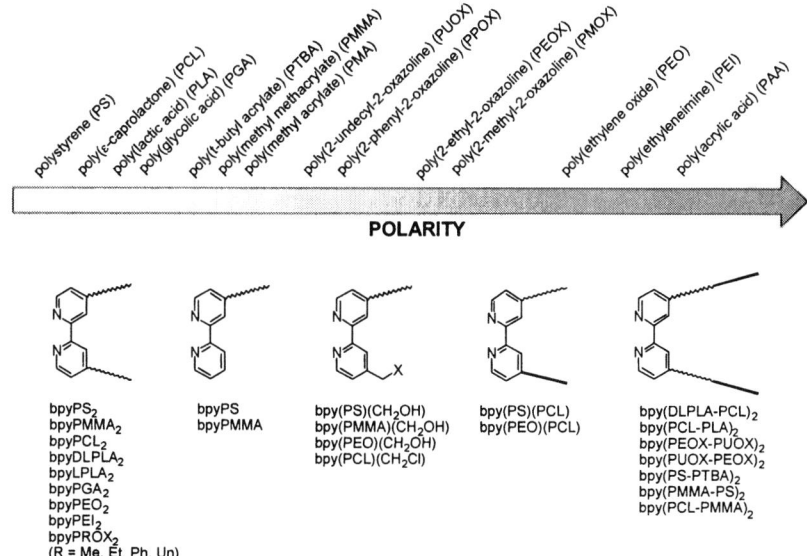

Figure 8. The bpy "macroligand toolkit" with estimated polarity scale.

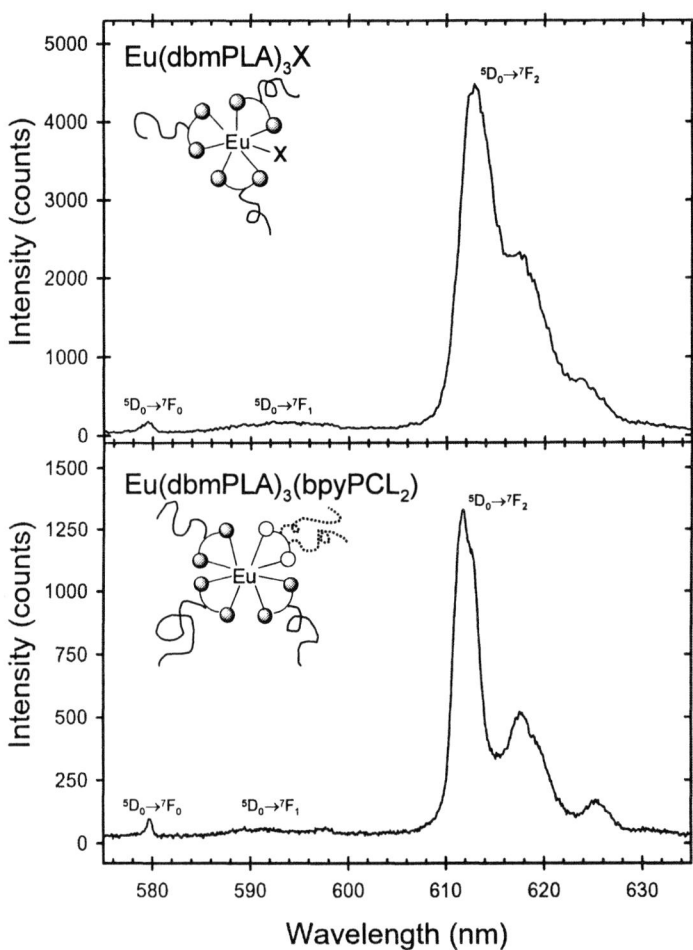

Figure 9. Emission spectra of europium tris dbmPLA and its bpyPCL$_2$ adduct (X = H$_2$O, solvent). (Adapted from reference 27. Copyright 2002 American Chemical Society).

Emission spectra are useful for verifying that adducts have formed. Comparison of the spectrum of a tris complex and that of its bpy adduct reveals a significant difference in the structural components of the $^5D_0 \rightarrow {}^7F_2$ feature, indicative of the change in geometry of the system upon binding to an additional ligand (Figure 9). Because Eu can function as a spectator and reporter of its chemical environment in this way, it is often exploited in sensors (*32*).

The synthesis of block copolymeric complexes however, is not always straightforward. In order for bdk and bpy macroligands of different compositions to form adduct PMCs, the benefits of chelation must outweigh the tendency of different polymers to phase separate. When dbmPLA and bpyPCL$_2$ were combined with EuCl$_3$, a one-pot synthesis in CH$_2$Cl$_2$, MeOH, and acetone yielded the desired block copolymer product, which appears homogeneous by luminescence lifetime measurement. But when dbmPLA was combined with bpyPEG$_2$ under the same conditions, more than one Eu species was observed by lifetime measurements. This observation could imply that unlike PLA and PCL polyesters, PLA and PEG have more contrasting properties and repel each other in the coordination sphere, resulting in the incomplete formation of the desired product. It is also possible that the second lifetime component of the mixture is a result of the polyether backbone or chain ends of the PEG acting as competitive donors to the metal center. Other polymer combinations too, may require careful screening of solvents and reaction conditions to determine whether reactions to form block copolymers can be driven to completion.

Block Copolymer Ordered Assemblies with Labile Metal Crosslinks: Morphological Changes upon Thermal Treatment

Block copolymers adopt diverse morphologies in concentrated solutions, thin films, and bulk materials (*33*). Metal-containing molecular structure translates into morphological positioning on the nanoscale. PMCs are ideally suited for selective positioning of chromophore placement in films. Depending on the type of star block copolymer architecture and the nature of the arms, metal cores can be localized to one domain (star blocks), or to the interface between domains (heteroarm stars).

Micrographs of Eu(dbmPLA)$_3$(bpyPCL$_2$) films prepared by simple casting from solution indicate a lamellar microstructure with a periodicity of 17.5 nm. The film shown in Figure 11 was swelled with water vapor prior to analysis to improve contrast. By virtue of its position at the block junction in the molecular architecture, the lanthanide resides at the intersection of microdomains in the ordered film.

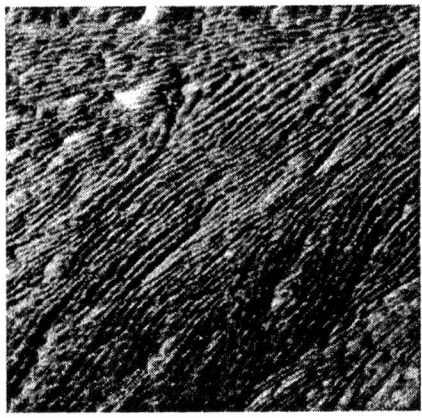

Figure 11. AFM image of Eu(dbmPLA)$_3$(bpyPCL$_2$). (Reproduced from reference 27. Copyright 2002 American Chemical Society).

For samples with labile crosslinks, including lanthanide complexes, ligand dissociation may be induced by heating. Evidence from variable temperature X-ray scattering indicates that elevated temperatures (~80 °C) effect a change in morphology, from an ordered microstructure to a largely macrophase separated material, consistent with Ln-bpy bond rupture (*34*). Although here, this thermal transformation appears to be irreversible, use of PMCs in other materials platforms may allow reversible ligand dissociation, leading to new kinds of switchable optical materials.

$$Eu(dbmA)_3(bpyB_2) \rightleftharpoons Eu(dbmA)_3X + bpyB_2$$

(X = solvent, H$_2$O)

The potential of template-assisted polymer synthesis to generate new chromogenic polymers for diverse applications is immense, and we and others (*35*) have only just begun to exploit the many possibilities.

References

(1) Special issue devoted to lanthanides: *Chem. Rev.* **2002**, *102*, 1271-1803.
(2) Caravan, P.; Ellison, J. J.; McMurray, T. J.; Lauffer, R. B. **1999**, *99*, 2293-2352.
(3) Wezel, T. J. *NMR Shift Reagents*; CRC Press: Boca Raton, FL, 1987.
(4) Selvin, P. R. *Annu. Rev. Biophys. Biomol. Struct.* **2002**, *31*, 275-302.
(5) Richardson, F. S. *Chem. Rev.* **1982**, *82*, 541-552.
(6) Blasse, G. *Luminescent Materials*; Springer-Verlag: New York, 1994.

(7) For a recent example, see: Male, N. A. H.; Salata, O. V.; Christou, V. *Synth. Met.* **2002**, *126*, 7-10.
(8) Graf, C.; von Blaadern, A. *Langmuir* **2002**, *18*, 524-534.
(9) Breen, M. L.; Dinsmore, A. D.; Pink, R. H.; Qadri, S. B.; Ratna, B. R. *Langmuir* **2001**, *17*, 903-907.
(10) Wang, W.; Asher, S. A. *J. Am. Chem. Soc.* **2001**, *123*, 12528-12535.
(11) Sabbatini, N.; Guardigli, M.; Lehn, J.-M. *Coord. Chem. Rev.* **1993**, *123*, 201-228.
(12) Becker, P. C.; Olsson, N. A.; Simpson, J. R. *Erbium-doped Fiber Amplifiers: Fundamentals and Technology*; Academic Press: San Diego, 1999.
(13) Moran, C. E.; Hale, G. D.; Halas, N. J. *Langmuir* **2001**, *17*, 8376-8379.
(14) Archer, R. D. *Inorganic and Organometallic Polymers*; New York: Wiley-VCH, 2001.
(15) Manners, I. *Science*, **2001**, *294*, 1664-1666.
(16) Nguyen, P.; Gomez-Elipe, P.; Manners, I. *Chem. Rev.* **1999**, *99*, 1515-1548 and references therein.
(17) Kido, J.; Okamoto, Y. *Chem. Rev.* **2002**, *102*, 2357-2368.
(18) Kawa, M.; Fréchet, J. M. J. *Chem. Mater.* **1998**, *10*, 286-296.
(19) Müller, M.; Zentel, R.; Maka, R.; Romanov, S. G.; Sotomayor Torres, C. M. *Chem. Mater.* **2000**, *12*, 2508-2522.
(20) Xu, L.; Zhou, W.; Kozlov, M. E.; Kyayrullin, I. I.; Udod, I.; Zakhidov, A. A.; Baughman, R. H.; Wiley, J. B. *J. Am. Chem. Soc.* **2001**, *123*, 763-764.
(21) Wijnhoven, J. E. G. J.; Bechger, L.; Vos, W. L. *Chem. Mater.* **2001**, *13*, 4486-4499.
(22) Chen, X. L.; Jenekhe, S. A. *Macromolecules* **2000**, *33*, 4610-4612.
(23) Edrington, A. C.; Urbas, A. M.; DeRege, P.; Chen, C.; Swager, T.; Hadjichristidis, M.; Xeridou, M.; Fetters, L. J.; Joannopoulos, J. D.; Fink, Y.; Thomas, E. L. *Adv. Mater.*, **2001**, *13*, 421-425.
(24) For e.g., see: Fraser, C. L.; Smith, A. P. *J. Polym. Sci, Part A: Polym. Chem.* **2000**, *38*, 4704-4716.
(25) Mecerreyes, D.; Jérôme, R.; Dubois, P. *Adv. Poly. Sci.* **1999**, *147*, 1-59.
(26) Matyjaszewski, K.; Xia, J. *Chem. Rev.* **2001**, *101*, 2921-2990.
(27) Bender, J. L.; Corbin, P. S.; Fraser, C. L.; Metcalf, D. H.; Richardson, F. S.; Thomas, E. L.; Urbas, A. M.; *J. Am. Chem. Soc.* **2002**, *124*, 8526-8527.
(28) Nederberg, F.; Connor, E. F.; Möller, M.; Glauser, T.; Hedrick, J. L. *Angew. Chem. Int. Ed.* **2001**, *40*, 2712-2715.
(29) Parker, D.; Dickins, R. S.; Puschmann, H.; Crossland, C.; Howard, J. A. K. *Chem. Rev.* **2002**, *102*, 1977-2010 and references therein.
(30) Batista, H. J.; de Andrade, A. V. M.; Longo, R. L.; Simas, A. M.; de Sa, G. F.; Ito, N. K.; Thompson, L. C. *Inorg. Chem.* **1998**, *37*, 3542-3547.

(31) Frey, S. T.; Gong, M. L.; DeW Horrocks, Jr. *Inorg. Chem.* **1994**, *33*, 3229-3234.
(32) Tsukube, H.; Shinoda, S. *Chem. Rev.* **2002**, *102*, 2389-2403.
(33) Fasolka, M. J.; Banerjee, P.; Mayes, A. M. *Macromolecules* **2000**, *33*, 5702-5712 and references therein.
(34) Urbas, A. M.; Thomas, E. L.; Bender, J. L.; Fraser, C. L. Unpublished results.
(35) For e.g., see: Schubert, U. S.; Heller, M. *Chem. Eur. J.* **2001**, *7*, 5252-5259.

Chapter 19

Synthesis, Photophysical Property, and Electroluminescent Applications of Silicon-Based Alternating Copolymers

H. K. Kim[1], N. S. Baek[1], K. L. Paik[1], Y. Lee[2], and J. H. Lee[3]

[1]Center for Smart Light-Harvesting Materials and Department of Polymer Science and Engineering, Hannam University, Daejeon 306–791, Korea
[2]Dongbu Research Council, Deajeon 103–2, Korea
[3]SAIT, Daejeon 103–12, Korea

Silicon-based alternating copolymers with a uniform π-conjugated segment regulated by alkyl/aryl-substituted distyrylsilanes units were synthesized by the Heck coupling reaction to use as full color electroluminescent (EL) materials. The EL color of the resulting copolymers was tuned by controlling π-conjugated length as well as by introducing various aromatic units into the polymer backbone. Both single and multilayered light-emitting diodes were fabricated by vacuum deposition of the Al or Ca onto a polymer film formed by spin-coating. Some of them exhibited a white EL color, due to the formation of a charge transfer complex. From photophysical and time-resolved transient decay studies, the formation of a charge separated complex was proposed. In this paper, the synthesis, characterization, photophysical properties and electroluminescence device (ELD) applications of silicon-based alternating copolymers are discussed.

Introduction

Electroluminescent (EL) devices based on polymeric thin layers have attracted much attention because of their academic interest and wide variety of applications such as flat-panel displays, light-emitting diodes, and lasers (*1-5*). EL polymeric materials offer a number of advantages, such as low operating voltages, easy accessibility of three primary R/G/B colors with the control of π-π^* energy gap through the manipulation of the molecular structure, fast response time, high display quality, and ease of device processability compared to inorganic EL materials and organic dye molecules (*6-9*).

Very recently, many efforts towards the main chain materials have been focused on developing blue and red light-emitting diodes capable of operating at ambient temperature, low voltages and easy processability. We also reported the development of a new type of processable silicon-based alternating copolymers having thiophene, carbazole, fluorene unit, etc in the polymer main chain by the well-known Pd-catalyzed Heck coupling reaction for blue and red light-emitting diodes (*10,11*). Instead of the Wittig method used in the early stage of this research (*12,13*), the Heck synthetic route to the preparation of silicon-based copolymers was used mainly for the following reasons. (1) The Heck route could overcome the problem of low quantum efficiency due to the formation of the triplet state arising from the unreacted or remained aldehyde functional groups in the copolymers obtained from the Wittig reaction. (2) To obtain quantitatively *trans*-double bond from *cis*-double bond, the Wittig reaction requires a further post-reaction of isomerization step, achieved by heating the crude polymers with a trace of iodine in toluene. However, the Heck reaction directly produces the desired polymers with trans configuration, which is important for the optimization of the luminescence efficiency and the emission wavelengths. Our results have shown that the introduction of organosilicon units with aromatic or flexible aliphatic group into π-conjugated systems could improve their processability and limit the π-conjugation length, resulting in blue light-emitting diodes (*10-18*). Surprisingly, the silicon-based copolymers with a relatively short π-conjugation length could be fabricated as blue light-emitting diodes operated at low voltages, due to the lowering of the LUMO level in luminescent polymers and the *d*-orbital participation of silicon atoms. In this chapter, we describe the direct synthesis, photophysical and EL properties of the silicon-based copolymers by the Pd-catalyzed Heck reaction of the distyrylsilane monomer with various aromatic or heteroaromatic dibromides.

Synthesis of Silicon-based Alternating Copolymers

The synthesis of monomers such as distyryl silane monomers, 3,6-dibromo-N-(2-ethylhexyl)carbazole, 2,5-bis-(4-bromophenyl)-[1,3,4]-oxadiazole and 3,8-

dibromo-1,10-phenanthroline is described elsewhere (11-18). Ru(II)-chelated complexes were prepared according to the method of Meyer et al. (19). The structural characteristics of the final monomers were provided by FT-IR, ^{1}H- & ^{13}C-NMR, elecmental analysis, UV-Vis absorption and emission spectroscopies. The synthesis of silicon-based copolymers was carried out according to scheme 1 and 2 using the well-known Pd-catalyzed Heck coupling reaction as described previously (10,11,16-18). The polymerization results, thermal and photophysical properties of silicon-based copolymers are summarized in Table 1. All the copolymers have the glass transition temperature (T_g) in the range of 94 to 127 °C. They did not show any definite melting points, implying that the silicon-based copolymers are likely amorphous. All of the polymers showed good thermal stability up to 300 ~ 315 °C in a nitrogen atmosphere.

Table 1. Summary of polymerization results, thermal and photophysical properties of silicon-based alternating copolymers.

Polymers	Yield (%)	$M_w \times 10^{-3}$	T_g (°C)	$UV(\lambda_{max})$ (nm)	$PL(\lambda_{max})$ (nm)	$EL(\lambda_{max})$ (nm)
SiHMPPV	52	8.5	94	356&365	470	450
SiPhPPV	48	6.3	127	355&365	470	450
SiHMPVK	51	2.3	106	322&356	440	460
SiPhPVK	43	3.7	102	325&356	440	467&620
SiHMFPV	70	9.8	96	378	470	475
SiPhFPV	77	12.7	109	380	476	485
SiHMThV	50	10.3	94	400	520	-
SiPhThV	55	13.3	112	407	526	-
SiPhThThV	50	18.6	116	416	512	-
SiHMOXD/Cz 10	52	5.8	105	355	435	460
SiHMOXD/Cz 91	48	4.3	115	354	428	-
SiHMOXD/Cz 55	57	8.1	116	356	430	-
SiHMOXD/Cz 19	65	5.1	108	345	422&442	464&644
Ru(II)-Chelated Polymer I	50	125	-	265, 392 & 460	495, 680 & 743	-
Ru(II)-Chelated Polymer II	52	116	-	289, 386 & 465	500, 678 & 740	-

SOURCE: Reproduced from Refercence 11a, 17 and 18.

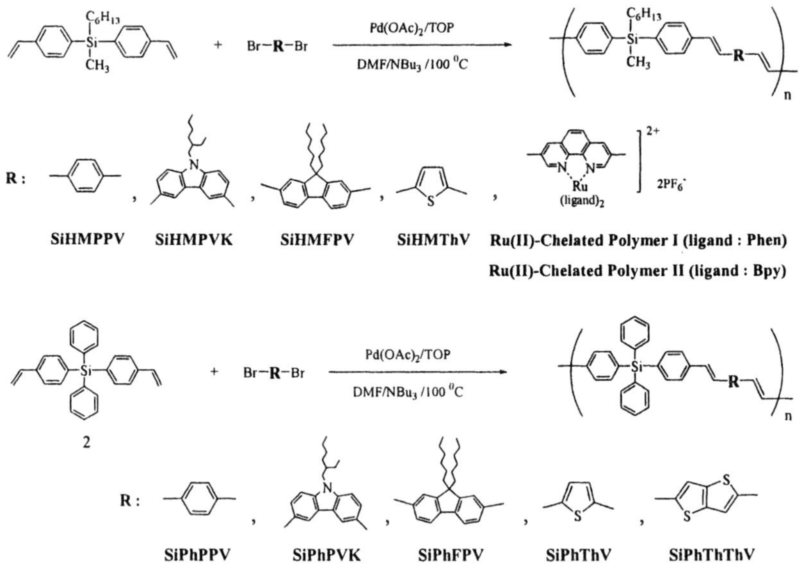

Scheme 1. Synthesis of silicon-based copolymers with flexible aliphatic and rigid aromatic group onto the organosilicon unit.

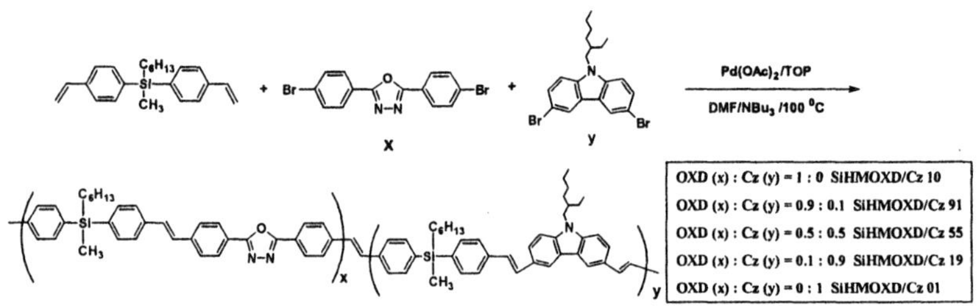

Scheme 2. Synthesis of silicon-based copolymers with various compositions.

Photophysical Properties

Their photophysical properties were obtained by means of UV-Vis absorption, steady-state, and time-resolved photoluminescence spectroscopies. Figure 1 shows the typical UV-Vis absorption and emission spectra of SiHM/SiPhPPV and SiHM/SiPhPVK in chloroform.

As shown in Figure 1, two maximum absorption wavelengths (λ_{max}) of SiHMPPV and SiPhPPV are at 355 and 365 nm, which is attributed to the π-π^* transition of the π-conjugated segment. Interestingly, SiPhPPV with high quantum efficiency, 26.4 % (photon/photon) could be candidates for luminescent materials in polymeric blue LEDs and blue lasers (14). When the excitation intensity of the nitrogen laser pulse at 337.1 nm was gradually increased, the broad photoluminescence (PL) spectrum of SiPhPPV changed at high excitation energy to a much narrower and stronger emission band peaked at 444 nm with the spectal width of 5 nm. This phenomenon observed at the input energy of about 10 µJ/pulse and emission intensity depends strongly on excitation intensities, accompanying by a nonlinear amplication.

Furthermore, we also introduced a carbazole unit into silicon-based copolymers to prepare copolymers containing both silyl groups and carbazole units, since carbazole-containing polymers have good electro- and photoactive properties due to their high hole-transporing mobility (12). The absorption spectra of SiHMPVK and SiPhPVK show two peaks of both a strong absorption band of the π-π^* transition of the carbazole segments around 322 ~ 325 nm and a strong absorption band of the π-π^* transition of the π-conjugated segment around 356 nm. Their PL spectra similarly shifts to those observed in the absorption spectra. With an excitation wavelength of 365 nm, the SiHMPPV and SiPhPPV spectra show a peak around 470 nm, indicating a blue emission. And, the PL spectra of both SiHMPVK and SiPhPVK give a peak in the blue emission region at 440 nm. On the other hand, the absorption spectra of SiHMFPV and SiPhFPV show a strong absorption band of the π-π^* transition of the π-conjugated segment around 380 nm. The maximum absorption wavelength (λ_{max}) of SiHMThV, SiPhThV and SiPhThThV appear at longer wavelengths of 400 nm, 407 nm and 416 nm, due to the strong delocalization of the π-conjugated thiophene units. With an excitation wavelength of 365 nm, the SiHMFPV and SiPhFPV spectra give a peak in the emission spectra at 470 nm for SiHMFPV and 476 nm for SiPhFPV. Also, with an excitation wavelength of 410 nm, the PL spectra of SiHMThV, SiPhThV and SiPhThThV give a peak in the green region at 520 nm, 526 nm, and 512 nm, respectively. Interestingly, the present silicon-based polymers have shown strong shifts relative to PPV. These results suggest that the regular π-conjugated system is effectively interrupted by the organosilicon units, reducing π-conjugated length.

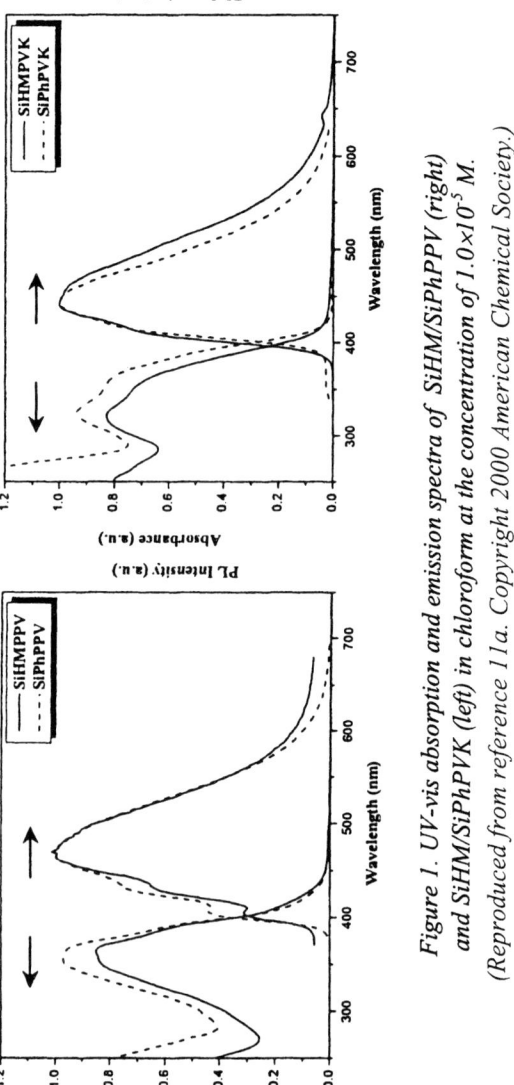

Figure 1. UV-vis absorption and emission spectra of SiHM/SiPhPPV (right) and SiHM/SiPhPVK (left) in chloroform at the concentration of 1.0×10^{-5} M. (Reproduced from reference 11a. Copyright 2000 American Chemical Society.)

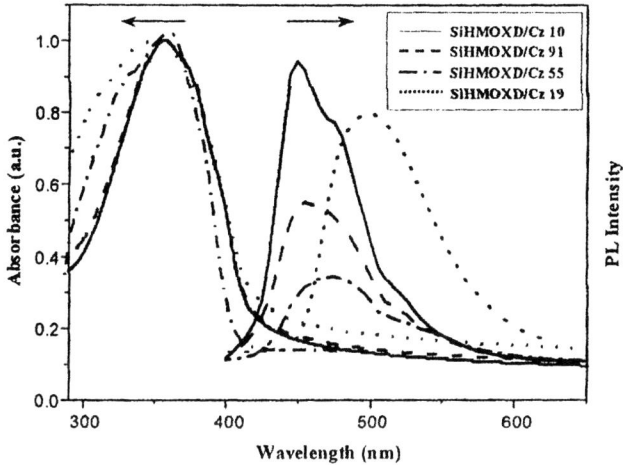

Figure 2. UV-visible absorption and emission spectra of SiHMOXD/Cz 10, SiHMOXD/Cz 91, SiHMOXD/Cz 55, and SiHMOXD/Cz 19 in thin films. (Reproduced from reference 17. Copyright 2002 American Chemical Society.)

Also, to balance the hole-electron charges injected and improve the quantum efficiency, we introduced the electron-transport oxadiazole units into the SiHMPVK (or SiHMOXD/Cz 01) with various molar ratios, yielding novel silicon-based copolymers containing both electron-transport oxadiazole and hole-transport carbazole moieties in the main chain (SiHMOXD/Cz xy). Figure 2 shows the UV-vis absorption and emission spectra of SiHMOXD/Cz xy. In the UV-visible spectra, the absorption maximum wavelength (λ_{max}) of SiHMOXD/Cz 1 0 appears a t 355 nm and 357 nm in chloroform solution and thin film, respectively, and its PL spectra show a strong band around 435 nm and 451 nm in chloroform and thin film, respectively. The maximum absorption wavelength (λ_{max}) of SiHMOXD/Cz 91, 55, and 19 appears at the wavelength of 357 nm, 360 nm, and 349 nm in film state, respectively. The PL spectra of these polymers exhibit a strong band around 455 nm, 483 nm, and 475 nm in the blue region, respectively.

To obtain red EL materials, also, we introduced the transition metal complexes into the polymer main chain. These chemical structures are shown in Scheme 1. In Figure 3, Ru(II)-chelated polymer I and II show one strong band around 265 ~ 289 nm for ligand (1,10-phenanthroline and 2,2'-bipyridine) units, one strong absorption band around 386 ~ 392 nm for π-conjugated backbones, and a broad shoulder MLCT band around 460 ~ 465 nm. Upon a photoexcitation with 325 nm, 394 nm and 423 nm, their PL spectra in a DMAc solution and thin

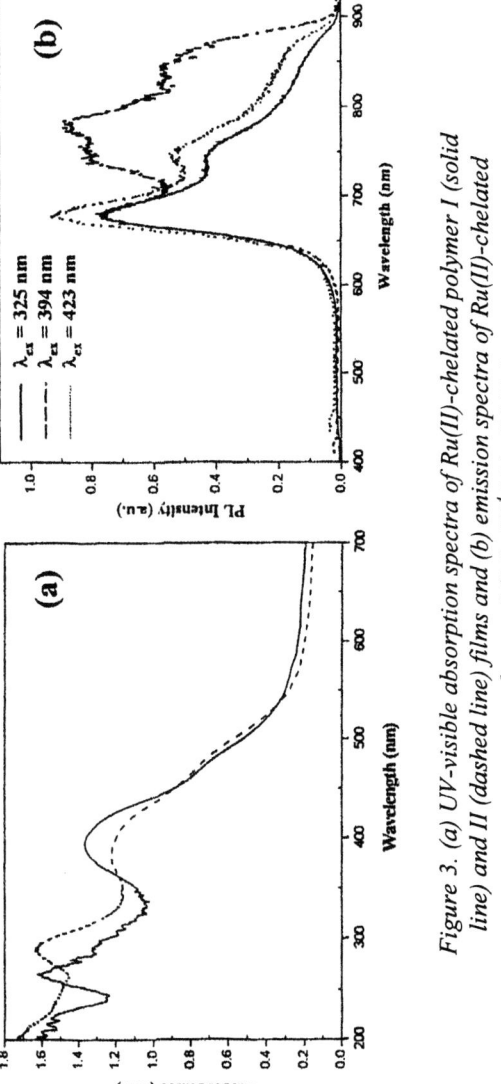

Figure 3. (a) UV-visible absorption spectra of Ru(II)-chelated polymer I (solid line) and II (dashed line) films and (b) emission spectra of Ru(II)-chelated copolymers I (2.5×10^{-4} M in DMAc).

films have similar features. They show two strong bands in the red region at room temperature. We couldn't obtain a strong greenish blue emission, corresponding to π-conjugated backbones, from the Ru(II)-chelated polymers at room temperature. It could be explained by the energy transfer of the excited state of π-conjugated backbones to metal-to-ligand charge transfer at room temperature in solution and film. As seen from Figure 3, the intensities of the red PL bands are moderate.

It indicates that its nonradiative process such as the thermal relaxation decay occurs at room temperature. Their PL intensity gradually increased with decreasing temperature, presumably due to the reduction of the thermal relaxation decay as nonradiative process (*18b*).

ELD Applications

The single and multilayered light-emitting diodes were fabricated by vacuum deposition of the Al or Ca onto a polymer film formed spin-coating technology. The single-layered LED of an Al/SiHMPPV or SiPhPPV/ITO showed a peak at 450 nm in the blue region when the operating voltage of 12 V was applied. Also, they show a typical rectifying characteristics. The threshold voltage is in the range 10 ~ 12 V from the J-V curve. Figure 4 shows the EL spectra of the multi-layered LEDs of a Al/LiAl/SiPhPVK/PANI/ITO glass as a function of applied voltages (*11a*). From the J-V curve, the threshold voltage of SiPhPVK is in the range of 6 ~ 12 V. The spectrum of the SiPhPVK gives a peak 467 nm when an operating voltage of 7 V was applied.

Unusually, when an operating voltage of higher than 11 V was applied, an additional strong EL emissive band showed the red emissive color from EL spectra. Therefore, the increase of an applied voltage was accompanied by a color tunning, and the two EL peaks may combined to produce a white EL color. From time-resolved PL measurements and photophysical studies, we proposed that the additional new PL band, unlikey SiHMPVK, might be attributed to the formation of a certain charge transfer complex, as shown in Figure 5. The stable resonance structure in the excited state was formed through stabilizing it with phenyl side groups in the SiPh unit of SiPhPVK. Phenyl side groups in the SiPh unit of SiPhPVK behave like electron-withdrawing group (*15,17*).

Also, to balance the hole-electron charges injected and improve the quantum efficiency, we have introduced the oxadiazole units into novel silicon-based EL copolymers (SiHMOXD/Cz xy) (*17*). The multi-layered light emitting diodes of Al (200nm)/Ca (50 nm)/ SiHMOXD/Cz 10 (80nm)/PEDOT (50nm)/ITO glass were fabricated. Their EL properties depend strongly on both the applied voltage and the loading amount of hole-transport carbazole moieties in the present copolymers. Its EL spectrum, as shown in Figure 6, exhibits a broad band around 460 nm in the blue region and a very broad, weak band in the red region at the

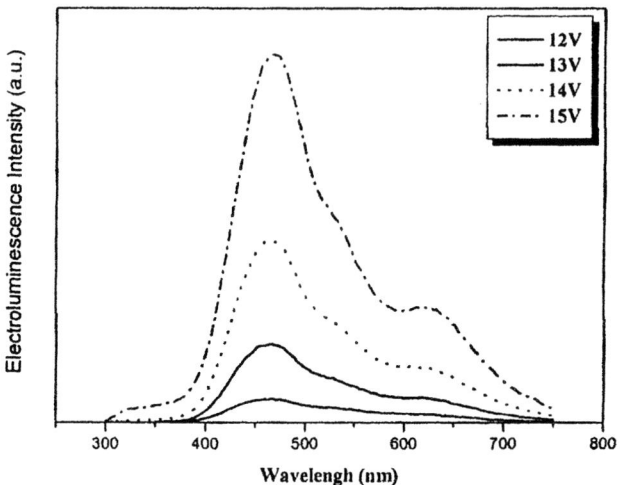

Figure 4. EL spectra of SiPhPVK as a function of applied voltages.
(Reproduced from reference 11a. Copyright 2000 American Chemical Society.)

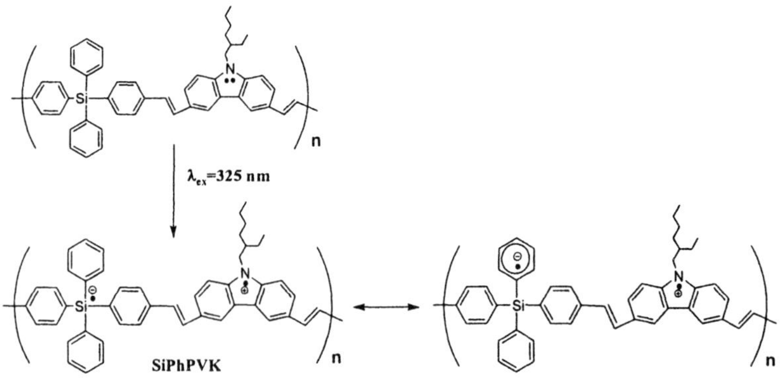

Figure 5. Proposed scheme for the formation of a charge transfer complex in SiPhPVK.
(Reproduced from reference 17. Copyright 2002 American Chemical Society.)

operating voltage of 12.5 V. The J-V curve shows the turn-on voltage of 6 - 7 V. This value is a little lower than any other silicon-based copolymers developed by our research group, due to the more electron affinity of the oxadiazole units.

With the applied voltage, the emissive EL bands were red-shifted from blue region to red region. When the higher voltage was applied, however, the broad band in the blue region decreases and a new broad band in red region increases. The new red EL peak is generated only by electric field, since this new red band is exhibited only in EL, but not in PL spectra. In other word, a certain charge complex, more like electroplex, is formed under a strong electric field inside the device a nd c annot b e p roduced b y p hotoexcitation (*17*). Therefore, t wo broad emissive bands combine to produce the white emissive color above 13 V. And, the maximum luminance of the white emissive color was 3.71 cd/m^2 at the applied voltage of 12 V. The EL spectra from the devices based on SiHMOXD/Cz 91 – 19 have a similar EL spectral feature with the applied voltage. The intensity of a blue EL band at the relatively high operating voltages increases with the loading amount of carbazole units.

Eventually, the LED device from the copolymer containing the mole ratio of electron-transport oxadiazole moiety to hole-transport carbazole moiety = 1/9 exhibits the almost same intensity of two bands. Surprisingly, the intensity of two emissive EL bands increases concomitantly with the applied voltage, showing two strong emissive bands, like two crests. They combine together and emit a strong white color. And, the maximum luminance of the white emissive color was 6.04 cd/m^2 at the applied voltage of 17 V.

Two peaks in the blue region and the new red peak at about 650 nm can be assigned to emissions from the individual lumophore and a certain charge complex like excimer, exciplex, or electroplex, respectively. Also, in order to gain further insight into the nature of the certain charge complex, we fabricated two LED devices from the blend systems of SiHMOXD/Cz 19 with polystyrene (PS). Even with the device from the blend system of SiHMOXD/Cz 19 with the large amount of PS (80%), in which intermolecular interaction between the polymer chains should be avoided, we still obtained two peaks similar to that of SiHMOXD/Cz 19.

This indicates that these unusual EL properties, originating from the formation o f a c ertain charge complex, do not exclude intra-chain interaction. Then, we measured the emission and the excitation spectroscopies for SiHMOXD/Cz 01, SiHMOXD/Cz 10, SiHMOXD/Cz 55, and SiHMOXD 19 at the various wavelengths. These two excitation wavelengths at 300 and 350 nm were chosen for the selective photoexcitation of the intrinsic carbazole unit (Cz units: D_{Cz}) or the intrinsic oxadiazole unit (OXD unit: A_{OXD}) and the π - conjugated segments with the carbazole unit (D_π) or the oxadiazole unit (A_π). The P L s pectra o f t he S iHMOXD/Cz 0 1 a nd t he S iHMOXD/Cz 1 0 s how the same emission maximum band around 430 nm and 455 nm, which is attributed to photoexcitation of the carbazole unit and the oxadiazole unit, respectively.

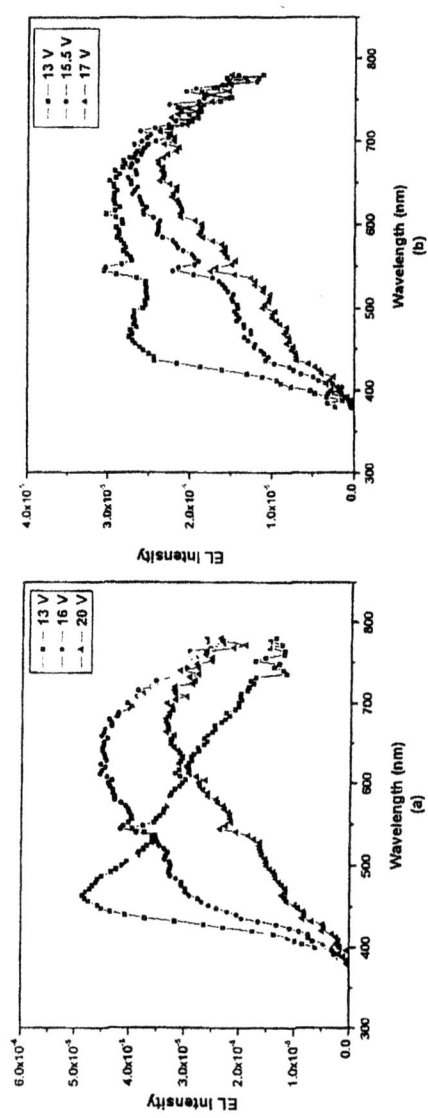

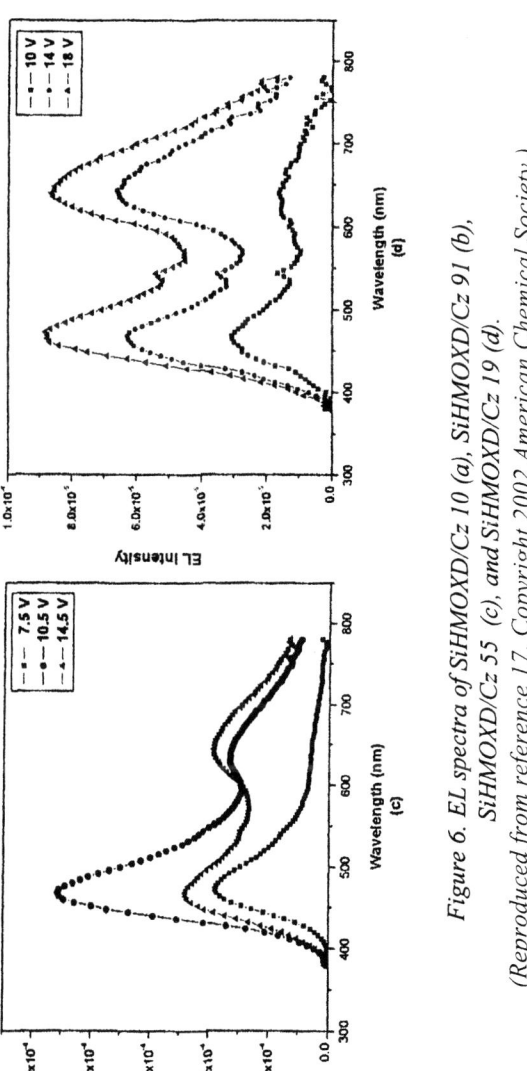

Figure 6. EL spectra of SiHMOXD/Cz 10 (a), SiHMOXD/Cz 91 (b), SiHMOXD/Cz 55 (c), and SiHMOXD/Cz 19 (d). (Reproduced from reference 17. Copyright 2002 American Chemical Society.)

However, these PL spectral features behave quite differently from SiPhPVK, while the stable resonance structure in the excited state was formed through stabilizing it with phenyl side groups in the SiPh unit of SiPhPVK (15,17). From these different results, we suggest the electron withdrawing power of oxadiazole units in the SiHMOXD is not enough to stabilize the excited state of SiHMOXD for the formation of the stable resonance structure in the excited state, since the oxadiazole units are connected indirectly to silicon atoms via styryl groups, unlike SiPhPVK (see Scheme 2). Their PL spectra of SiHMOXD/Cz 55 and SiHMOXD/Cz 19 show the same emission maximum band around 465 nm, which is attributed to photoexcitation of the intrinsic oxadiazole unit (OXD unit: A_{OXD}) as well as the π-conjugated segements with the oxadiazole unit (A_π). Even though the carbazole units (D_{Cz}) as well as the π-conjugated segments with the carbazole unit (D_π) were photoexcited at 300 nm, the emission maximum band around 430 nm, corresponding to carbazole moieties, was not observed. It might due to the energy transfer of the excited state of carbazole moieties to the ground state of the oxadiazole units. From these emission studies, we could not observe the red PL spectra, ascribed to the formation of a certain intramolecular charge complex. Therefore, we could propose the formation of a certain intramolecular charge complex for the silicon-based copolymers containing both electron-transport oxadiazole and hole-transport carbazole moieties, corresponding to a red color, as shown in Figure 7.

The proposed scheme can be deduced as follows: In the LED device of SiHMOXD/Cz 10, this new red band is exhibited only in EL, but not in PL spectra. And, in the LED device of SiHMOXD/Cz 01, the blue EL band is exhibited in both EL and PL spectra. So, the blue EL color and the new red EL color come from the silicon-based copolymer with only the carbazole moiety and the silicon-based copolymer with only the oxadiazole moiety, respectively. Thus, a certain charge complex, more like excimer, exciplex, or electroplex, is formed under a strong electric field inside the device and cannot be produced by photoexcitation. From the blending study, the charge complex can be formed intramolecularly between the oxadiazole units and the carbazole units.

When the voltage was applied, the electrons can inject into the oxadiazole units (A_{OXD}) and π-conjugated segments with the oxadiazole unit (A_π), yielding a certain charge complex of the negative polarons in the oxadiazole units (A_{OXD}) and π-conjugated segments with the oxadiazole unit (A_π). Alternatively, the holes can inject into the carbazole units (D_{Cz}) as well as the π-conjugated segments with the carbazole unit (D_π), forming a certain charge complex of the positive polaron in the the carbazole units (D_{Cz}) as well as the π-conjugated segments with the carbazole unit (D_π). As seen from Figure 7, a certain charge transfer complex emits the longer emissive bands, which

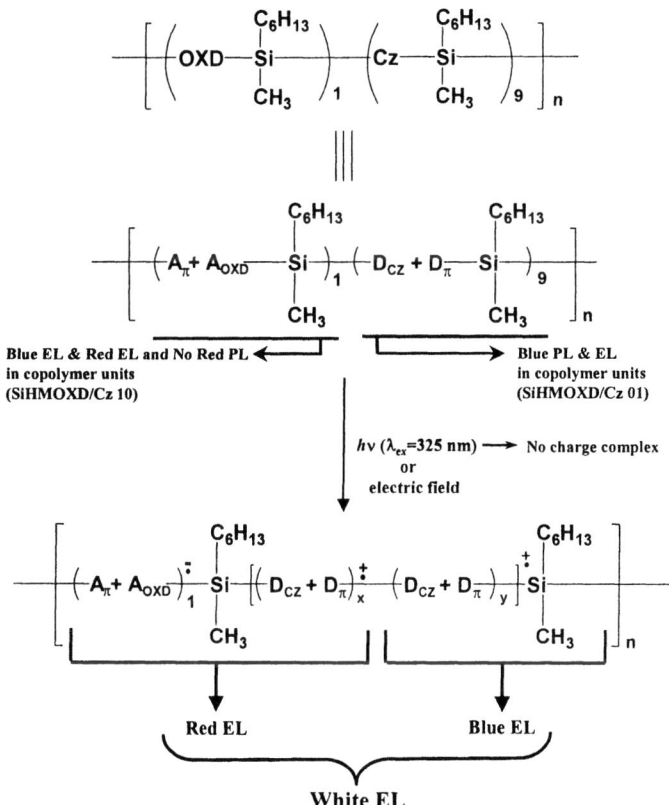

Figure 7. Proposed scheme for the formation of a charge transfer complex in SiHMOXD/Cz 19.
(Reproduced from reference 17. Copyright 2002 American Chemical Society.)

corresponds to the red EL color in EL spectra. This kind of charge transfer complexes from photoinduced charge separation occurring in donor-acceptor substituted silanes (A_π-SiMe$_2$-D_π) has been reported (20). Also, some of a certain charge complex of the positive polaron in the the carbazole units (D_{Cz}) as well as the π-conjugated segements with the carbazole unit (D_π) emits a blue EL color. Therefore, two broad EL bands combine to produce the white emissive color.

Summary

Silicon-based alternating copolymers with a uniform π-conjugated segment regulated by alkyl/aryl-substituted distyrylsilanes units were synthesized by means of Heck c oupling r eaction t o u se a s f ull c olor e lectroluminescent (EL) materials. The EL color of the resulting copolymers was tuned by controlling π-conjugated length as well as by introducing various aromatic units into the polymer backbone. The single and multilayered light-emitting diodes were fabricated by vacuum deposition of the Al or Ca onto a polymer film formed by spin-coating. Some of them exhibited a white EL color, possibly due to the formation of a certain charge complex. From photophysical and time-resolved transient decay studies, the formation of a certain charge complex was proposed. In this paper, the synthesis, characterization, photophysical properties and electroluminescence device (ELD) applications of silicon-based alternating copolymers were discussed.

Acknowledgement. The authors acknowledge that this research was financially supported by National Creative Research Initiative Program of the Ministry of Science and Technology of Korea at Hannam University.

References

(1) Tang, C. W.; VanSlyke, S. A. *Appl. Phys. Lett.* **1987**, *51*, 913.
(2) Burroughes, J. H.; Bradley, D. D. C.; Brown, A. R.; Marks, R. N.; Mackay, K.; Friend, R. H.; Burns, P. L.; Holmes, A. B. *Nature* **1990**, *347*, 539.
(3) Gustafasson, G.; Cao, Y.; Treacy, G. M.; Klavetter, F.; Colaneri, N.; Heeger, A. J. *Nature* **1992**, *357*, 477.
(4) Greenham, N. C.; Moratti, S. C.; Bradley, D. D. C.; Friend, R. H.; Holmes, A. B. *Nature* **1993**, *365*, 628.
(5) Tessler, N.; Denton, G. J.; Friend, R. H. *Nature* **1996**, *382*, 695.

(6) (a) Zhang, C.; Seggern, H. Von.; Parbaz, K.; Kraabel, B.; Schmidt, H. W.; Heeger, A. J. *Synth. Met.* **1994**, *62*, 35. (b) Fisher, T. A.; Lidzey, D. G.; Pate, M. A.; Weaver, M. S.; Whittaker, D. M.; Skolnick, M. S.; Bradley, D. D. C. *Appl. Phys. Lett.* **1995**, *67*, 1355. (c) Zhang, X.; Jenekhe, S. A. *Macromolecules* **2000**, *33*, 2069.
(7) (a) Chang, S. C.; Bharathan, J.; Yang, Y.; Helgeson, R.; Wudl, F.; Ramey, M. B.; Reynolds, J. R. *Appl. Phys. Lett.* **1998**, *73*, 2561. (b) Lupton, J. M.; Samuel, I. D.; Beavington, W. R.; Burn, P. L.; Bassler, H. *Adv. Mater.* **2001**, *13*, 258.
(8) Hesemann, P.; Vestweber, H.; Pommerehne, J.; Mahet, R. F.; Greiner, A. *Adv. Mater.* **1995**, *7*, 388.
(9) Caccialli, F.; Li, X.-C.; Friend, R. H.; Moratti, S. C.; Holmes, A. B. *Synth. Met.* **1995**, *75*, 161.
(10) Paik, K. L.; Baek, N. S.; Kim, H. K.; Lee, J. H.; Lee, Y. *Opt. Mater.* **2002**, *21*, 135.
(11) (a). Jung, S. H.; Kim, H. K.; Kim, S. H.; Jeoung, S. C.; Kim, Y. H.; Kim, D. *Macromolecules* **2000**, *33*, 9277. (b) Jung, S. H.; Kim, H. K. *J. Luminescence* **2000**, *87-89*, 51. (c) Baek, N. S.; Jung, S. H.; Oh, D. J.; Kim, H. K.; Hwang, G. T.; Kim, B. H. *Synth. Met.* **2001**, *121*, 1743.
(12) (a) Kim, H. K.; Ryu, M.-K.; Lee, S.-M. *Macromolecules* **1997**, *30*, 1236.; (b) Kim, H. K.; Ryu, M.-K.; Kim, K. D.; Lee, S. M.; Cho, S. W.; Park, J. W. *Macromolecules* **1998**, *31*, 1114. (c) Kim, K. D.; Park, J. S.; Kim, H. K.; Lee, T. B.; No, K. T. *Macromolecules* **1998**, *31*, 7267.
(13) (a) Ryu, M.-K.; Lee, J. H.; Lee, S. M.; Zyung, T.; Kim, H. K. *ACS Polym. Mater. Sci. Eng.* **1996**, *75(2)*, 408. (b) Kim, H. K.; Ryu, M.-K.; Kim, K. D.; Lee, J. H. *Synth. Met.* **1997**, *91(1-3)*, 297.
(14) Yoshida, Y.; Nichihara, Y.; Ootake, R.; Fujii, A.; Ozaki, M.; Yoshino, K.; Kim, H. K.; Baek, N. S.; Choi, S. K. *J. Appl. Phys.* **2001**, *90*, 6061.
(15) Lee, S. N.; Baek, N. S.; Kim, H. K.; Shim, S.D.; Joo, T. *Mol. Cryst. & Liq. Cryst.* **2002**, *377*, 121.
(16) Paik, K. L.; Baek, N. S.; Kim, H. K.; Lee, J.-H. *ACS Polym. Prepr.* **2002**, *43(1)*, 77.
(17) Paik, K. L.; Baek, N. S.; Kim, H. K.; Lee, J.-H.; Lee, Y. *Macromolecules* **2002**, *35*, 6782.
(18) (a) Baek, N. S.; Kim, H. K.; Hwang, G. T.; Kim, B. H. *Mol. Cryst. & Liq. Cryst.* **2001**, *370*, 387. (b) Baek, N. S.; Kim, H. K.; Lee, Y.; Hwang, G. T.; Kim, B. H. *Thin Solid Films* **2002**, *417*, 111.
(19) Sullivan, B. P.; Salmon, D. J.; Meyer, T. J. *Inorg. Chem.* **1978**, *17*, 3334.
(20) Walree, C. A. V.; Roest, M. R.; Schuddeboom, W.; Jenneskens, L. W.; Verhoeven, J. W.; Warman, J. M.; Kooijman, H.; Spek, A. L. *J. Am. Chem. Soc.* **1996**, *118*, 8395.

Chapter 20

Tunable Photoluminescence of Poly(quinoline)s in Polymer Blend Films and Silica

S. W. Ho, W. Y. Huang, T. K. Kwei, and Y. Okamoto*

Polymer Research Institute, Polytechnic University, 6 Metrotech Center, Brooklyn, NY 11201

Poly(2,6-[4-phenylquinoline]), PPQ, and poly(2,6-[p-phenylene]-4-phenylquinoline), PPPQ, were synthesized and their photoluminescence was investigated in solution and in different host materials: blends and composites. The polymers form aggregates/excimers in concentrated solutions and gradually dissociate upon dilution. When PPQ were blended with poly(acrylic acid), three different emission peaks at 460, 565, and 605 nm were obtained with increasing poly(acrylic acid) concentration, resulted from the protonation capability and steric hindrance effect of poly(acrylic acid). A sol-gel process was employed to incorporate different concentrations of quinoline polymers into silica. At high polymer concentrations, multiple aggregates were formed within the silica network, leading to the emission of red light. At low polymer concentrations, the polymer chains are isolated and trapped individually in the silica channels, resulting in the emission of blue light. For concentrations in between, moderate interchain interaction leads to the emission of green, yellow and orange colors. These results demonstrate that color tunability can be achieved by simply varying the concentration of quinoline polymers in host materials.

Introduction

Emissive conjugated polymers with their potential applications in light emitting devices, photodetectors, and lasers have drawn a lot of attentions over the past 10 years. Blue light emitting diodes using poly(pyridine) as the emitting layer have been reported (*1, 2*). The polymer is soluble in acidic solvents such as formic and dichloroacetic acids, thereby allowing solvent or spin casting. The excitation and emission wavelengths were varied by quaternization and methylation of the nitrogen atom (*3*). Other related basic aromatic π-conjugated polymeric systems are the polyquinolines, which have excellent thermal stability and high mechanical strength (*4,5*). Recently, the optoelectronic properties such as photoluminescence, photoconductivity, and nonlinearity of polyquinolines have been extensively investigated by Jenekhe and co-workers (*6-10*). We have synthesized two polyquinolines, poly(2,6-[4-phenylquinoline]), PPQ, and poly(2,6-[p-phenylene]-4-phenylquinoline), PPPQ, as shown in Scheme 1. These poly(quinoline)s were readily dissolved in acids such as formic and dichloroacetic acids. Thus the photoluminescence properties of these polymers in solution and solid state were investigated in detail.

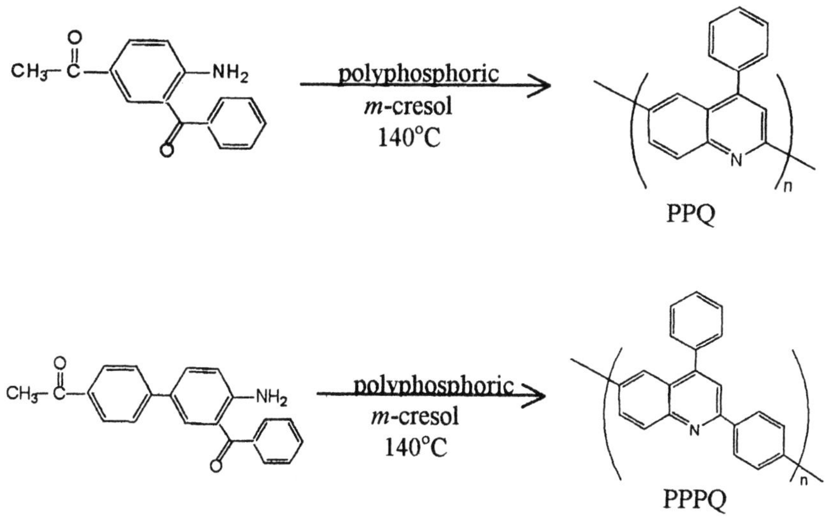

Scheme 1. Reactions and chemical structures of poly(2,6-[4-phenylquinoline]), PPQ, and poly(2,6-[p-phenylene]-4-phenylquinoline), PPPQ.

Experimental

Most of the chemicals used in this study were purchased from Aldrich Chemical Co. HCOOH (95-97%) and Cl$_2$CHCOOH (99%+) were used without further purification. Poly(acrylic acid) (M.W. 450,000) was purchased from Polysciences, Inc. Poly(2,6-[4-phenylquinoline]) and poly(2,6-[p-phenylene]-4-phenylquinoline) were synthesized by the self-condensation of 5-acetyl-2-aminobezophenone and 4-amino-4'-acetyl-3-benzoylbiphenyl by the method reported by Stille and co-workers (4,5). The self-condensation polymerization was carried out in a mixture of polyphosphoric acid and m-cresol, as shown in Scheme 1. The molecular weights of the polymers obtained were found to be dependent on the reaction times. Typical intrinsic viscosities, [η], of PPQ and PPPQ in concentrated H$_2$SO$_4$ were 3.6 dL/g (70 hrs reaction time) and 11.6 dL/g (24 hrs reaction time), respectively. From these results, the number polymerization, n, was estimated in the range of 8 to 10 (2,4).

The polymer solutions were characterized with a Shimadzu UV-Vis UV 2401 Spectrometer and a Perkin-Elmer LS-50B Luminescence Spectrophotometer with 1 cm and 0.1 cm quartz cells in the 30°/60° geometry. The quantum yields of the polymer solutions were determined by using quinine sulfate (~ 1.3 × 10^{-5} M in 0.1 N H$_2$SO$_4$, ϕ = 55%) (11,12) as a standard.

$$\phi_x = \phi_r \left(\frac{I_x \varepsilon_r c_r n_x^2}{I_r \varepsilon_x c_x n_r^2} \right) = \phi_r \left(\frac{I_x A_r n_x^2}{I_r A_x n_r^2} \right)$$

Eq. 1

where the x and r subscripts refer to the sample of unknown quantum yield and the standard, respectively. ϕ is the quantum yield, I is the fluorescence intensity, ε is the molar extinction coefficient, c is the molar concentration, A is the UV absorbance and n is the refractive index.

Results and Discussion

Fluorescent properties in solutions

Acidic solvents assist the dissolution of polyquinolines by protonating or hydrogen bonding with the nitrogen atoms. The UV absorption maximum of the PPQ/formic acid solution shifts to shorter wavelengths from 385nm to 375nm upon dilution, and the fluorescence emission peak becomes sharper at around 450nm with increasing intensity upon dilution, as shown in Figure 1. These results suggest that poly(quinoline)s form aggregates/excimers in concentrated solutions. It is well known that excimer formation reduces the quantum yield of conjugated polymers (13). Thus we have measured the quantum yield of PPQ in formic acid and the quantum yield is found to be increasing with decreasing PPQ concentration, reaching 15 % at 0.001 g/L.

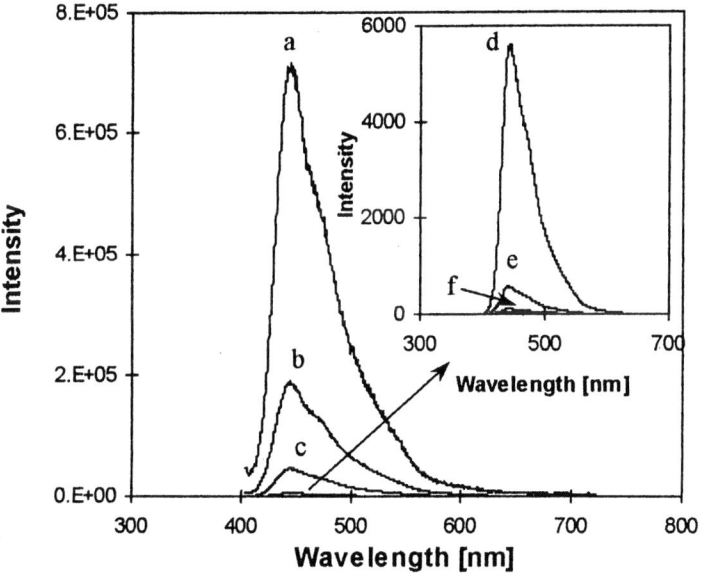

Figure 1. Fluoresence spectra of PPQ in formic acid:
(a) 1.25×10^{-4}, (b) 1.25×10^{-3}, (c) 1.25×10^{-2}, (d) 1.25×10^{-1}, (e) 1.25, and (f) 6.25 g/L. The measurements were taken in a 1 mm quartz cell at room temperature.

When poly(acrylic acid) is added into the high concentration PPQ or PPPQ acidic solution, the mixture becomes viscous and the fluorescence intensity is dramatically increased with the addition of increasing amount of poly(acrylic acid). Since poly(acrylic acid) has a limited solubility in formic acid, we used dichloroacetic acid as the solvent. The emission spectra of PPQ and poly(acrylic acid) in dichloroacetic acid solutions are shown in Figure 2. There are two emission peaks at 525 and 460 nm in all conditions, where the 525 nm peak is indicative of excimer formation and the 460 nm peak is considered the outcome of "isolated polymer chains". When the molar ratio of poly(acrylic acid) to PPQ equals to 385, the intensity of the emission peak at 525 nm becomes smaller while the intensity of the 460 nm peak greatly increases, reaching about 100 times larger than the original PPQ acidic solution without poly(acrylic acid). The poly(acrylic acid) is competing with dichloroacetic acid for the association with PPQ. As the ratio of poly(acrylic acid) to PPQ becomes large, the PPQ chains are more separated from each other through the intervention of poly(acrylic acid), resulting in decreased aggregation and increased emission intensity at 460 nm.

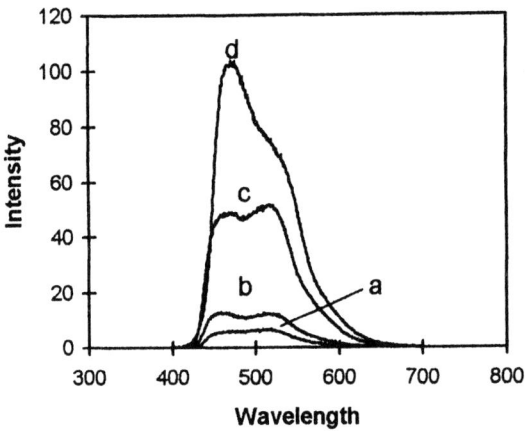

Figure 2. Fluorescence spectra of PPQ/poly(acrylic acid) mixtures in dichloroacetic acid. The measurements were taken in 1 cm quartz cell. (a) 1 g/L PPQ/dichloroacetic acid solution. Solution (a) is used as the base PPQ solution for the addition of poly(acrylic acid). The molar ratio of poly(acrylic acid) to PPQ in the mixtures are: (b) 32, (c) 82, and (d) 384, respectively.

Fluorescent properties in blended polymer films

A dilution effect on the fluorescence intensity was observed when poly(quinoline)s were blended with nonfluorescent poly(vinyl alcohol) (PVA). Although these two polymers were not completely miscible over a wide range of the concentration ratios, visually clear thin films were obtained for many compositions by spin coating. Typical fluorescence spectra of PPQ/poly(vinyl alcohol) blend systems are shown in Figure 3. At high concentrations (> 50 wt% PPQ/PVA), broad emission peaks around 550-600 nm are seen whereas at concentrations of 5 wt% or less, the emission peaks shift to a shorter wavelength (~ 450 nm). However, no intermediate peak is found at any concentration level.

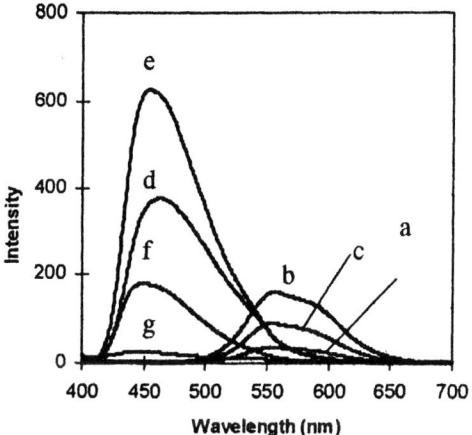

Figure 3. Fluorescence emission spectra of PPPQ/PVA blend films (excited at 380 nm): (a) pristine PPPQ, (b) 80 wt% PPPQ, (c) 50 wt% PPPQ, (d) 20 wt% PPPQ, (e) 5 wt% PPPQ, (f) 1 wt% PPPQ, and (g) 0.1 wt% PPPQ. (Reproduced from *Macromolecules*, **1999**, *32(24)*, 8091. Copyright 1999 ACS.)

When poly(acrylic acid) is added instead of poly(vinyl alcohol), it shows a different phenomenon. The poly(acrylic acid) films were prepared by adding a very small amount of concentrated PPQ/dichloroacetic acid solutions at different concentrations into a stock of 300 g/L poly(acrylic acid) solution in ethanol. The volume ratio of the PPQ acidic solution to the poly(acrylic acid)/ethanol stock solution was estimated to be 0.2. The mixtures were stirred vigorously and then dried in ambient overnight to remove the solvents. Most of the resulting films are clear but brittle. Figure 4 is the fluorescence spectra of PPQ/poly(acrylic acid) blended films at different molar ratios. As the molar ratio of poly(acrylic acid) to PPQ decreases from 9×10^6 to 9×10^2, the films show the emission color of blue/green, orange and red, respectively, under UV radiation at 365 nm, corresponding to the emission peaks at 460, 565 and 605 nm. The difference between the additions of poly(acrylic acid) and poly(vinyl alcohol) is that poly(acrylic acid) is capable of competing with the acidic solvent to protonate the nitrogen atoms on PPQ. By association with PPQ, the large excess of poly(acrylic acid) may wrap around the PPQ chains and prevent the interchain interaction, resulting in a better molecular level separation and three distinguished emission peaks.

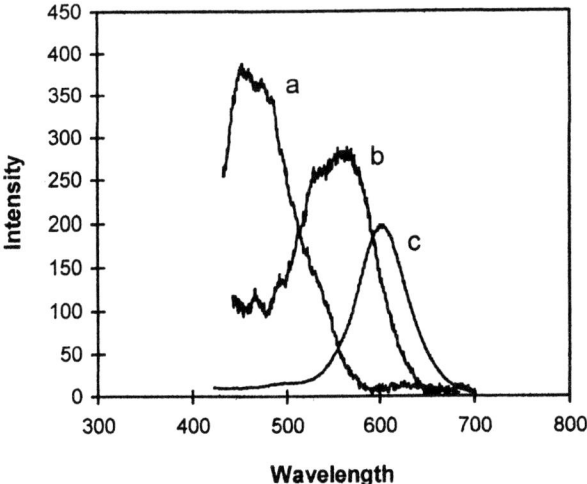

Figure 4. Fluorescence spectra of PPQ/ poly(acrylic acid) blend films. The molar ratio of poly(acrylic acid) to PPQ in the blends are: (a) 9×10^6, (b) 5×10^4, and (c) 9×10^2, respectively.

Fluorescent properties in silica

The sol-gel technique was employed to prepare a composite of poly(quinoline)s and silica. A large number of organic-inorganic nanocomposites have been successfully synthesized by performing sol-gel condensation in organic polymers that are inert or reactive with respect to the sol-gel chemistry (14,15). Prasad and his co-workers reported a π-conjugated optically active polymer, poly(p-phenylene vinylene), incorporated in silica glass via the sol-gel process. The resulting glass was porous though the average pore diameter was typically much smaller (1.5-10 nm) than the wavelength of near UV or visible radiation. Thus the composite exhibited low scattering, good optical quality, chemical and thermal stability (16).

In the sol-gel process, various concentrations of quinoline polymers in formic acid solutions were mixed with tetraethyl orthosilicate (TEOS) at room temperature. The gelation of the solutions occurred rapidly (< 15 minutes in most cases). The gels were then heated at 40-50°C to complete the TEOS polymerization. When the resulting composite materials were excited by UV, full-colored emissions spanning the entire visible range with high photoluminescence intensities were observed. The typical emission colors under UV of PPQ/silica composites are shown in Plate 1 and the fluorescence spectra are shown in Figure 5.

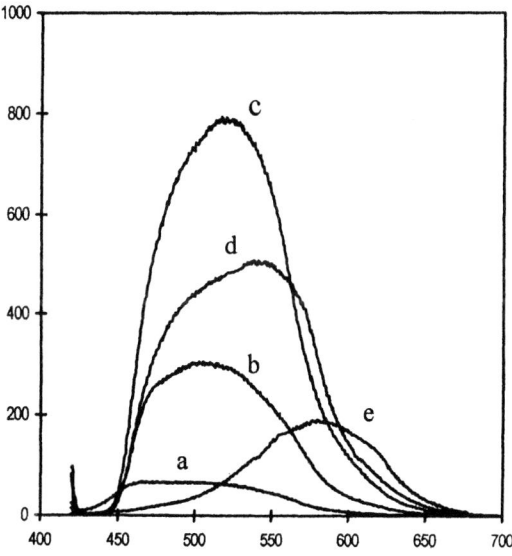

Figure 5. Fluorescence spectra of PPQ-silica gels (excited at 380nm): (a) 0.0001 wt% PPQ/TEOS; (b) 0.001 wt%; (c) 0.1 wt%; (d) 1 wt%; (e) 5 wt%. (Reproduced with permission from *Appl. Phys. Lett.* **2000**, *80(7)*, 163. Copyright 2000.)

This result is caused by the combination effect of excimer formation and the gelation of inorganic silica. Excimer formation among small molecules are often reported as the interaction of an excited chromophore A^* with an unexcited chromophore A, $A^* + A \longrightarrow (AA^*)$. Such an excited couple is stable as a resonance contribution, resulting in the shifting of the emission peak to the red region. π-Conjugated polymers such as poly(quinoline)s are generally stiff chain molecules with relatively planar geometries and very strong intermolecular interactions. It is therefore reasonable to expect that excimers efficiently form among quinoline polymer chains. The possibility may also exist that multiple excimers, $(RRRn)^*$, form among the trapped polymers in silica channels.

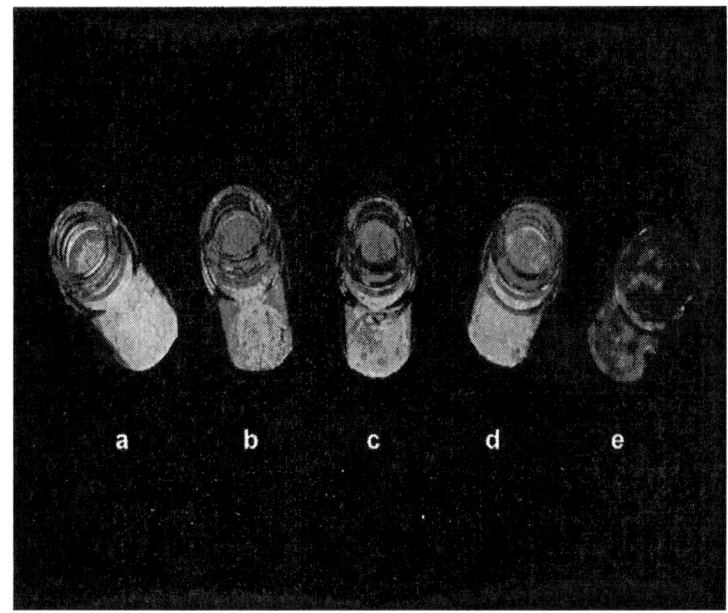

Plate 1. Emitting colors under UV (365 nm) radiation of PPQ-Silica gels.
(a) 0.0001 wt% PPQ/TEOS; (b) 0.001 wt%; (c) 0.1 wt%; (d) 1 wt%;
(e) 5 wt%. A small amount of solvent was still remained in the gel.
(See Page 6 of color insert.)

$$R \xrightarrow{h\nu} R^* \quad (1)$$

$$R^* + R \longrightarrow (RR)^* \quad (2)$$

$$(RR)^* + nR \longrightarrow R_{n+2}^* \quad (n=1, 2, 3, \ldots) \quad (3)$$

$$R_{n+2}^* \xrightarrow{Decay} R_{n+2} \quad (4)$$

where R is a polymer chain, R* an excited polymer chain, (RR)* an excimer and R_{n+2}^* a multiple excimer.

The interchain interaction among these polymer chains results in the shift of emission peaks progressively to longer wavelengths with increasing polymer concentration. When the amount of the polymer in the sol-gel solutions is small, $< 10^{-3}$wt%, the majority of the chains in solution are in the "isolated" state. The isolated polymer chains may be trapped individually in a channel of the silica domains upon gel formation, resulting in the emission of a blue color (~ 440 nm) upon UV radiation. At a higher concentration (~5 wt%), the emission peak is around 600 nm and the red emitting color is indicative of possibly high degree of cofacial chain interaction and stacking of the chromophores. For concentrations in between, less extensive cofacial chain interaction and fewer multiple excimers trapped inside the silica domains than those at 5 wt% lead to the emission of green, yellow and orange colors. A sketch illustrating the interaction between PPQ chains and silica matrix is shown in Scheme 2. Thus the emission color can be tailored simply by varying the concentration of quinoline polymers in the sol-gel composites.

Under certain conditions, cf., fast mixing, white luminescent silica upon UV radiation was obtained as its emission spectrum shown in Figure 6. The white emission may be due to the trapping of different amounts of polymer chains in silica domains, causing different degrees of chain interactions.

Scheme 2. A sketch of the interaction between PPQ and silica. The black thick lines symbolize the silica matrix and the gray short lines are PPQ. (a) at very low PPQ concentration, and (b) at high PPQ concentration.

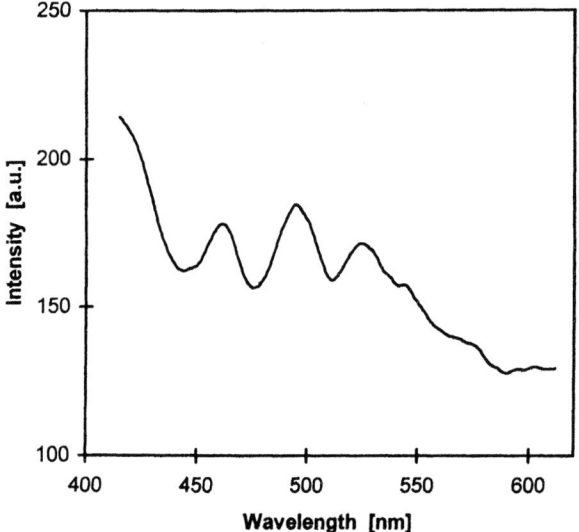

Figure 6. Fluorescence spectra of PPPQ/TEOS glass under fast mixing. The weight ratio of PPPQ to TEOS is 0.00001%. (Reproduced with permission from *Appl. Phys. Lett.* **2000**, *80(7)*, 163. Copyright 2000.)

Ongoing Work and Conclusions

Recently we have studied the kinetics of aggregation and the conformation of the PPQ aggregates in solution and solid state using dynamic light scattering and atomic force microscopy. We obtained results showing an equilibrium between aggregate association and dissociation in solution and a cylindrical conformation of aggregates. These phenomena are being investigated in detail (17).

Photoluminescence properties of PPQ in solution and solid states were investigated in this study. The emission spectra of the PPQ/formic acid solutions show an increasing intensity of the emission peak at 450 nm, as the PPQ concentration decreases and less excimer formation is formed in the system. The quantum yields of solutions at different concentrations are reported. When poly(acrylic acid) is added to the solutions, the fluorescence intensities increase dramatically due to reduction of the interchain interactions by having poly(acrylic acid) wrap around quinoline polymer chains.

When PPQ is blended with nonfluorescent poly(vinyl alcohol) or poly(acrylic acid), visually clear films were obtained. The emitting colors of PPQ/ poly(acrylic acid) blends upon UV radiation show three colors of blue/green (460 nm), orange (565 nm) and red (605 nm), while the PPQ/poly(vinyl alcohol) blends have only two peaks at 450 and 550 nm. By employing the silica sol-gel technique, the emitting colors of the composites can be further tailored by varying the PPQ concentration in the silica matrix. A full color display with its spectrum covering the entire visible color range is obtained.

Acknowledgement

We wish to thank the National Science Foundation, Division of Materials Research under Grant No. DMR 9802108 for financially supporting this research.

References

(1) Gebler, D. D.; Wang, Y. Z.; Blatchford, J. W.; Jessen, S. W.; Lin, L. B.; Gustafson, T. L.; Swager, T. M.; MacDiarmid, A. G.; Epstein, A. J. *J. Appl. Phys.* **1995**, *78*, 4264.
(2) Yamamoto, T.; Lee, B.; Saitoh, Y.; Inoue, T. *Chem. Lett.* **1996**, 679.
(3) Yun, H.; Kwei, T. K.; Okamoto, Y. *Macromolecules* **1997**, *30*, 4633.

(4) Sybert, P. D.; Beever, W. H.; Stille, J. K. *Macromolecules* **1981**, *14*, 493.
(5) Pelter, M. W.; Stille, J. K. *Macromolecules*, **1990**, *23*, 2418.
(6) Jenekhe, S. A.; Zhang, X.; Chen, X. L.; *Chem. Mater.* **1997**, *9*, 409.
(7) Zhang, X.; Shetty, A. S.; Jenekhe, S. A. *Pro. SPIE* **1997**, 3148, 89
(8) Agrawal, A. K.; Jenekhe, S. A.; Vanherzeele, H. V.; Meth, J. S. *J. Phys. Chem.* **1992**, *96*, 2837.
(9) Agrawal, A. K.; Jenekhe, S. A. *Macromolecules* **1993**, *26*, 895.
(10) Zhang, X.; Kale, D. M.; Jenekhe, S. A. *Macromolecules* **2002**, *35*, 382.
(11) Calvert, J. G.; Pitts, J. N. *Photochemistry*; John Wiley & Sons, Inc.; New York, 1966.
(12) Lakowicz, J. R.; Lakowicz, J. *Principles of Fluorescence Spectroscopy*, 2nd Ed.; Kluwer Academic Publishers, 1999.
(13) Jenekhe, S. A.; Osaheni, J. A. *Science*, **1994**, *265*, 765.
(14) Wen, J.; Wilkos, G. L.; *Chem. Mater.*, **1996**, *8,* 1667.
(15) Ellsworth, M. W.; Gin, D. L. *Polymer News*, **1999**, *24*, 331.
(16) Wung, C. J.; Pang, Y.; Prasad, P. N.; Karasz, F. E. *Polymer*, **1991**, *32*, 605.
(17) Ho, S. W.; Spagnoli, C.; Kwei, T. K.; Cowman, M.K.; Teraoka, I.; Okamoto, Y.; *unpublished results*.

Tunable Reflection and Photonic Band-Gap Structures

Chapter 21

Electrically Switchable Reflectors of Chiral Gels

Rifat A. M. Hikmet

Philips Research, Polymers and Organic Chemistry, Prof. Holstlaan 4, 5656AA Eindhoven, The Netherlands (email: rifat.hikmet@philips.com)

In-situ polymerisation of LC reactive molecules in the presence of non-reactive LC molecules leads to the formation of anisotropic networks containing free molecules (anisotropic gels). In this chapter gels obtained by polymerisation in the cholesteric phase will be described. In the cholesteric phase the optical rotary dispersion shows extremely high values and a band of light is split into two opposite circularly polarised components. These properties of cholesterics make them suitable to be used in passive optical components such as reflectors, polarizers, bandpass and notch filters. Cholesteric gels can be switched fast and in a controllable way. Using combination of heat and ultraviolet light in a pattern-wise manner, a single switchable gel layer reflecting various colours are produced. Homogeneity of the gels can be also be manipulated to increase the width reflection band to obtain electrically switchable silver coloured mirrors.

Introduction

Cholesteric liquid crystal phase is obtained when a nematic phase is doped with chiral molecules. Chiral molecules are optically active and are known to show optical rotary dispersion in the order of 1°/cm. However in the cholesteric phase they induce rotation of the long axes of the liquid crystal molecules (the director **n**) about a helix as shown in Figure 1.

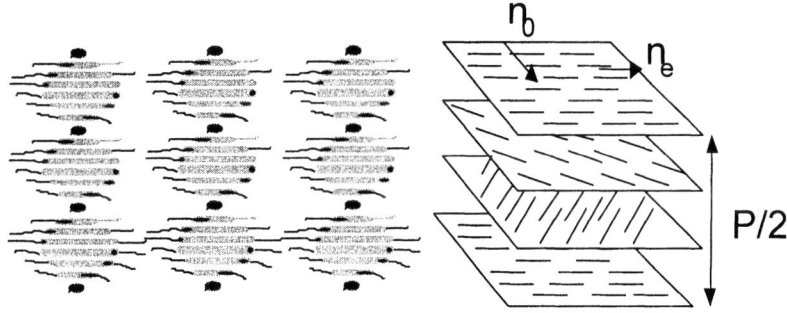

Figure 1 Schematic representation of a cholesteric Phase.

Such a macromolecular arrangement leads to optical effects unique to this phase. For example, the optical rotary dispersion shows a dramatic increase and reaches values in the order of 100°/cm. Furthermore, a band of circularly polarized light having the same sense as the cholesteric helix is reflected while the band with the opposite sense is transmitted. The upper (λ_{max}) and lower (λ_{min}) boundaries of the reflected band are $\lambda_{min}=p*n_o$ and $\lambda_{max}=p*n_e$ respectively where p is the cholesteric pitch corresponding to the length over which the director rotates 360°, n_e and n_o are the extraordinary and the ordinary refractive indices of a uniaxially oriented phase respectively. The reflected bandwidth ($\Delta\lambda$) is given by $\Delta\lambda = \lambda_{min}-\lambda_{max} = p*(n_e-n_o)$.

It has already been shown that such a cholesteic structure can be frozen-in by cooling a liquid crystal polymer below its glass transition temperature (*1*) or polymerisation of liquid crystal molecules with reactive end groups (*2*) to be used in passive optical applications. There has also been much interest in the use of cholesteric materials in display applications where switching can be realized between various misaligned states. Switching characteristics of cholesterics, for example the presence of hysteresis has been used in addressing schemes to produce passive matrix displays (*3*). The scanning speed of the voltage has also shown to influence the switching characteristic of cholesterics. This effect has been used in gels where bistable switching took place between weakly scattering focal-conic texture and planar texture with misaligned helix. In our earlier attempts to produce cholesteric gels, which could be switched between defect free planar Grandjean texture and homeotropic texture we used liquid crystal diacrylates which were polymerized in the presence of non-reactive cholesteric liquid crystal (*4*). However fast switching to the defect free state could not be obtained. For this purpose new types of gels, which can be used in defect free switching of cholesteric liquid crystals were developed (*5*). In this review production, various properties, patterning of such gels in lateral directions (*6*) in order to obtain multi colour displays, and creating a pitch gradient along the cell thickness to broaden the reflection band to produce broad band switchable reflector will be described.

Materials

The reactive liquid crystals were synthesized at Philips research. Conventional liquid crystals were purchased from Merck. The polymerisation of the mixtures was initiated by means of UV radiation using photoinitiator Irgacure 651 purchased from Ciba Speciality Chemicals. The structures of the acrylates are shown in Figure 2.

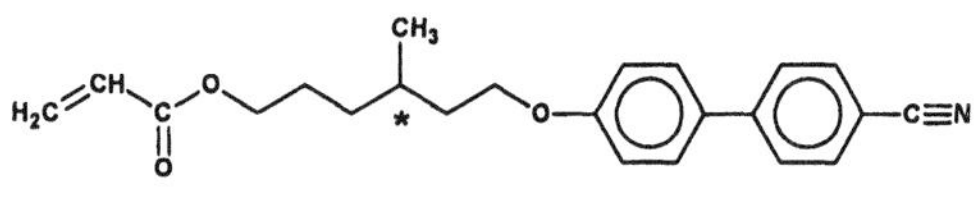

CCB6

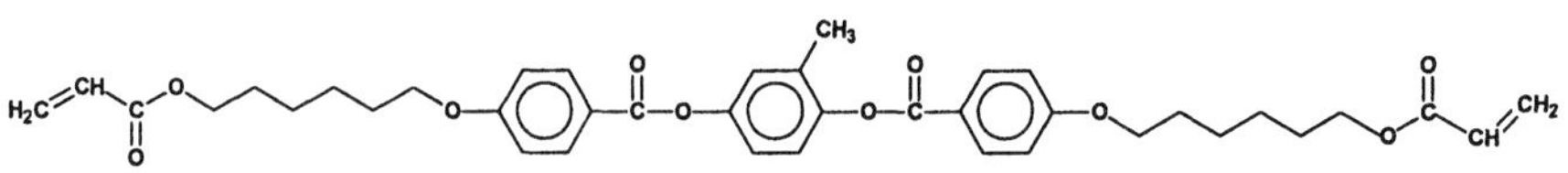

C6M

Figure 2. Structures of the acrylates.

Results and Discussion

In conventional LC cells, long range orientation of LC molecules is induced by orientation layers at the cell. Switching is induced by applying an electric field across transparent electrodes present on the cell surfaces underneath the orientation layers. Upon removal of the field, the LC molecules revert to the initial orientation state under the influence of these orientation layers. As shown in Figure 1 cholecterics have a complicated helical structure and in order to obtain sufficient reflection the cell gap needs to be at least ten pitches thick. When such a structure is switched from a defect free planar orientation it is almost impossible for the system to reorient itself back to the initial state under the influence of the surface orientation layers. Instead, the cholesteric helix becomes oriented in various directions and the cell shows a scattering texture. In order to obtain fast switching a memory state needs to be built into such a cholesteric system. We tried to do this by creation of a lightly cross-linked network dispersed within the non-reactive LC molecules. This is done by in-situ polymerisation of a liquid crystal monoacrylate and diacrylate mixture in the presence of non-reactive LC molecules. The planar orientation of the cholesteric mixtures containing monomers with reactive groups is obtained in cells containing uniaxially rubbed polymer layers. The polymerisation of the reactive molecules is induced by UV radiation freezing-in the cholesteric configuration and orientation by creating a network containing non-reactive LC molecules (anisotropic gel). The gel structure is schematically represented in Figure 3.

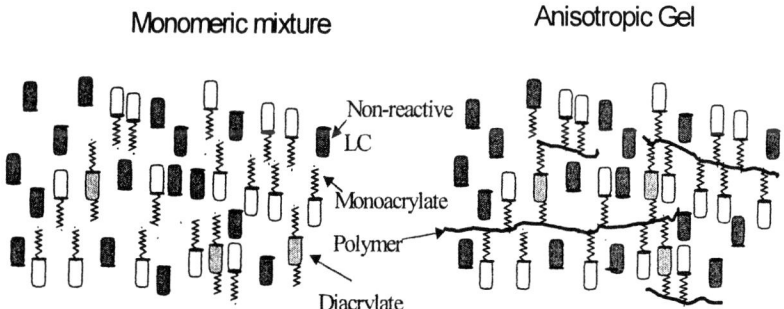

Figure 3. Schematic representation of gel formation.

The network in these gels consists of monoacrylate molecules forming the side-chain polymers which are cross-linked by the diacrylates. The network is in strong interaction with non reactive LC which can be switched together with the side chain polymer upon application of an electric field. In these gels the functions of the diacrylate molecules which are present at fractions of a percent is two folds: i) They form the cross-links thus provide system memory function ii) They preserve the polymer structure and its distribution within the system preventing its diffusion. The second function is especially important in producing broadband reflectors and patterned gels described below.

The switching behavior of the gels was studied using UV-Vis spectroscopy. In the gels, two different types of switching have been characterized. Figure 4 shows the first mode of switching in these gels. In this mode the reflection band shifts gradually to low wavelengths with increasing voltage before decreasing in magnitude and shifting back to higher wavelengths and disappearing at higher voltages.

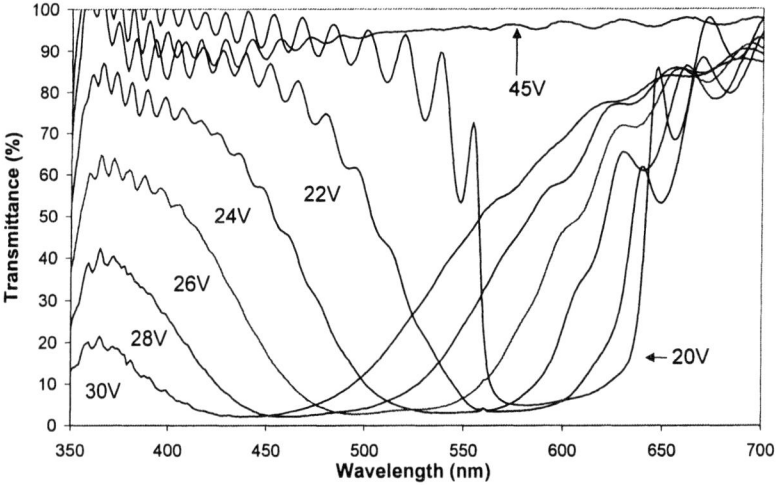

Figure 4. Transmission as a function of voltage.

Shifting of the reflection band to lower wavelengths is a well understood effect associated with the cholesteric layers getting tilted with respect to the incident beam of light followed by helical unwinding. The threshold voltage associated with the helical deformation has been previously described as grid-like deformations or lattice dislocations in the literature. This transition, which is also referred to as Helfrich's deformation, leads to the formation of the fingerprint texture which occurs above the critical voltage $V_{c,h}$ given by (7)

$$V_{c,h} = (6k_3k_2)^{0.25} \left(\frac{2\pi^2 d}{p\varepsilon_o \Delta\varepsilon} \right)^{0.5} \quad (1)$$

where d is the cell thickness, p the cholesteric pitch, ε_o the permittivity of free space, $\Delta\varepsilon$ dielectric anisotropy and k_3 and k_2 are the bend and twist elastic constants respectively. Following this transformation at higher voltages the helical unwinding is initiated above the threshold voltage $V_{c,n}$ given by

$$V_{c,n} = \frac{d\pi^2}{p} \left(\frac{k_2}{\varepsilon_o \Delta\varepsilon} \right)^{0.5} \quad (2)$$

In the second mode shown in Figure 5 the reflection band becomes narrower and decreases in magnitude as only the lower limit of the reflection band retain its position with increasing voltage.

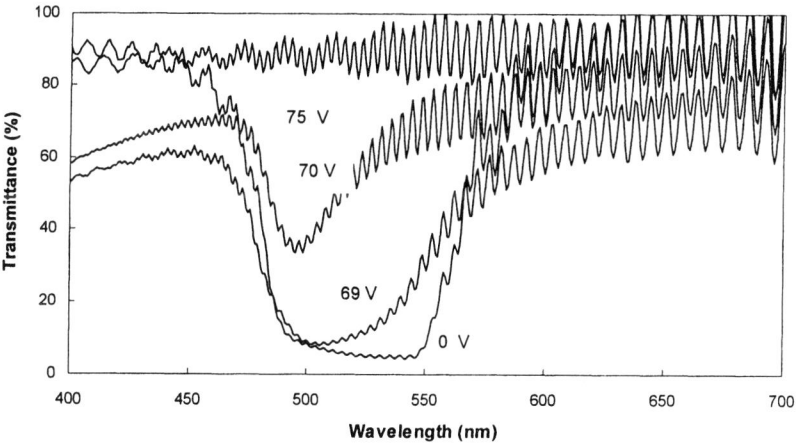

Figure 5. Transmission as a function of voltage.

The fact that the lower limit of the reflection band remains the same while the upper limit decreases indicate that the cholesteric structure and the helical pitch remains the same while molecules tilt to become oriented in the direction of the applied field. This mode of switching is associated with conical deformation and the equation describing the threshold voltage is given by

$$V_c = \left(\frac{dk_1 \pi^2}{p\varepsilon_0 \Delta\varepsilon} \right)^{0.5} \left(1 + \frac{4d^2 k_3}{k_1 p^2} \right)^{0.5} \tag{3}$$

where k_1 is the splay elastic constant. As can be deduced from equations 1 and 3 desired switching mode could be obtained by adjusting parameters such as the elastic constants. This could be done by changing the network structure (8) as well as the pitch and the cell thickness.

We also produced monomeric mixtures showing strong temperature dependent reflection colours. Using the same mixture polymerisation could be carried out at various temperatures and gels reflecting various colours could be produced. In these gels the pitch is frozen-in by the network thus the reflection colours are not temperature sensitive. This effect is clearly illustrated in Figure 6 where the λ_{max} is plotted as a function of temperature for the monomeric mixture together with a gel obtained by polymerization at 35 °C (6).

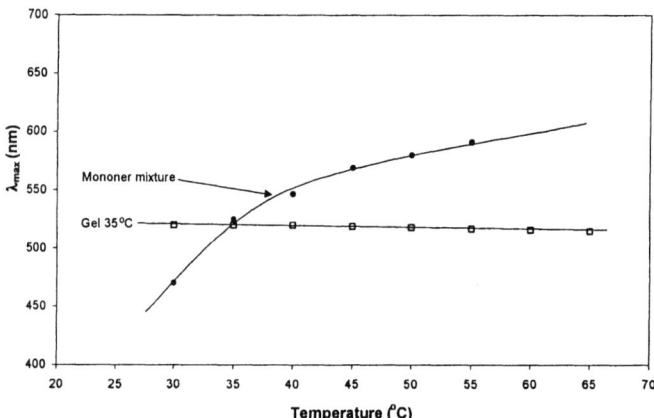

Figure 6. Position of the reflection band as a function of temperature.

Having shown that a single cholesteric mixtures can be used to produce gels reflecting different colours by means of polymerisation at various temperatures we patterned gels in the following way. A cell containing indium tin oxide transparent electrodes and orientation layers was filled with the cholesteric mixture. The temperature of cell was first raised to 55°C so that the cell was reflecting red light. A mask was placed on top of the cell and it was irradiated using a UV lamp. Subsequently the cell was cooled to 35°C. The areas which did not get irradiated during the previous exposure became green. Another mask was placed on top of the cell and the cell was irradiated. Subsequently the cell was cooled down to 30°C. At this temperature the unirradiated areas turned blue and the cell was exposed to flood exposure. Plate 1 shows photographs of the cell at room temperature (6). At zero voltage each colour indicates the temperature of polymerisation. The switching behavior of the patterned gel is also shown in plate 1. It can be seen that the application of an electric field the red areas start switching followed by the green and blue areas. This behavior can be related to pitch-length dependent threshold voltage for switching.

During our study, the tendency of some of the mixtures to show phase separation upon polymerisation was observed. This was evident from the increase in the reflection band of the gels, which in some cases doubled in width while the birefringence of the system showed only a slight increase upon polymerisation. This tendency could be influenced by various parameters. For example factors determining the kinetic chain length such as the initiator concentration and the UV intensity had a profound influence on the width of the reflection band. As known with increasing molecular weight of the polymer, its miscibility with a monomer decreases. In the gels during polymerisation such a phase separation leading to concentration fluctuations occurs. These fluctuations are fixed by the presence of the cross-links and the system further remains kinetically stable. As a function of time and temperature, no homogenization or change in the structure of the network is observed. Such a phase separation has also been observed for gels containing only diacrylate molecules. In that case, two distinct phases of liquid crystal and polymer network swollen by liquid crystal could be observed. It is also found that when compounds referred to as

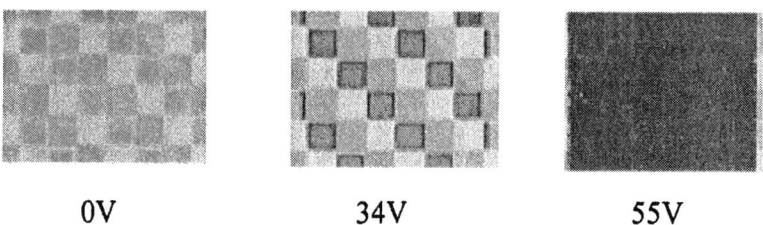

0V   34V   55V

Plate 1. Photographs of a patterned cholesteric gel at various applied voltages.
(See Page 6 of color insert.)

excited state quenchers were added to the monomeric mixtures further increase in the bandwidth was observed. Such quenchers work by capturing the energy of the excited state of the initiator thus radical formation can be prevented. The change in the width was followed as function of time for a system containing excited state quencher and is plotted in Figure 7. It can be seen that after an incubation time the width of the reflection band starts increasing before reaching a certain value where after it remains the same.

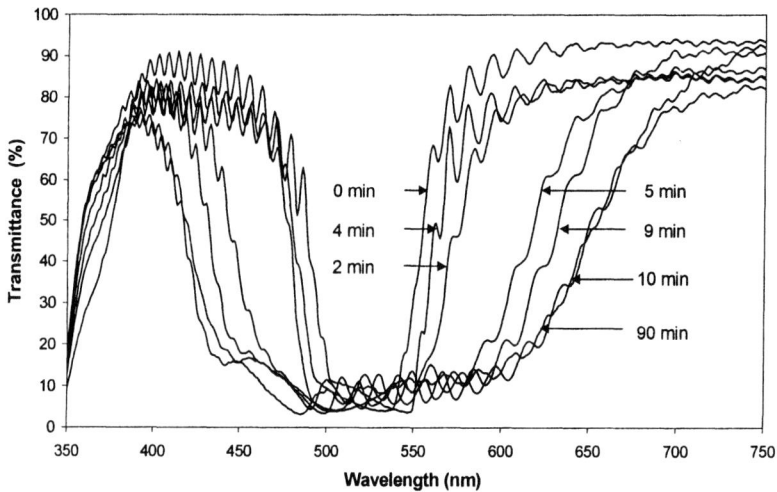

Figure 7. Reflection band a s a function of time.

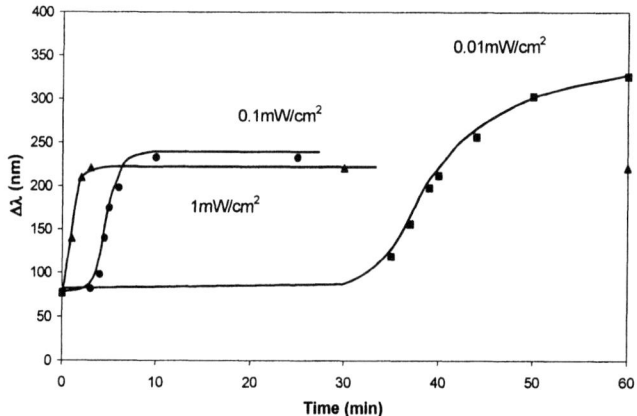

Figure 8. Width of the reflection band as a function of time during polymerisation with various UV intensities.

The effect of the UV intensity on this process is shown in Figure 8. It can be seen that with decreasing UV intensity the incubation time and the final width of the reflection band increases.

The effect of the excited state quencher (Tinovin P) concentration on the band width is shown in Figure9. It can be seen that with increasing intensity of UV light the width of the reflection band decreases. In these gels formation of a pitch gradient as a result of light intensity variation across the cell due to absorption of UV light is the likely cause of the broadening of the reflection band. Due to the absorbance of the initiator and the excited state quencher UV light is absorbed within the cell which means a difference in the intensity of light at the two surfaces of cell.

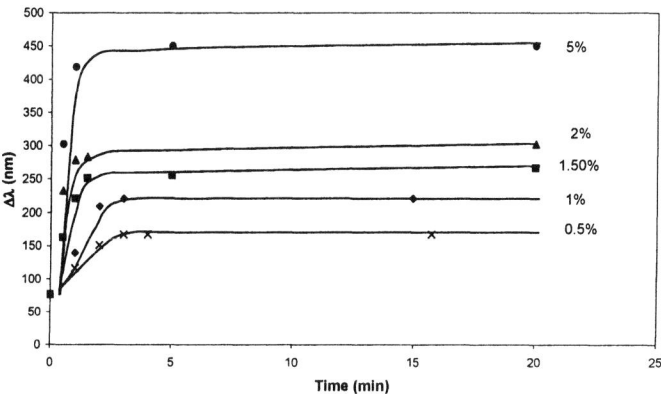

Figure 9. Width of the reflection band as a function of time for samples containing various amounts of excite state quencher during polymerisation at UV intensity of 1mW/cm².

Figure 9 shows that increase in the excited state quencher concentration gives rise to an increased width of the reflection band. The fact that intensity plays a very important role indicates the competition between the phase separation process which causes the increase as opposed to the polymerisation speed which tends to freeze in the structure and the pitch present within the system. Such a pitch gradient has been observed in polymerized acrylate networks (9). In the case of studies of other gels, the band broadening occurred as the system phase separated into two regions containing liquid crystal and the liquid crystal swollen network.

The temperature dependence of the reflection band was also ion studied as shown in Figure 10. It can be seen that with increasing temperature the positof the reflection band remains almost constant and only a slight decrease can be

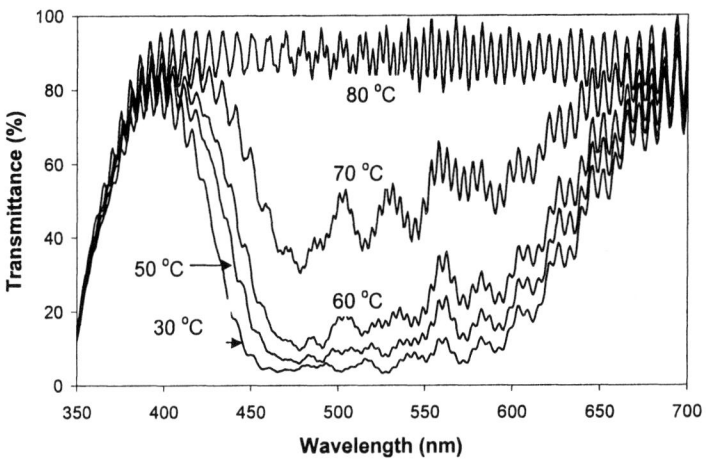

Figure 10. Reflection band as a function of temperature.

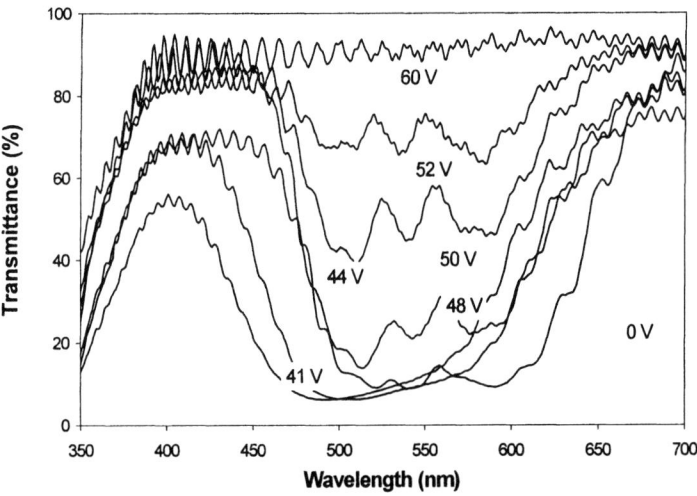

Figure 11. Reflection band as a function of voltage.

observed in the width of the reflection band. At higher temperatures the intensity of the reflection band decrease before the system becomes isotropic.

These broad band cholesteric gels could be switched reversibly between silver colored reflecting and non-reflecting transparent states. Upon application of the electric field, the cholesteric structure disappears and the cell becomes transparent as shown in Figure 11. Upon removal of the voltage, the cell reverts to the silver colored reflecting state very rapidly.

Conclusions

New kinds of gels, which can be used in the production of defect free reversible switchable cholesteric mirrors, have been developed. These gels can be patterned with ease to show various reflection colours. By manipulating the structure of the gels the reflection band of the switchable cholesteric, mirrors can also be increased to cover the whole visible range.

Acknowledgements

The author would like to thank J. Lub for the synthesis of acrylates.

References

(1) Maurer, R.; Anderejewski, D.; Kreuzer, F.H.; Miller A. *SID Digest* **1990**, 110.
(2) Lub, J.; Broer, D.J.; Hikmet, R.A.M.; Nierop, K.G.J. *Liq. Cryst.* **1995**, *18*, 319.
(3) Yang, D.K.; Doane, J.W. *SID Digest* **1992**, 759.
(4) Hikmet, R.A.M.; Zwerver, B.H. *Liq. Cryst.* **1992**, *12*, 319.
(5) Hikmet, R.A.M.; Kemperman, H. *Nature* **1998**, *392*, 476
(6) Hikmet, R.A.M.; Polesso, R. *Adv. Mater.* **2002**, 14(7), 502.
(7) Kelker, H.; Hatz, R. *Handbook of Liquid Crystals*, **1980**, Verlag Chemie.
(8) Hikmet, R.A.M.; Kemperman *Liq. Cryst.* **1999**, *11*, 1645.
(9) Broer, D.; Lub J.; Mol, G.N. *Nature* **1995**, *378*, 467.

Chapter 22

Glassy Liquid Crystals for Tunable Reflective Coloration

Shaw H. Chen, Philip H. M. Chen, Dimitris Katsis, and John C. Mastrangelo

Department of Chemical Engineering and Laboratory for Laser Energetics, Center for Optoelectronics and Imaging, University of Rochester, 240 East River Road, Rochester, NY 14623-1212

Novel glassy liquid crystals possessing elevated phase transition temperatures and excellent morphological stability have been developed for photonic applications. The concepts of high-performance circular polarizers, optical notch filters and reflectors with a tunable spectral range have been demonstrated with glassy cholesterics. The design principle governing glassy liquid crystals has been generalized to photoresponsive systems containing diarylethene moieties for reversible tunability of reflective coloration with superior thermal stability and fatigue resistance.

Amorphous Molecular Glasses

Glasses have been in existence for thousands of years, and yet they still represent one of the frontiers in materials science today (*1*). They are traditionally classified as amorphous solids exhibiting isotropic properties. Cooled at a sufficiently rapid rate, all liquids should bypass crystallization to form glass. It is well known that liquid viscosity increases exponentially with decreasing temperature. Phenomenologically, the glass transition temperature, T_g, is defined on the basis of viscosity reaching a value of 10^{13} poise (*2*). Despite intensive efforts over several decades, understanding of molecular processes accompanying glass transition has remained largely qualitative (*1*, *2*).

Glassy films are characterized by their superior optical quality over a large area with no grain boundaries, and hence are ideally suited for electronics, optics, photonics, and optoelectronics (3). Glass formation is ubiquitous among polymeric materials. To take advantage of the ease of film processing due to low melt viscosity and the feasibility of vacuum deposition, considerable efforts have been devoted in recent years to developing low-molar-mass organic materials. Existing amorphous molecular glasses can be categorized as follows: (i) bulky, odd-shaped or twin molecules (4-6); (ii) starburst molecules and dendrimers (7-9); (iii) spiro-linked molecules (10-12); and (iv) tetrahedrally configured molecules (13). In general, an elevated T_g relative to the application temperature is desired for stability against recrystallization. Representative structures are presented in *Figure 1*, where G x°C I expresses a T_g at x°C.

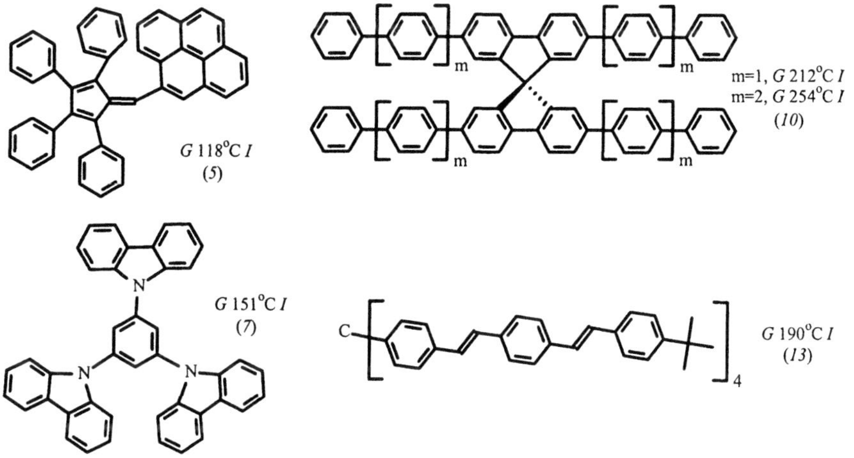

Figure 1. Representative amorphous molecular glasses.

Glassy Liquid Crystals (or Supercooled Liquid Crystals)

It is arguable that glasses are not necessarily amorphous or isotropic. In principle, mesomorphic organic glasses can be realized by vitrifying liquid crystals through thermal quenching. Liquid crystals are a class of self-organizing fluids characterized by a uniaxial, lamellar, helical or columnar arrangement in nematic, smectic, cholesteric, and discotic liquid crystalline order (14), respectively, as depicted in *Figure 2*:

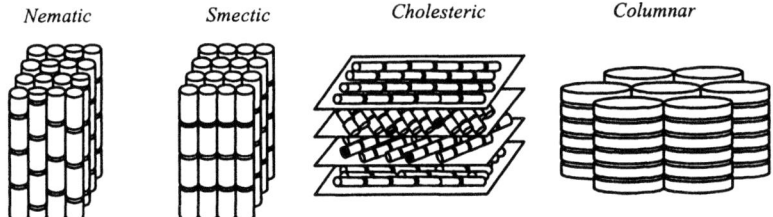

Figure 2. Liquid crystalline order via molecular self-organization.

Each type of liquid crystal has found its respective niche in optics, photonics, or optoelectronics. With these molecular arrangements frozen in the solid state, glassy liquid crystals (GLCs), or supercooled liquid crystals, represent a novel material class that combines properties intrinsic to liquid crystals with those common to polymers, such as glass transition and film- and fiber-forming abilities. The preparation of defect-free GLC films requires slow cooling from mesomorphic melts without encountering crystallization, a challenge to thermal quenching as a conventional means to vitrification. From a fundamental perspective, transition of liquid crystal into mesomorphic solid adds a new dimension to the traditional view of transition from isotropic liquid into isotropic solid. The differential scanning calorimetric thermograms compiled in *Figure 3* serve to distinguish a GLC from a conventional liquid crystal.

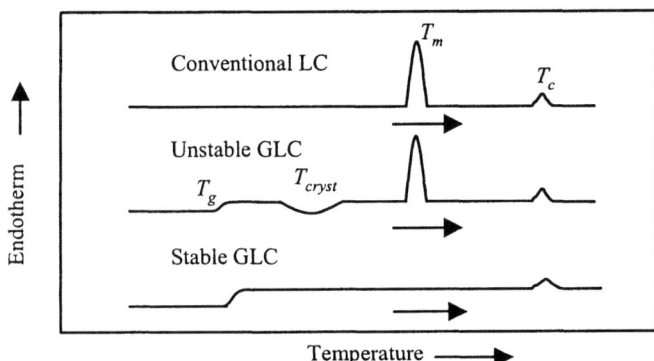

Figure 3. DSC thermograms of liquid crystals.

Heating a conventional liquid crystal causes a first-order transition from crystalline solid to liquid crystal at its melting point, T_m, followed by a transition into isotropic liquid at its clearing point, T_c. In contrast, a stable GLC undergoes a second-order transition from mesomorphic solid into liquid crystal at T_g, without modifying the molecular order, followed by a transition into isotropic liquid at T_c. Intermediate between a conventional liquid crystal and a stable GLC lies an unstable GLC, which tends to recrystallize from mesomorphic melt above T_g upon heating with subsequent melting to liquid crystal at T_m and clearing at T_c. Empirically, T_g/T_m was found to fall between 2/3 and 3/4 on an absolute temperature scale.

The very first attempt to synthesize GLCs in 1971 yielded materials with a low T_g and poor morphological stability (15). In parallel to low-molar-mass GLCs, liquid crystalline polymers have been explored for the past three decades (16-18). In essence, GLCs are advantageous in their superior chemical purity and favorable rheological properties (19). Existing GLCs can be categorized into: (i) laterally or terminally branched, one-string compounds with a T_g mostly around room temperature (20); (ii) twin molecules with an above-ambient T_g, but generally lacking morphological stability (21-24); (iii) cyclosiloxanes functionalized with mesogenic and chiral pendants (25-27); (iv) carbosilane dendrimers exhibiting a low T_g (28-30), and (v) macrocarbocycles with mesogenic segments as part of the ring structure (31).

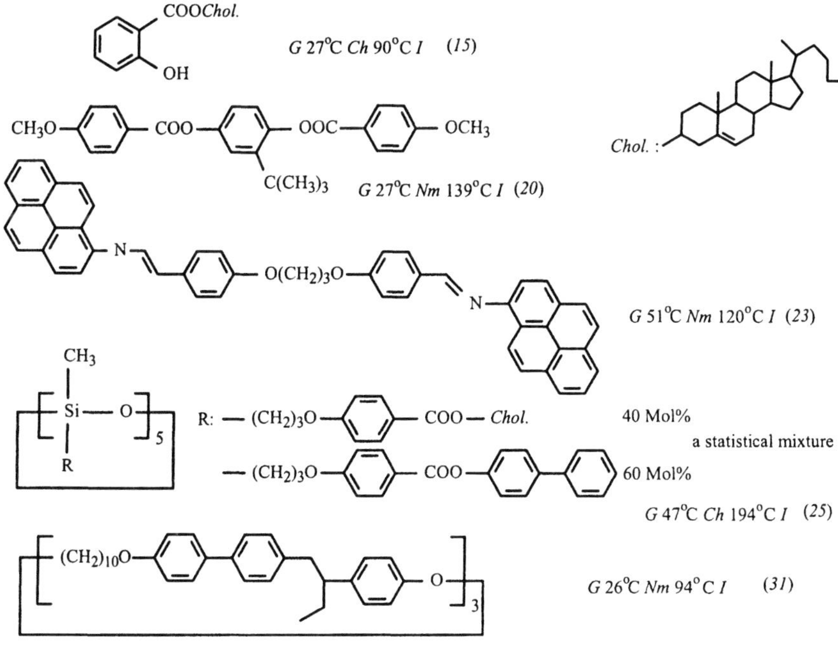

Figure 4. Representative glassy liquid crystals reported previously.

Representative structures are presented in *Figure 4*, where G x°C Nm or Ch y°C I expresses a T_g and a T_c (for a nematic or cholesteric to isotropic transition) at x and y°C, respectively. Based on the previously reported structures, there does not seem to be a systematic approach to the design of glassy liquid crystals. Specifically, the structural factors determining the type of mesomorphism, T_g and T_c, and the stability against recrystallization from the glassy state have remained elusive. Glassy cholesteric liquid crystals capable of selective wavelength reflection are of particular interest because of the relevance to tunable reflective coloration.

Optical Properties of Cholesteric Liquid-Crystal Films

A cholesteric liquid crystal contains both nematic and chiral moieties in a single molecular entity or as a binary mixture. Consisting of a helical stack of quasinematic layers, a well-aligned cholesteric film can be characterized by handedness and helical pitch length, p, as depicted in *Figure 5*:

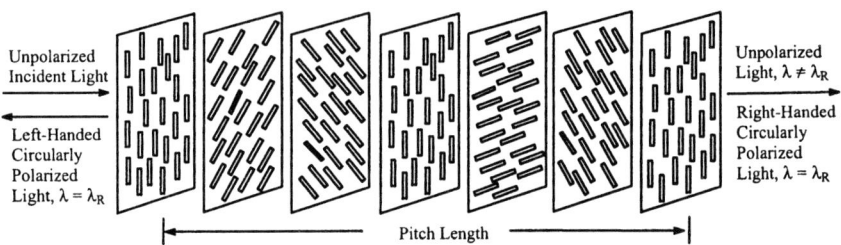

Figure 5. Selective reflection by a left-handed cholesteric film (32).

Handedness describes the direction in which twisting of the nematic director occurs from one layer to the next, and p is defined as the distance over which the director rotates by 360°. The property of selective reflection can be described in terms of $\lambda_R = p(n_e + n_o)/2$, in which n_e and n_o are the extraordinary and ordinary refractive indices of the quasinematic layer, respectively. The other parameter of interest is the helical twisting power, a measure of the ability of a chiral moiety to generate a given helical pitch length. Let us consider incident unpolarized white light propagating through a left-handed film as illustrated in *Figure 5*. Natural (*i.e.* unpolarized) light consists of equal amounts of left- and right-handed (LH and RH) circularly polarized components. The LH circularly polarized component in the neighborhood of

λ_R is selectively reflected, while the RH component is completely transmitted. The selective reflection bandwidth is determined by optical birefringence, $\Delta n = n_e - n_o$. A sufficiently thick, single-handed cholesteric film is capable of reflecting 50% of incident unpolarized light within the selective reflection band. Outside the selective reflection band, incident light is transmitted regardless of its polarization state. It follows that a stack of RH and LH films tuned at the same λ_R will reflect 100% of incident unpolarized light within the selective reflection band without attenuating the rest of the spectrum.

A New Approach to Glassy Liquid Crystals with Elevated Phase Transition Temperatures and Superior Morphological Stability

Most existing liquid crystals tend to crystallize upon cooling to below their melting points, thus losing the desired molecular order characteristic of liquid crystals and resulting in polycrystalline films that scatter light or limit charge transport. As an emerging class of advanced materials, GLCs preserve varied forms of molecular order characteristic of liquid crystals in the solid state. To prevent spontaneous crystallization, we have implemented a molecular design strategy in which mesogenic and chiral pendants are chemically bonded to a volume-excluding core. While the core and the pendant are crystalline as separate entities, the chemical hybrid with a proper flexible spacer connecting the two readily vitrifies into a GLC on cooling. A definitive set of GLCs has been synthesized and characterized (*33-43*) as summarized in *Figure 6*, where G x°C (*Nm*, S_A, or *Ch*) y°C *I* expresses a T_g and a T_c (for a nematic, smectic A or cholesteric to isotropic transition) at x and y°C, respectively. Cyclohexane, bicyclooctene, adamantane and cubane serve as the cores to which nematic and chiral pendants are attached, a manifestation of the versatility of our design concept. A spacer length of two to four methylene units was found to be optimum for vitrification with an elevated T_g.

Major advances have been made recently using a cyanoterphenyl nematogen, with an exceptionally high T_m and T_c, in the construction of GLCs with substantially elevated transition temperatures without compromising morphological stability. Note the glassy nematics with a T_g close to 130°C and a T_c close to 350°C (*44*), the highest values ever achieved in GLCs. As shown in *Figure 7*, a linear nematogen is superior to an angular one in terms of phase transition temperatures. Glassy cholesterics have been synthesized in the past via a statistical approach, which requires intensive purification of a multicomponent reaction product (*45*). A deterministic synthesis strategy, as described in *Reaction Scheme 1*, produced enantiomeric glassy cholesterics with an identical molecular structure except opposite chirality (*46*).

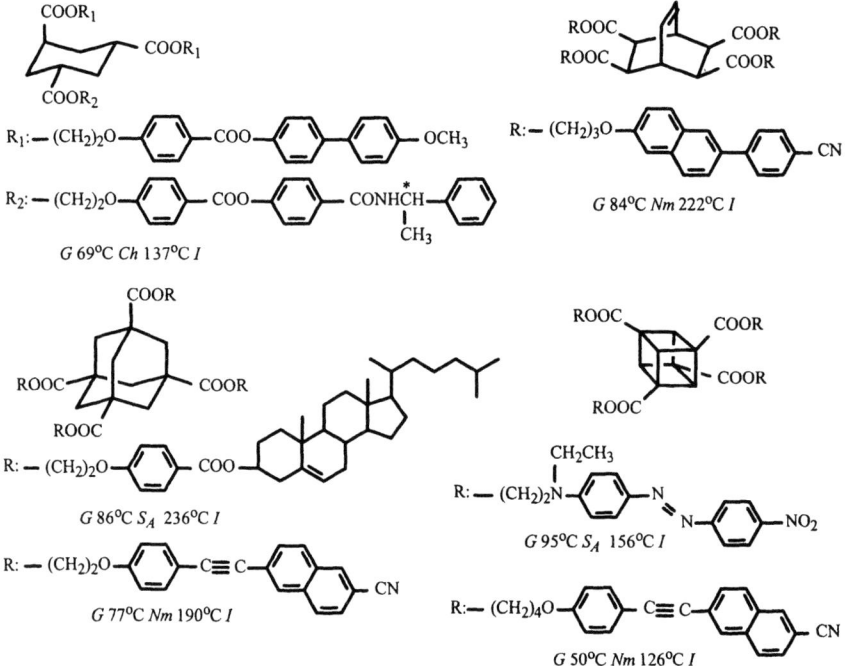

Figure 6. Representative morphologically stable glassy liquid crystals.

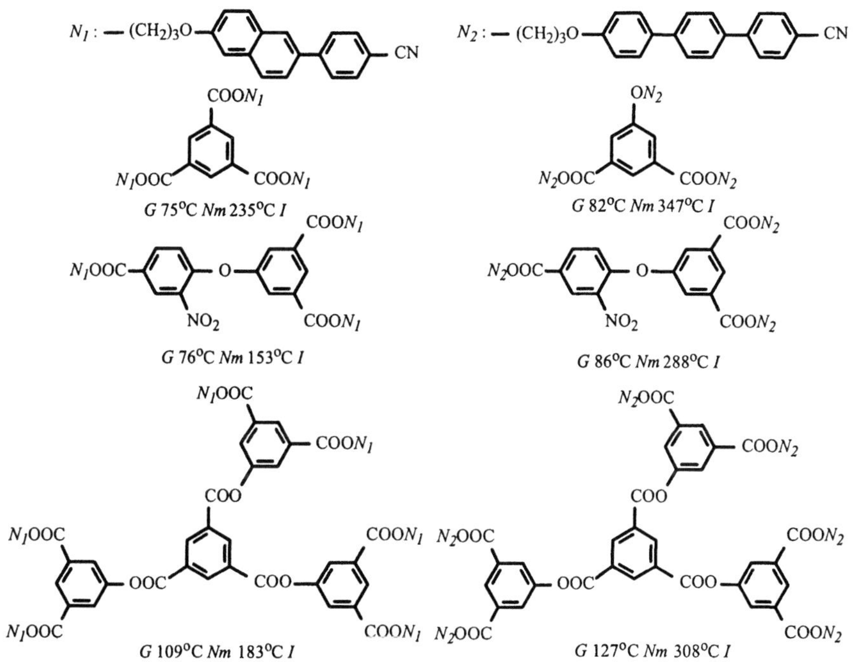

Figure 7. High-temperature glassy nematics.

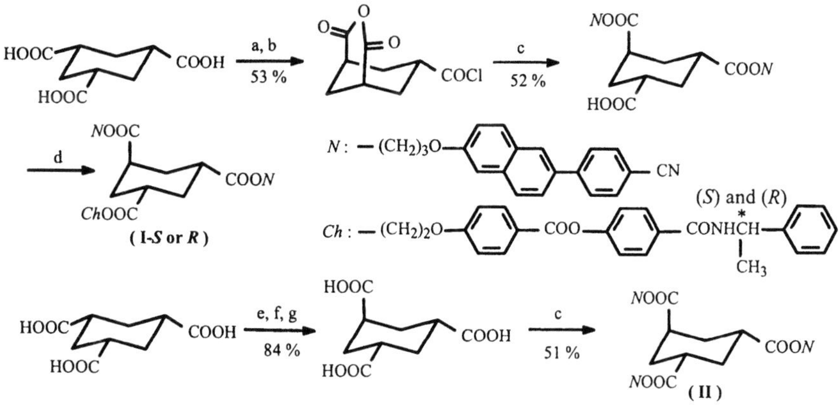

a. $(CH_3CO)_2O$ / reflux; b. $SOCl_2$ / reflux; c. NOH / DMAP / Et_3N; d. ChOH / DEADC / PPh_3
e. $(CH_3CO)_2O$, CH_3CO_2Na; f. CH_3COCl; g. H_2O

DMAP: 4-(dimethylamino)pyridine; DEADC: diethyl azodicarboxylate; PPh_3: triphenylphosphine

Reaction Scheme 1. Deterministic synthesis of glassy cholesterics.

The mixture of (I) and (II) at a mass ratio of 42 to 58 showed a T_g at 67 and a T_c at 131°C. The polarization spectra of single-handed glassy cholesteric films are shown in *Figure 8*. An unpolarized incident beam is decomposed into two circularly polarized components of equal intensities propagating in opposite directions. Since handedness of circularly polarized light can be reversed via reflection from a specular surface, essentially 100% circular polarization of an unpolarized light source can be accomplished. An optical notch filter consisting of a stack of glassy cholesteric films of opposite handedness is shown in *Figure 8* to yield an attenuation of 3.75 optical density units, equivalent to a contrast ratio of better than 5,000 to 1, representing the best performance of organic materials to date. The spectral range intended for polarization and reflection can be readily tuned by varying the chemical composition.

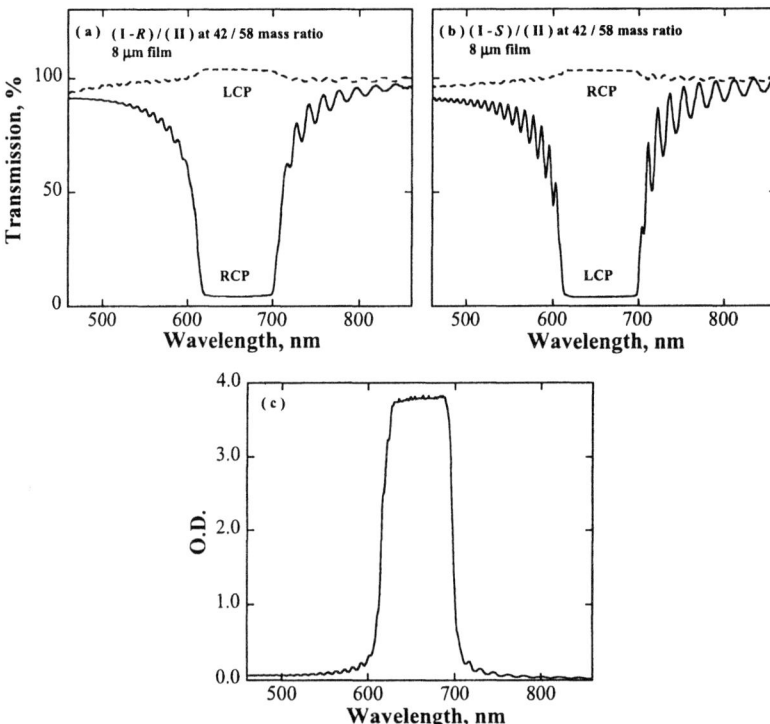

Figure 8. Optical spectra of an unpolarized beam through (a) a right-handed, (b) a left-handed glassy cholesteric film, and (c) a notch filter comprising the two single-handed films. (Reproduced with permission from reference 46. Copyright 2000 Wiley-VCH.)

The cyanoterphenyl group has also been successfully implemented in high-temperature glassy cholesterics synthesized in a deterministic fashion using the 5-oxyisophthalate linking unit, as shown in *Figure 9*. Note that G x°C Ch y°C I expresses a T_g and a T_c (for cholesteric to isotropic transition) at x and y°C, respectively, determined from a DSC heating scan. The DSC cooling scan is presented as I w°C Ch z°C G, indicating a T_c (for isotropic to cholesteric transition) and a T_g at w and z°C, respectively.

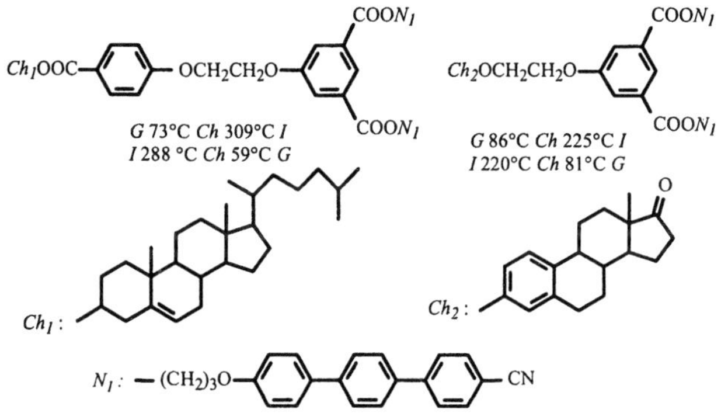

Figure 9. High-temperature glassy cholesterics.

Tunable Reflective Coloration by Glassy Cholesteric Films

There are two distinct modes of coloration in nature: *pigmentary*, involving electronic transitions of chromophores underlying light absorption and emission; and *structural*, involving interference, diffraction, or scattering of ambient light by nanostructures (*47*). Examples of structural colors include butterfly wings, bird feathers, and beetle cuticles. In particular, beetles' exocuticles resemble cholesteric liquid crystalline films capable of selective wavelength reflection with simultaneous circular polarization, giving rise to long-lasting brilliant colors. A wide variety of cholesteric liquid crystalline materials have been developed, such as low viscosity liquids, lyotropic and thermotropic polymers, liquid crystal/polymer composites, and glassy liquid crystals. Of particular interest are glassy liquid crystals that resist spontaneous crystallization through heating-cooling cycles, such as mixtures of Compounds (III) and (IV) depicted in *Figure 10*. Although (IV) crystallizes upon heating to 95°C, both (III) and its binary mixtures with (IV) form stable glassy cholesteric films (*45*). Also shown in *Figure 10* are the selective reflection bands ranging from blue to green, red and the infrared region with mixtures at an increasing ratio of (IV) to (III).

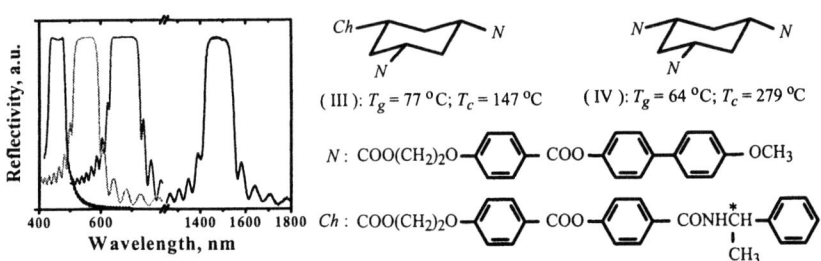

Figure 10. Reflective coloration by glassy cholesteric films.

Furthermore, the selective reflection band and its width were shown to be tunable, albeit irreversibly, via photoracemization of a bridged binaphthyl dopant (*48*). In particular, phototunability is demonstrated in *Figure 11*, where UV-irradiation of a cholesteric film at a temperature above its T_g (i.e. 120 vs 68°C) over an increasing time period followed by cooling to room temperature is shown to result in an increasing selective reflection wavelength. Morphologically stable GLCs that resist crystallization upon heating and cooling are the key to the successful implementation of this device concept.

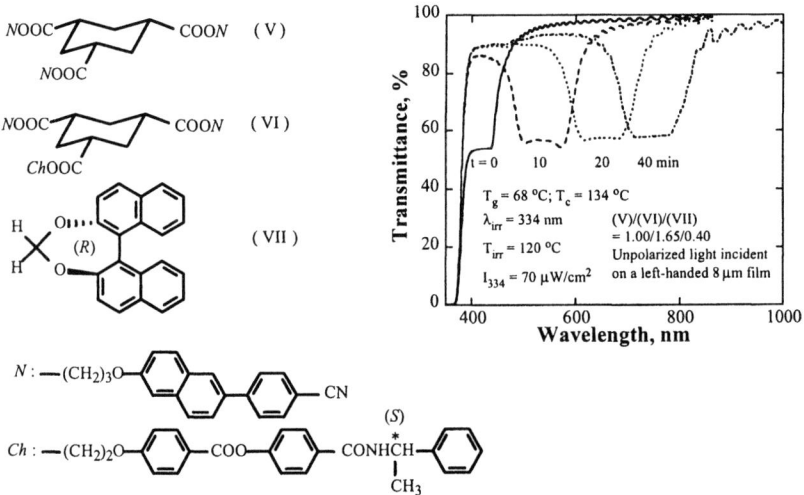

Figure 11. A phototunable glassy cholesteric film.

Reversible Tunability of Reflective Coloration

Reversible tunability of reflective coloration has been extensively explored with temperature, pressure, electric field, and light as the external stimuli (*49-57*). The approach based on photoisomerization appears to be the most promising. The concept was first demonstrated by Sackman (*54*), and revisited recently by Ikeda et al. (*55*), based on *cis-trans* isomerization of azobenzene dopants in cholesteric fluid films. Shibaev et al. (*56*) employed a chiral azobenzene, both as a dopant and as a comonomer, in a polymer system with an ambient T_g and a response time of tens of minutes. Tamaoki et al. (*57*) used a glassy cholesteric matrix containing an azobenzene dopant, a material system allowing for photomodulation of pitch length at temperatures above T_g followed by cooling to below T_g to preserve the modified pitch in the solid state. All these approaches employing azobenzenes to tune reflective coloration suffer from fatigue and thermally activated *cis*-to-*trans* isomerization. Photoinduced interconversion between nematic and cholesteric mesomorphism and that between the right- and left-handed cholesteric mesomorphism have also been reported (*58-63*), where a helical pitch length on the order of 10 µm was observed.

Of all the photoresponsive moieties that have been explored, diarylethenes (*64*) appear to be the most promising in terms of thermal stability and fatigue resistance. The premise is that the two interconvertible isomers, *viz.* the open and closed forms, present disparate helical twisting powers in a liquid-crystal matrix. Indeed, the closed form of chiral diarylethenes was found to have a stronger helical twisting power than the open form in nematic liquid-crystal hosts with a helical pitch length on the order of 10 µm in all cases (*62, 63*). To accomplish tunable reflective coloration in the visible region, a cholesteric liquid crystal with a short pitch length must be used as the host to define the base case. Moreover, glassy liquid-crystal films are much preferred over liquid-crystal films in practical application. Isomerization of diarylethenes was found to take place not only in liquid but also in single-crystalline (*64*) and amorphous (*65, 66*) solids without altering the morphology. However, modulation of pitch length requires a relative rotation between quasinematic layers as depicted in *Figure 5*, a macroscopic rearrangement allowable only in the liquid state. Therefore, we aim at cholesteric GLCs containing diarylene moieties.

A Novel Class of Photoresponsive GLCs

Reversible tunability of reflective coloration using diarylethene-containing GLCs is envisioned in *Figure 12*.

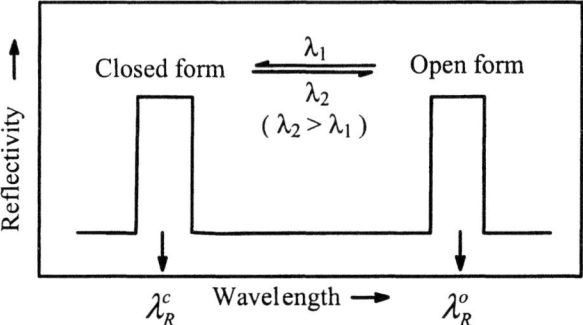

Figure 12. Reversible photowunability of reflective coloration.

A glassy cholesteric film comprising the open form of a diarylethene moiety with a predetermined reflection wavelength, λ_R^o, is heated to above its T_g, where irradiation at λ_1 is performed to afford the closed from. This treatment results in a shorter reflection wavelength, λ_R^c, which can be preserved in the solid state with subsequent cooling to below T_g. The more extended conjugation of the closed form could cause light absorption in the visible region. With λ_R^c frozen in the solid state, irradiation of the closed form at λ_2 will regenerate the open form, thereby bleaching the undesired absorptive color without altering λ_R^c, a clear advantage over the use of liquid films. As dictated by the open form, subsequent heating to above T_g will recover λ_R^o, which can then be frozen in the solid state by cooling through T_g. To attain the envisioned phototunability, we have developed the first diarylethene-containing nematic and chiral glasses. The molecular structures are depicted in *Figure 13*, where N and Ch denote a nematic and a chiral moiety, respectively. Work is in progress to take advantage of these novel glassy materials together with the high-temperature GLCs shown in *Figures 7 and 9* for reversible tunability of reflective coloration with superior thermal stability and fatigue resistance.

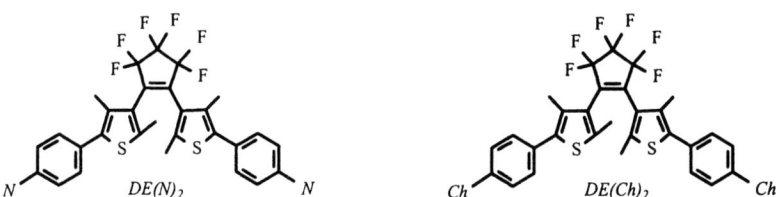

Figure 13. Novel glassy materials containing photoresponsive diarylethenes.

Conclusions

Glassy liquid crystals with elevated glass transition and clearing temperatures, a broad mesomorphic fluid temperature range, and excellent morphological stability have been developed for a diversity of photonic applications. In particular, glassy cholesteric films have been demonstrated for high-performance circular polarizers, optical notch filters and reflectors across a spectral range that can be tuned by varying the chemical composition or by UV-irradiation albeit irreversibly. The molecular design concept governing glassy liquid crystals has been generalized to photoresponsive material systems containing diarylethenes, thereby enabling reversible tunability of reflective coloration with superior thermal stability and fatigue resistance.

Acknowledgment

The authors wish to thank S. D. Jacobs and K. L. Marshall at the Laboratory for Laser Energetics, University of Rochester, and P. J. Hood of Cornerstone Research Group, Inc. in Dayton, Ohio for technical advice and helpful discussions. They are grateful for the financial support provided by the Multidisciplinary University Research Initiative, administered by the Army Research Office, under DAAD19-01-1-0676, the National Science Foundation under Grants CTS-9500737, 9818234 and 0204827, Eastman Kodak Company, and Kaiser Electronics. Glassy liquid crystals have served as the foundation for SBIR and STTR programs funded by AFOSR, BMDO, DTRA, NASA, NSF, and ONR in partnership with Cornerstone Research Group. Additional funding was provided by the Department of Energy Office of Inertial Confinement Fusion under Cooperative Agreement No. DE-FC03-92SF19460 with the Laboratory for Laser Energetics and the New York State Energy Research and Development Authority. The support of DOE does not constitute an endorsement by DOE of the views expressed in this article.

References Cited

(1) Angell, C. A. *Science* **1995**, *267*, 1924.
(2) Debenedetti, P. G.; Stillinger, F. H. *Nature* **2001**, *410*, 259.
(3) *Organic Materials for Photonics: science and technology*; G. Zerbi, Ed.; Elsevier: Amsterdam, 1993.
(4) Braun, D.; Langendorf, R. *J. Prakt. Chem.-Chem. Ztg.* **1999**, *341*, 128.
(5) Alig, I.; Braun, D.; Langendorf, R.; Wirth, H. O.; Voigt, M.; Wendorff, J. H. *J. Mater. Chem.* **1998**, *8*, 847.

(6) O'Brien, D. F.; Burrows, P. E.; Forrest, S. R.; Koene, B. E.; Loy, D. E.; Thompson, M. E. *Adv. Mater.* **1998**, *10*, 1108.
(7) Naito, K.; Miura, A. *J. Phys. Chem.* **1993**, *97*, 6240.
(8) Katsuma, K.; Shirota, Y. *Adv. Mater* **1998,** *10*, 223.
(9) Halim, M.; Pillow, J. N. G.; Samuel, I. D. W.; Burn, P. L. *Adv. Mater.* **1999**, *11*, 371.
(10) Johansson, N.; Salbeck, J.; Bauer, J.; Weissörtel, F.; Broms, P.; Andersson, A.; Salaneck, W. R. *Adv. Mater.* **1998**, *10*, 1136.
(11) Bach, U.; De Cloedt, K.; Spreitzer, H.; Gratzel, M. *Adv. Mater.* **2000**, *12*, 1060.
(12) Schartel, B.; Damerau, T.; Hennecke, M. *Phys. Chem. Chem. Phys.* **2000**, *2*, 4690.
(13) Wang, S.; Oldham, W. J., Jr.; Hudack, R. A., Jr.; Bazan, G. C. *J. Am. Chem. Soc.* **2000**, *122*, 5695.
(14) *Handbook of Liquid Crystals Research*; Collings, P. J.; Patel, J. S., Eds.; Oxford University Press: New York, **1997**.
(15) (a) Tsuji, K.; Sorai, M.; Seki, S. *Bull. Chem. Soc. Jap.* **1971**, *44*, 1452; (b) Sorai, M.; Seki, S. *Ibid.*, 2887.
(16) *Side Chain Liquid Crystal Polymers*; McArdle, C. B., Ed.; Chapman and Hall: New York, **1989**.
(17) *Liquid-Crystal Polymers*; Platé, N. A., Ed.; Plenum Press: New York, **1993**.
(18) *Liquid Crystallinity in Polymers: Principles and Fundamental Properties*; Ciferri, A., Ed.; VCH: New York, **1991**.
(19) Shi, H.; Chen, S. H.; De Rose, M. E.; Bunning, T. J.; Adams, W. W. *Liq. Cryst.* **1996**, *20*, 277.
(20) Wedler, W.; Demus, D.; Zaschke, H.; Mohr, K.; Schafer, W.; Weissflog, W. *J. Mater. Chem.* **1991**, *1*, 347.
(21) Attard, G. S.; Imrie, C. T. *Liq. Cryst.* **1992**, *11*, 785.
(22) Dehne, H.; Roger, A.; Demus, D.; Diele, S.; Kresse, H.; Pelzl, G.; Weissflog, W. *Liq. Cryst.* **1989**, *6*, 47.
(23) Attard, G. S.; Imrie, C. T.; Karasz, F. E. *Chem. Mat.* **1992**, *4*, 1246.
(24) Tamaoki, N.; Kruk, G.; Matsuda, H. *J. Mater. Chem.* **1999**, *9*, 2381.
(25) Kreuzer, F. H.; Andrejewski, D.; Haas, W.; Haberle, N.; Riepl, G.; Spes, P. *Mol. Cryst. Liq. Cryst.*, **1991**, *199*, 345.
(26) Kreuzer, F. H.; Maurer, R.; Spes, P. *Makromol. Chem., Macromol. Symp.* **1991**, *50*, 215.
(27) Gresham, K. D.; McHugh, C. M.; Bunning, T. J.; Crane, R. J.; Klei, H. E.; Samulski, E. T. *J. Polymer Sci.: Part A: Polymer Chem.* **1994**, *32*, 2039.
(28) Lorenz, K.; Hölter, D.; Stühn, B.; Mülhaupt, R.; Frey, H. *Adv. Mater.* **1996**, *8*, 414.

(29) Ponomarenko, S. A.; Boiko, N. I.; Shibaev, V. P.; Richardson, R. M.; Whitehouse, I. J.; Rebrov, E. A.; Muzafarov, A. M. *Macromolecules* **2000**, *33*, 5549.
(30) Saez, I. M.; Goodby, J. W.; Richardson, R. M. *Chem. Eur. J.* **2001**, *7*, 2758.
(31) Percec, V.; Kawasumi, M.; Rinaldi, P. L.; Litman, V. E. *Macromolecules* **1992**, *25*, 3851.
(32) Dreher, R.; Meier, G. *Phys. Rev. A* **1973**, *8*, 1616.
(33) Shi, H.; Chen, S. H. *Liq. Cryst.* **1994**, *17*, 413.
(34) Shi, H.; Chen, S. H. *Liq. Cryst.* **1995**, *18*, 733.
(35) Mastrangelo, J. C.; Blanton, T. N.; Chen, S. H. *Appl. Phys. Lett.* **1995**, *66*, 2212.
(36) Shi, H.; Chen, S. H. *Liq. Cryst.* **1995**, *19*, 785.
(37) Chen, S. H.; Mastrangelo, J. C.; Shi, H.; Bashir-Hashemi, A.; Li, J.; Gelber, N. *Macromolecules* **1995**, *28*, 7775.
(38) Shi, H.; Chen, S. H. *Liq. Cryst.* **1995**, *19*, 849.
(39) De Rosa, M. E.; Adams, W. W.; Bunning, T. J.; Shi, H.; Chen, S. H. *Macromolecules* **1996**, *29*, 5650.
(40) Chen, S. H.; Mastrangelo, J. C.; Blanton, T. N.; Bashir-Hashemi, A. *Liq. Cryst.* **1996**, *21*, 683.
(41) Chen, S. H.; Shi, H.; Conger, B. M.; Mastrangelo, J. C.; Tsutsui, T. *Adv. Mater.* **1996**, *8*, 998.
(42) Chen, S. H.; Mastrangelo, J. C.; Blanton, T. N.; Bashir-Hashemi, A. *Macromolecules* **1997**, *30*, 93.
(43) Chen, S. H.; Katsis, D.; Mastrangelo, J. C.; Schmid, A. W.; Tsutsui, T.; Blanton, T. N. *Nature* **1999**, *397*, 506.
(44) Fan, F. Y.; Culligan, S. W.; Mastrangelo, J. C.; Katsis, D.; Chen, S. H.; Blanton, T. N. *Chem. Mater.* **2001**, *13*, 4584.
(45) Katsis, D.; Chen, H. P.; Mastrangelo, J. C.; Chen, S. H.; Blanton, T. N. *Chem. Mater.* **1999**, *11*, 1590.
(46) Chen, H. P.; Katsis, D.; Mastrangelo, J. C.; Chen, S. H.; Jacobs, S. D.; Hood, P. J. *Adv. Mater.* **2000**, *12*, 1283.
(47) Srinivasaro, M. *Chem. Rev.* **1999**, *99*, 1935.
(48) Chen, S. H.; Mastrangelo, J. C.; Jin, R. J. *Adv. Mater.* **1999**, *11*, 1183.
(49) Pollman, P.; Wiege, B. *Mol. Cryst. Liq. Cryst.* **1987**, *150b*, 375.
(50) Kimura, H.; Hosino, M.; Nakano, H. *J. Phys. Coll.* **1979**, *40*, C3-174.
(51) (a) Meyer, R. B. *Appl. Phys. Lett.* **1969**, *14*, 208; (b) Baessler, H.; Labes, M. M. *Phys. Rev. Lett.* **1968**, *21*, 1791.
(52) Hikmet, R. A. M.; Kemperman, H. *Nature* **1998**, *392*, 476.
(53) Nishio, Y.; Kai, T.; Kimura, N.; Oshima, K.; Suzuki, H. *Macromolecules* **1998**, *31*, 2384.
(54) Sackmann, E. *J. Am. Chem. Soc.* **1971**, *93*, 7088.

(55) Lee, H.-K.; Doi, K.; Harada; H.; Tsutsumi, O.; Kanazawa, A.; Shiono, T.; Ikeda, T. *J. Phys. Chem. B* **2000**, *104*, 7023.
(56) Bobrovsky, A. Y.; Boiko, N. I.; Shibaev, V. P.; Springer, J. *Adv. Mater.* **2000**, *12*, 1180.
(57) (a) Tamaoki, N.; Song, S.; Moriyama, M.; Matsuda, H. *Adv. Mater.* **2000**, *12*, 94; (b) Tamaoki, N. *Ibid.* **2001**, *13*, 1135.
(58) Feringa, B. L.; Huck, N. P. M.; van Doren, H. A. *J. Am. Chem. Soc.* **1995**, *117*, 9929.
(59) Suraez, M.; Schuster, G. B. *J. Am. Chem. Soc.* **1995**, *117*, 6732.
(60) Denekamp, C.; Feringa, B. L. *Adv. Mater.* **1998**, *10*, 1080.
(61) Feringa, B. L.; van Delden, R. A.; Koumura, N.; Geertsema, E. M. *Chem. Rev.* **2000**, *100*, 1789.
(62) Yamaguchi, T.; Inagawa, T.; Nakazumi, H.; Irie, S.; Irie, M. *Chem. Mater.* **2000**, *12*, 869.
(63) Yamaguchi, T.; Inagawa, T.; Nakazumi, H.; Irie, S.; Irie, M. *J. Mater. Chem.* **2001**, *11*, 2453.
(64) Irie, M. *Chem. Rev.* **2000**, *100*, 1685.
(65) Kawai, T.; Koshido, T.; Yoshino, K. *Appl. Phys. Lett.* **1995**, *67*, 795.
(66) Kawai, T.; Fukuda, N.; Gröschl, D.; Kobatake, S.; Irie, M. *Jpn. J. Appl. Phys.* **1999**, *38*, L1194.

Chapter 23

Tunable Near-IR Optical Properties from Trialkoxysilane-Capped Poly(methyl methacrylate)–Silica Waveguide Materials

Ming-Hsin Wei, Chia-Hua Lee, and Wen-Chang Chen*

Department of Chemical Engineering, National Taiwan University, Taipei 10617, Taiwan

A class of trialkoxysilane-capped poly(methyl methacrylate)(PMMA)-silica hybrid materials with tunable near infrared (NIR) optical properties was studied. The incorporation of the silica segment into the PMMA matrix induced an increase in the anharmonicity of the C-H bond. Thus, the overtone vibration bands of the C-H bond were red shifted and resulted in adjustment of the optical window of the NIR region. Optical planar waveguides were fabricated from the hybrid materials on silicon wafer using thermal oxide as the cladding layer. The optical loss of the prepared optical waveguides was reduced from 0.652 to 0.264 dB/cm as the silica content increased. This is due to both the increase in anharmonicity of the C-H bond and reduction of the C-H bonding density. The prepared waveguides have potential applications as passive components for optical communication.

Introduction

Organic-inorganic hybrid materials have attracted extensive fundamental and practical interest.[1-5] Molecular design of hybrid materials could produce materials with tunable properties for optoelectronic devices, including optical waveguides,[6-12] electroluminescent devices,[13-15] nonlinear optical materials,[16] photorefractive materials,[17] and semiconductors.[18]

The primary interest of this study is in hybrid materials for optical waveguide applications. Inorganic oxides, organic polymers, and organic-inorganic hybrid materials represent three different classes of materials for optical waveguides. Optical polymers such as acrylates or polyimides have been recognized as waveguide materials for optical communication because of ease of processing, low cost and tunable properties. These advantages are not be easily achieved by conventional inorganic oxides. However, the thermal and mechanical properties of organic polymers limits their practical applications. Furthermore, the C-H vibration overtones of organic polymers produce a large optical losses in the near infrared (NIR) region important for optical communication.[19] Organic-inorganic hybrid materials provide a possible solution for replacing organic polymers as waveguide materials.[6-12] The incorporation of inorganic moieties into polymer matrix could improve the thermal and mechanical properties. Besides, the incorporation of inorganic moieties reduces the C-H bonding density in the waveguide materials and thus the optical loss resulting from the C-H bond is reduced. Furthermore, the large polarizability difference between the organic and inorganic moieties possibly induces an increase of the anharmonicity of the C-H bond and thus the NIR spectra could be shifted to the optical window of the light source. The anharmonicity is affected by the bond type, bond angle, and polarizability. It has been shown in the literature [20,21] that the nonlinear optical properties were largely modified by bond anharmonicity. However, it has not been investigated in the linear optical absorption of organic-inorganic hybrid materials.

There are two possible drawbacks for using hybrid materials as waveguide materials: the inhomogeneous phases and the O-H residue, which might result in scattering loss and vibrational absorption loss, respectively. Our laboratory has successfully prepared highly uniform trialkoxysilane-capped poly(methyl methacrylate)(PMMA)-inorganic oxide hybrid optical thin films by the combination of in-situ acid-catalyzed sol-gel process followed by spin-coating, and multi-step curing.[22-25] The thermal and mechanical properties of the prepared PMMA-silica materials were superior than those of the parent PMMA.[22] The refractive index could be tuned through the silica content or the side group on the silica moiety. However, the NIR optical properties and their corresponding waveguides have not been investigated yet. Besides, the previously used reaction scheme[22] was unable to produce homogeneous thick films for optical waveguide applications.

In this study, a modification of the previous reaction scheme[22] was applied to prepare the trialkoxysilane-capped PMMA-silica waveguide materials, as shown in Figure 1. 3-(Trimethoxysilyl) propyl methacrylate (MSMA) was reacted with tetraethoxysilane (TEOS) first, coupled with methyl methacrylate (MMA), and then polymerized to form a precursor solution. The precursor was then spin coated on top of thermal oxide using silicon wafer as the substrate to form an optical planar waveguide. The structures and properties of the resulting hybrid materials were investigated, by FTIR, scanning electron microscopy (SEM), atomic force microspocy (AFM), refractive index, and NIR absorption. The optical loss of the optical planar waveguide was studied. The effects of the silica content on the optical properties are discussed.

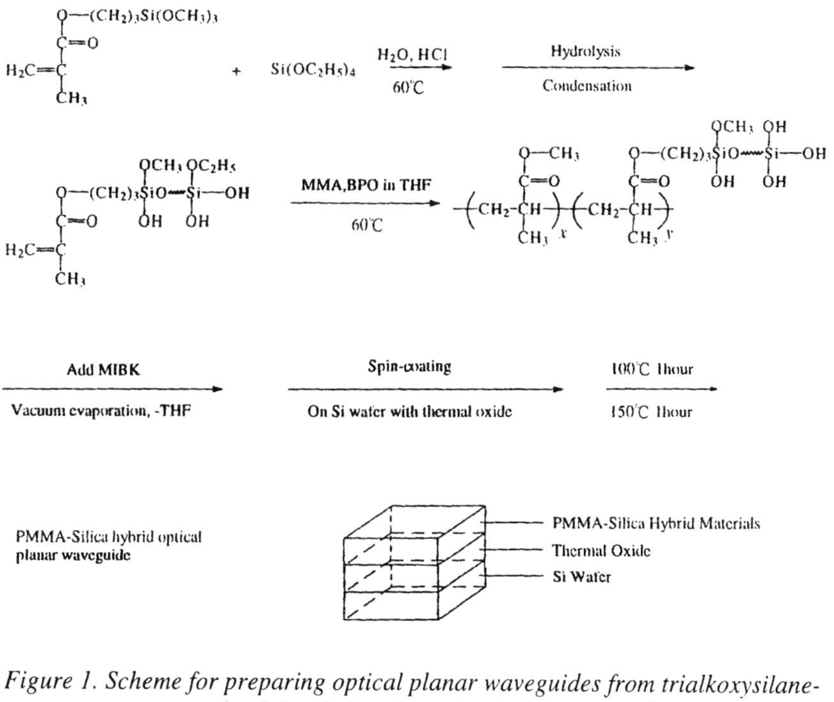

Figure 1. Scheme for preparing optical planar waveguides from trialkoxysilane-capped poly(methyl methacrylate)-silica materials.

Experimental

Materials

Methyl methacrylate (MMA, 99 wt%, Lancaster), 3-(trimethoxysilyl) propyl methacrylate (MSMA, 98 wt%, Aldrich), benzoyl peroxide (BPO, 75 wt%, Aldrich), tetrahydrofuran (THF, 99.9 wt%, Acros), methyl isobutylketone (MIBK, 99.5wt%, Acros), hydrogen chloride (HCl, Yakuri Pure Chemical), and tetraethoxysilane (TEOS, 98 wt%, Aldrich) were used to prepare precursor solutions for the hybrid materials. A PMMA sample for comparing the optical properties with the hybrid materials was purchased from Aldrich (M_w = 75,000).

Synthesis of Precursor Solutions

MSMA and TEOS were dissolved in THF, and a hydrolysis-condensation reaction was carried out for 1 hr to form silica segments with the existence of water (the molar ratio of $H_2O/\{MSMA+TEOS\}$ = 0.55) and HCl (the molar ratio of HCl/TEOS = 0.015). Then, MMA (the molar ratio of MSMA/{MSMA+MMA } = 0.25) and BPO (the mole ratio of BPO to monomer is 0.0375) dissolved in THF were added into the reaction mixture to form a polymeric moiety via copolymerization between MMA and MSMA for two hours. In order to form high quality films for optical waveguides, low boiling point THF solvent and alcohol generated by the hydrolysis and condensation reaction must be removed from the precursor solution to avoid film defects by spin coating. Hence, a high boiling point solvent, MIBK, was added to the precursor solution by the weight ratio of 50% THF. Then, low boiling point species were removed to form a precursor solution by vacuum evaporation. The denotation of **SiX** means the amount (wt%) of TEOS in the reaction mixture for preparing precursor solutions. In this study, the compositions of the prepared hybrid materials were from **Si10** to **Si30**.

Preparation of Optical Planar Waveguides

For the optical planar waveguides, the prepared precursor solution was spin coated on top of thermal oxide on silicon wafer to form the planar waveguide, as shown in Figure 1. Then, it was cured at 100℃ and 150℃ for 1 hr, respectively, to form planar optical waveguides. In this case, the guiding and cladding layers were the prepared hybrid films, and thermal oxide and air, respectively.

Chacterization

Infrared spectra of hybrid thin films were measured by using a Jasco Model FTIR 410 spectrophotometer. An atomic force microscope (Digital Instrument, Inc., Model DI 5000 AFM) was used to probe the surface morphology of the coated films. The microstructure of the hybrid materials was further examined by a field emission scanning electron microscopy (FE-SEM, Hitachi, Model-4000). Thermal properties of hybrid materials such as thermal-decomposition temperature (T_d) and glass-transition temperature (T_g) were obtained by a thermogravimetric analyzer (TGA, TA, Model-951) and a differential Scanning Calorimeter (DSC, TA, Model DSC-910S) under nitrogen atmosphere, respectively. The refractive indices of the prepared hybrid optical films were measured by a Prism Coupler (Metricon, model No.: 2010) at 1319 nm.

For measuring the NIR absorption spectra of the hybrid materials, a sample with a thickness of a few mm was obtained by evaporating the precursor solution in a 20 mL vial under vacuum. The NIR absorption spectra of the hybrid films were obtained using a UV-Visible-NIR spectrophotometer (Jasco, model No.:V-570, resolution : 0.5 nm in the NIR region) in the wavelength range of 1000 to 1600 nm. The interpretation of the NIR spectra was based on the following equation:[19]

$$v_v = \frac{v_1 v[1-x(v+1)]}{1-2x} \qquad (1)$$

Where v_1, v_v, and χ are the fundamental and vth harmonic of the C-H stretching vibration($v = 2,3,4...$) and χ is the anharmonicity, which represents the deviation of the vibration from the harmonic behavior. From the FTIR and NIR spectra, v_1 and v_v were obtained, respectively. Then, χ was calculated from eq.(1). The optical losses of the planar optical waveguides were measured by a designed optical system (manufactured by Center of Measurement and National Standards, Industrial Technology Research Institute, Hsinchu, Taiwan) using a cutback method at 1310 nm.

Results and Discussion

Structure of The Prepared Hybrid Thin Films

Figure 2 shows the FTIR spectra of PMMA and the prepared hybrid materials **Si10** to **Si30** in the range from 400 to 4000 cm^{-1}. Assignment of stretching vibration bands of the C-O-C or Si-O-Si, Si-C, C=O, C-H, and O-H bonds are made at 1110 cm^{-1}, 1270 cm^{-1}, 1730 cm^{-1}, 2943 cm^{-1}, and 3482 cm^{-1}, respectively. These bands are similar to those reported previously.[22] The C=C band at 1650 cm^{-1} completely disappears after curing at 150°C, which suggests that the polymerization of the MMA monomer is complete. The intensity ratio of the Si-O-Si absorption peak at 1078 cm^{-1} over that of C=O (1730 cm^{-1}) enhances with increasing TEOS amount, which suggests increasing silica moiety in the hybrid materials. The absorption peak at 3486 cm^{-1} is a resulte of the incomplete condensation of Si-OH in the hybrid materials due to the insufficient curing temperature of 150^{0}C. The position of the Si-OH band is smaller than 3520 cm^{-1}, which suggests the formation of hydrogen bonding. [22] Although the Si-OH group could not be completely polymerized, the hydrogen bonding of the Si-OH residue group with the carbonyl group makes hybrid materials with high optical transparency.

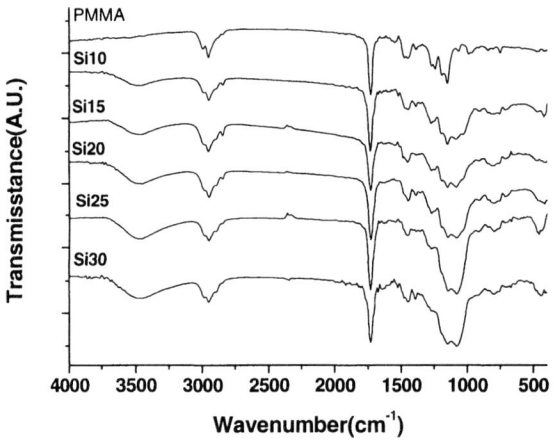

Figure 2. FTIR spectra of PMMA and the prepared hybrid materials, Si10~Si35.

Figure 3 shows AFM image of the prepared hybrid thin film, **Si15**. Highly uniform surface is observed in this figure with the roughness of R_a and R_q equal to 0.354 and 0.457 nm, respectively. It indicates the excellent coupling between the PMMA moiety and silica segment. This can be further supported by the SEM microgragh of Figure 4. There is no apparent silica domain observed in Figure 4. The coated gold particle is below 20 nm for the SEM measurement. Therefore, the domain size of silica segment should be smaller than 20 nm. The AFM and SEM results suggest the homogeneous structure of the prepared hybrid materials.

Properties of Hybrid Thin Films

Figures 5 and 6 illustrate the DSC and TGA curves of the PMMA and the prepared hybrid material, **Si15**, respectively. As shown in Figure 5, the T_g of PMMA is about 108°C but there is no observable T_g in the case of **Si15**. The disappearance of the T_g may be due to the restriction of motion of polymeric chain segments in the crosslinked structure of the prepared hybrid material. The incorporation of the silica segment into PMMA also raises the T_d from 251 °C to

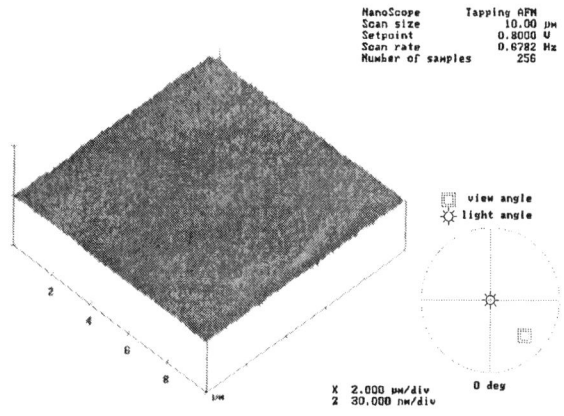

*Figure 3. AFM image of the prepared hybrid material, **Si15**.*

*Figure 4. SEM micrograph the prepared hybrid material, **Si15***

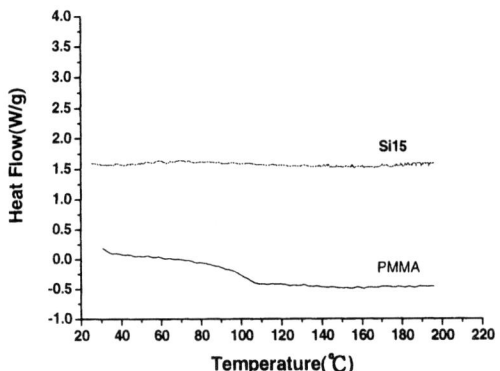

*Figure 5. DSC curves of PMMA and the prepared hybrid material, **Si15**.*

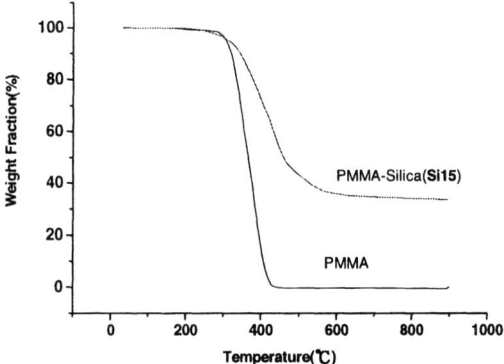

*Figure 6. TGA curves of PMMA and the prepared hybrid material, **Si15**.*

267 °C, as shown in Figure 6. The improvement of thermal properties in the hybrid materials compared to the parent PMMA is mainly a result of the incorporation of the inorganic silica moiety.

The hybrid films had a film thickness ranging from 1.53 to 2.61 μm, as shown in Table 1. The refractive index of the hybrid film at 1319 nm decreases with increasing the silica content since the refractive index of silica segment is lower than that of PMMA. However, it is higher than that of thermal oxide (refractive index = 1.458) and thus core-cladding waveguide structure can be formed in the prepared optical planar waveguides. The optical properties of the hybrid films are shown in Table 1.

Table 1. Properties of PMMA, the hybrid materials, and their planar optical waveguides.

	Silica (wt%)	$h(\mu m)$	n_{TE} at 1319 nm	$\nu_1(nm)$	$\nu_3(nm)$	χ	α_{1310} (dB/cm)
PMMA	0.00	1.53	1.4791	3388	1175	0.0187	0.652
Si10	3.35	2.31	1.4771	3389	1187	0.0231	0.371
Si15	5.22	1.70	1.4743	3389	1192	0.0249	0.299
Si20	7.24	2.29	1.4741	3389	1193	0.0253	0.274
Si25	9.43	2.61	1.4733	3388	1195	0.0261	0.281
Si30	11.81	1.80	1.4718	3388	1195	0.0261	0.264

Figure 7 shows the characteristic NIR absorption spectrum of the prepared hybrid material, **Si15**. There are three major absorption bands shown in Figure 7. The absorption bands in the range of 1100 ~1260 and 1600~1800 nm are assigned to be the third (ν_3) and second (ν_2) harmonic stretching vibration absorption bands of the C-H bond. The absorption band in the range of 1320 to 1530 nm is a result of the combination of the second harmonic stretching vibration (ν_2) and bending vibration (δ) of the C-H bond combined with the second harmonic stretching vibration of O-H bond (ν_2). The positions of the overtone absorption bands shown in Figure 7 are similar to those reported in the literature.[26] Figure 8 shows the NIR absorption spectra of the PMMA and hybrid materials, **Si10 ~ Si30**, in the wavelength range of 1000 to 1600 nm. As shown in the figure, there is red-shifting on both ν_3 and $\nu_2+\delta$ of the C-H bond as the silica content increases from 0 (PMMA) to 11.81 wt% (**Si30**). From Table

1, the fundamental C-H absorption band (ν_1) is almost unchanged with the variation of the silica content. However, the position of ν_3 of the C-H bond increases from 1175 to 1195 nm as the silica content increased. We propose that the red-shifting of the C-H overtone position is due to the variation of anharmonicity (χ in eq.(1)) due to the incorporation of silica moiety into the PMMA. Since the large difference on the polarizability between the PMMA and silica segments, the anharmonicity of the C-H bond increases with increasing silica content. The anharmonicity calculated from eq.(1) increases from 0.0187 to 0.0261 with increasing silica content from 0 to 11.81 wt%, as shown in Table 1. The same red shifting phenomenon also appears in the case of $\nu_2 + \delta$ of the C-H bond. However, the shifting of this absorption band could not be estimated quantitatively by eq.(1) due to the complicated combinational absorption bands.[27] Here, it is proposed that the red-shifting of the $\nu_2 + \delta$ band is also due to the increase of the anharmonicity of the C-H bond. Such shift of optical spectra is very important for the optical window of the light source at 1310 nm, as shown in the optical loss results of next paragraph. There are O-H residues in the prepared hybrid materials, as shown in Figure 2. However, the red shifting of the NIR optical absorption spectra further moves the O-H vibration overtone around 1430 nm, which is out of the optical window of the light source at 1310 nm.

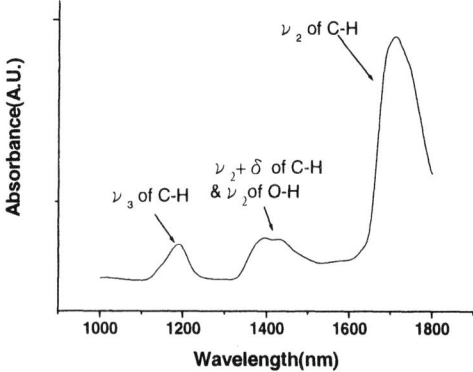

Figure 7. The NIR absorption spectrum of the prepared hybrid materias, Si15, in the wavelength range of 1000 to 1600 nm.

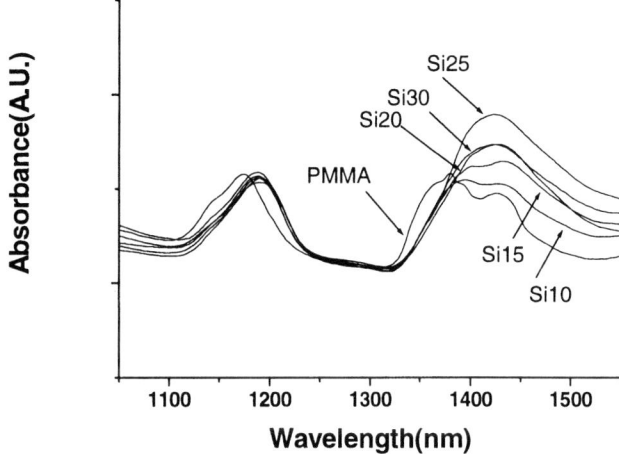

Figure 8. The NIR absorption spectra of PMMA and the prepared hybrid materials in the wavelength of 1000 to 1600 nm.(Reproduced from reference 28. Copyright 2003 Tamkang Journal of Science and Engineering)

Optical Losses of the Hybrid Planar Optical Waveguides

Figure 9 shows the variation of the optical losses of the planar optical waveguides at 1310 nm, which were fabricated from PMMA and the hybrid films of **Si10~Si30**. The optical loss reduces from 0.652 to 0.264 dB/cm with increasing silica content from 0 to 11.81 %. The loss of the hybrid film based waveguide is less than the parent PMMA-based planar waveguide. As shown in Figure 8, the optical transparency at 1310 nm is increasing with the silica content due to the red shifting of the absorption spectra and thus results in reducing optical loss. Another factor in the optical loss result is the reduction in the C-H bonding density in the material structure by increasing the silica content. It also suggests that increasing the silica content does not result in a decrease of optical quality and thus scattering loss is not significant in the prepared optical planar waveguides. The conventional approach for reducing optical loss in organic polymers is replacing the C-H bond with C-F or C-D bond. The current hybrid materials approach can also achieve the same goal of reducing optical loss. The optical losses of the present waveguides are among the lowest optical loss found in planar waveguide prepared from sol-gel materials.[7-11] Hence, the hybrid materials have potential applications as waveguide materials for optical communication.

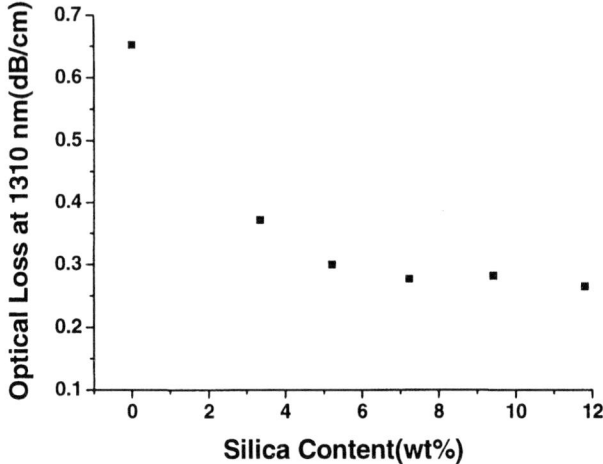

Figure 9. Variation of Optical loss of the prepared optical planar waveguide at 1310 nm with the silica content.

Conclusions

Tunable NIR properties were obtained from the trialkoxysilane-capped poly (methyl methacrylate)-silica hybrid materials. Highly homogeneous and good thermal resistant hybrid materials. The incorporation of the silica segment into the PMMA matrix induces an increase of the anharmonicity of the C-H bonds and results in the red-shift of the absorption spectra in the NIR region. Hence, the optical transparency at 1310 nm was adjusted by the silica content in the hybrid material. The optical loss of the prepared optical waveguides was reduced from 0.652 to 0.264 dB/cm with increasing silica content. Shifting of the optical spectra is due to both increase in the anharmonicity of C-H bond and reduction the C-H bonding density. The prepared waveguides have potential applications in optical communication.

Acknowledgement

The authors thank National Science Council (NSC), Ministry of Economic Affairs, and Department of Education of Taiwan, R.O.C. for the financial supports of this work.

References

(1) Novak, B. M. *Adv. Mater.* **1993**, *5*, 422.
(2) Allcock, H. R. *Adv. Mater.* **1994**, *6*, 106.
(3) Jiang, H.; Kakkar, A. K. *Adv. Mater.* **1998**, *10*, 1093.
(4) Sanchez, C.; Lebeau, B. *MRS Bull.* **2001**, *26*, 377.
(5) Wen, J.; Wilkes, G. L. *Chem. Mater.* **1996**, *8*, 1667.
(6) Yoshida, M.; Lal, M.; Kumar, N. D.; Prasad, P. N. *J. Mater. Sci.* **1997**, *32*, 4047.
(7) Yoshida, M.; Prasad, P. N. *Chem. Mater.* **1996**, *8*, 235.
(8) Xu, C.; Eldada, L.; Wu, L. C.; Norwood, R. A.; Schacklette, L. W.; Tardley, J. T. *Chem. Mater.* **1996**, *8*, 2701.
(9) Yamada, N.; Yoshinaga, I.; Katayama, S. *J. Appl. Phys.* **1999**, *85*, 2423.
(10) Ribeiro, S. J. L.; Messaddeq, Y.; Goncalves, R. R.; Ferrari, M.; Montagna, M.; Aegerter, M. A. *Appl. Phys. Lett.* **2000**, *77*, 3502.
(11) Motakef, S. ; Suratwala, T.; Roncone, R. L.; Boulton, J. M.; Teowee, G.; Neilson, G. F.; Uhlmann, D. R. *J. Non-Cryst. Solids* **1994**, *178*, 31.
(12) Chu, A. K.; Lee, K. M.; Pong, B. J.; Lin, C. J.; Ho, W. J.; Shih, T. T. *Electron. Lett.* **2000**, *36*, 1539.
(13) Lee, T. W.; Park, O.; Yoon, J.; Kim, J. J. *Adv. Mater.* **2001**, *3*, 211.
(14) Huang, W. Y.; Ho, S. W.; Kwei, T. K.; Okamoto, Y. *Appl. Phys. Lett.* **2002**, *80*, 1162..
(15) Tang, J.; Wang, C.; Wang, Y.; Sun, J.; Yang, B. *J. Mater. Chem.* **2001**, *11*, 1370.
(16) Huynh, W. U.; Dittmer, J. J.; Alivisatos, A. P. *Science* **2002**, *295*, 2425.
(17) Guloy, A. M.; Tang, Z.; Miranda, P. B.; Srdanov, V. I. *Adv. Mater.* **2001**, *13*, 833.
(18) Kagan, C. R.; Mitzi, D. B.; Dimitrakopoulos, C. D. *Science* **1999**, *286*, 945.
(19) Groh, W. *Makromol. Chem.* **1988**, *189*, 2861.
(20) Marder, S. R.; Beratan, D. N.; Cheng, L. T. *Science* **1991**, *252*, 103.
(21) Risser, S. M.; Beratan, D. N.; Marder, S. R. *J. Am. Chem. Soc.* **1993**, *115*, 7719.
(22) Chen, W. C.; Lee, S. J.; Lee, L. H.; Lin, J. L. *J. Mater. Chem.* **1999**, *9*, 2999.
(23) Chen, W. C.; Lee, S. J. *Polym. J.* **2000**, *32*, 67.
(24) Chang, C. C.; Chen, W. C. *J. Polym. Sci. Polym. Chem.* **2001**, *39*, 3419.
(25) Lee, L. H.; Chen, W.C. *Chem. Mater.* **2001**, *13*, 137-1142.
(26) Ando, S.; Matsuura, T.; Sasaki, S. *J. Am. Chem. Soc.* **1994**, *20*, 304.
(27) Burneau, A.; Carteret, C., *Phys. Chem. Chem. Phys.* **2000**, *2*, 3217.
(28) Lee, C.-H.; Chen, W.-C. *Tamkang J. Sci, Eng.* **2003**, *6*, 73.

Chapter 24

Photochemical Control of Reflection Colors of Glass-Forming Non-Polymeric Cholesterics with Azobenzene Chromophore

Nobuyuki Tamaoki and Masaya Moriyama

Molecular Function Group, National Institute of Advanced Industrial Science and Technology (AIST), Central 5, Higashi 1–1–1, Tsukuba, Ibaraki, Japan

A new series of non-polymer photochromic liquid crystalline compounds, **2, 3** and **4**, which form glassy solids maintaining the liquid crystalline molecular alignment has been synthesized. Their optical properties such as cholesteric reflections are controlled by the E-Z photoisomerization of the azobenzene unit in liquid crystalline phases. Upon quenching, the optical properties are fixed in the glassy state. Because the glassy state of some compounds is stable at room temperature, the compounds can be utilized for the construction of rewritable information-storing materials.

Introduction

The photochromic reaction in liquid crystals (LCs) is one of the effective stimuli inducing modification of the LC optical properties. For example, photochromic reactions in cholesteric liquid crystals (CLCs) can induce changes in the cholesteric iridescent reflections (reflected colors) due to the helical molecular alignment.(*1, 2*) Since the molecular alignment of the CLC is dependent on the molecular composition, the photochromic reactions transforming the molecular structure can change the cholesteric reflection. This photo-induced modification can be utilized for optical information recordings. Meanwhile, compounds showing the glassy solid state in LC phases have advantages of stable information storage compared to those with just a normal LC phase having fluidity. In general, polymeric LC materials can easily form glassy solids that maintain the LC molecular alignment, though they are inferior to low molecular weight materials in their response time to stimuli.

Recently, it was reported that dicholesteryl compounds with ca. 1000 molecular weight such as **1** have properties showing a CLC phase with good response to stimuli, forming a stable glassy solid that maintains the cholesteric helical alignment of the molecules (*3*). The mixtures of these kinds of CLC compounds and photochromic compounds can be effective materials for color information recordings (*2*).

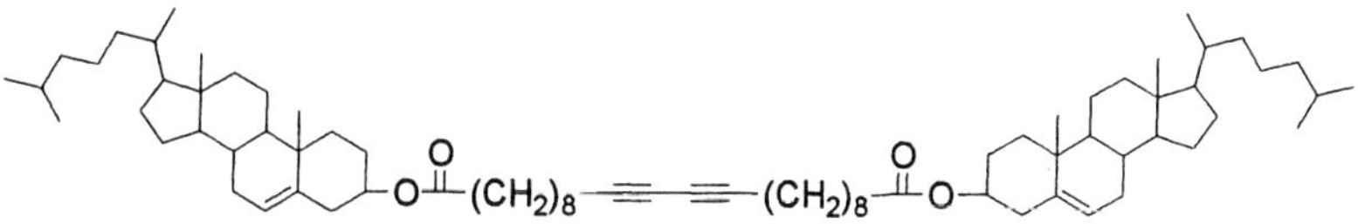

1

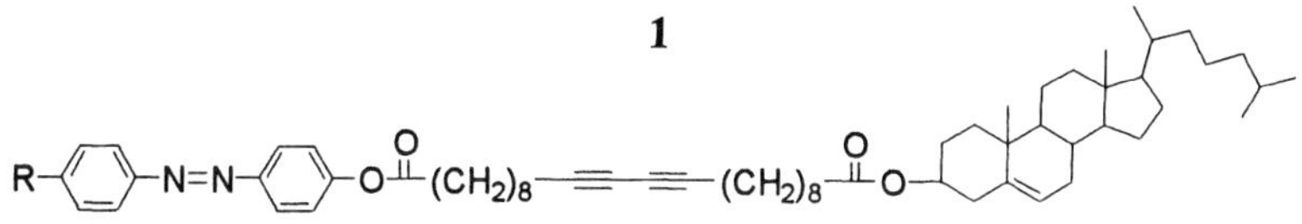

2 (R = H), 3 (R = C$_7$H$_{15}$)

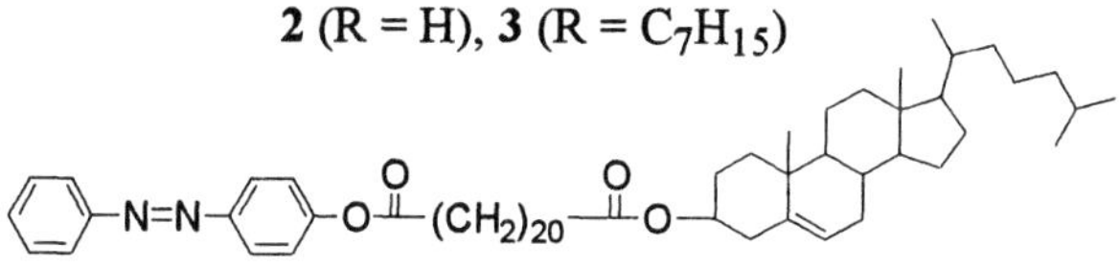

4

In this study, photochromic, non-polymeric glass-forming LC compounds, **2, 3** and **4**, consisting of a cholesteryl group and an azobenzene moiety connected with a long flexible aliphatic spacer, were prepared and their properties were studied. Although analogous compounds, whose central parts are shorter, have been extensively investigated from the viewpoints of sol-gel formations (*4*), nonlinear optics (*5*) and photoswitching dopnats (*1i*), the study of the glass-forming LC analogues have never been reported. Additionally, the repetitive photo-control and storage of cholesteric iridescent reflections of single-component materials (not mixtures) have been first achieved in this study.

Experimental

All the compounds were obtained by esterification of the corresponding dicarboxylic acids with cholesterol and the corresponding hydroxyazobenzenes. The phase transition temperature was observed by DSC measurements (±2 °C min^{-1} rate of temperature change). The reflection spectra were measured for the films by dipping the sample in ice water before or after irradiation at the liquid crystalline temperature.

Results and Discussion

Compounds **2** and **4** showed CLC phases with iridescent reflection, and **3** showed only a smectic phase with no iridescent reflection. DSC thermograms of the compounds are shown in Figures 1-3. All the materials formed glassy solids which maintained the LC structure by rapid cooling from the LC phase. Their phase transition temperature observed by DSC measurements (±2 °C min^{-1} rate of heating or cooling) and λ_{max} of the cholesteric reflection band are summarized in Table 1. For cholesterics **2** and **4**, the λ_{max} shifted to a small value with an increase in the temperature ($d\lambda_{max}/dT<0$). The reflection region of **2** having a diyne unit was wider than that of **4** having a long n-alkyl chain. Azobenzenes substituted at the 4,4' positions with appropriate alkyl chains show a smectic phase (*6*), accordingly, it is reasonable that **3** shows a smectic phase. The glass transition temperature of all compounds is 30-45 °C.

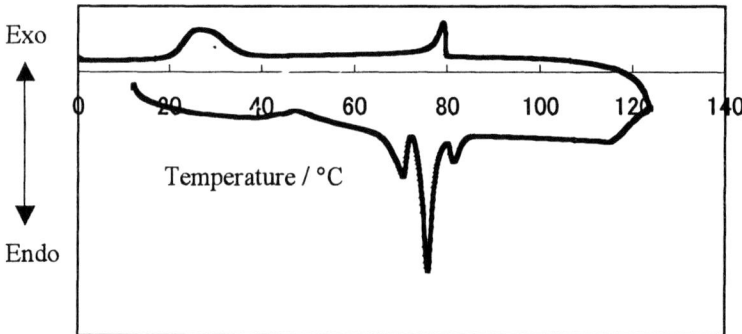

Figure 1. DSC thermogram of compound **2**.

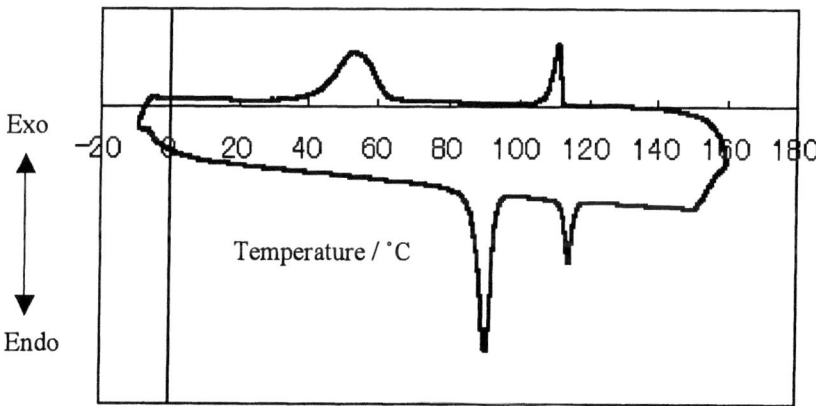

Figure 2. DSC thermogram of compound **3**.

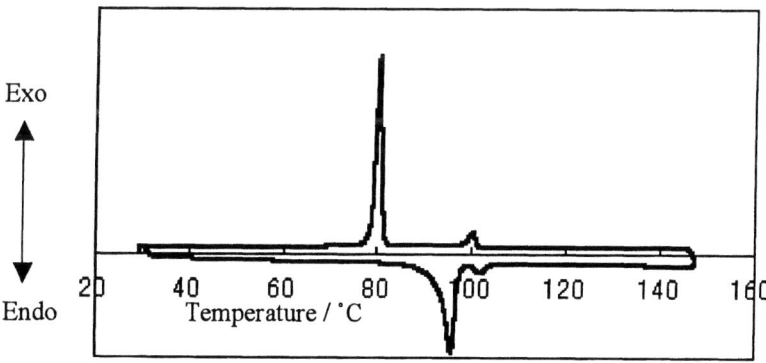

Figure 3. DSC thermogram of the compound **4**.

Table 1. Phase transition temperatures and λ_{max} of cholesteric reflection of the photochromic LC compounds

Compound	Phase transition Temperature/°C[a]	λ_{max}/nm (Temperature/°C)
2	Cr 70 Sm 76 Ch 82 I	696(50), 595(60)
	I 80 Ch and Sm 36[b] Cr	514(70), 463(80)
3	Cr 90 Sm 114 I	-
	I 111 Sm 31 Cr[c]	
4	Cr 97 Ch 104 I	533(88), 515(90)
	I 101 Ch 81 Cr	485(95), 457(100)

[a]I: isotropic phase, Ch: cholesteric phase, Sm: smectic phase, Cr: crystal. [b]A broad signal was observed. Ch-Sm and Sm-Cr transitions could not be separated. [c]The crystallization temperature is much changed depending on the cooling rate. Reproduced with permission from reference 8. Copyright 2001 The Chemical Society of Japan.

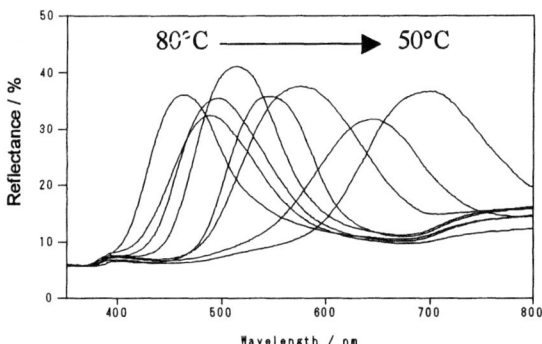

Figure 4. Change of reflection spectra of the thin film of the dimesogenic compound 2 between a pair of glass plates.

In the cholesteric phase of the compounds 2 and 4, a reflection band due to the helical molecular superstructure was observed, the position of which depends on the temperature. Decreasing the temperature shifted the reflection band toward longer wavelength (Table 1 and Figure 4). The trend of the shift of the reflection band, $d\lambda_{max}/dT<0$, is usually explained by the formation of the cybotactic smectioc domain at lower temperature.

Upon irradiation at 366 nm, which induced E/Z isomerization of the azobenzene units, the λ_{max} of the cholesteric reflections of **2** and **4** shifted to a shorter wavelength as shown in Figure 5 (*7*). The *cis* isomer would cause disorder in the cholesteric molecular arrangement, with the result that the λ_{max} shifted to a disorder state which is the same as the thermally disordered state of the CLC compounds (*1g*). The cholesteric reflections including the changed ones due to the photoisomerization were fixed in the glassy solid obtained by the rapid cooling from the CLC phase. The fixed reflection did not change upon irradiation of the glassy solid state. Due to the rigidity of the glassy solid matrix, the molecular alignment would not be disturbed by the isomerization of the azobenzenes. However, raising the temperature above the isotropic one induced Z/E thermal back isomerization within 1 min and the CLC alignment returned to its initial state. Meanwhile, prolonged irradiation induced a phase transition from the cholesteric to isotropic phase (Figure 5(c) and Figure 6(b)). Cooling from the isotropic phase produced crystals which showed no reflection but optical light scattering. For the smectic LC compound **3**, although the difference in the reflection could not be produced, switching of light transmission was possible. As compound **3** spontaneously aligned homeotropically, the transparent and colorless films could be prepared. In the films, the E/Z photoisomerization induced a phase transition from the homeotropic smectic to the isotropic phase and a contrast formed and was fixed after rapid cooling (Figure 6(c)).

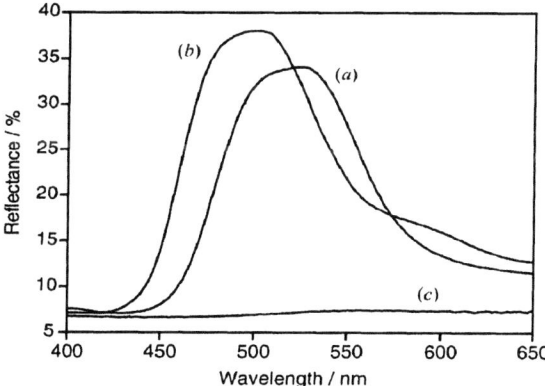

Figure 5. Reflection spectra of the films of **2** obtained by dipping the samples in ice water before and after irradiation at 70 °C. (a) Before irradiation. (b) After irradiation with 366-nm light for 10 s. (c) After irradiation for 30 s. (Reproduced with permission from reference 8. Copyright 2001 The Chemical Society of Japan.)

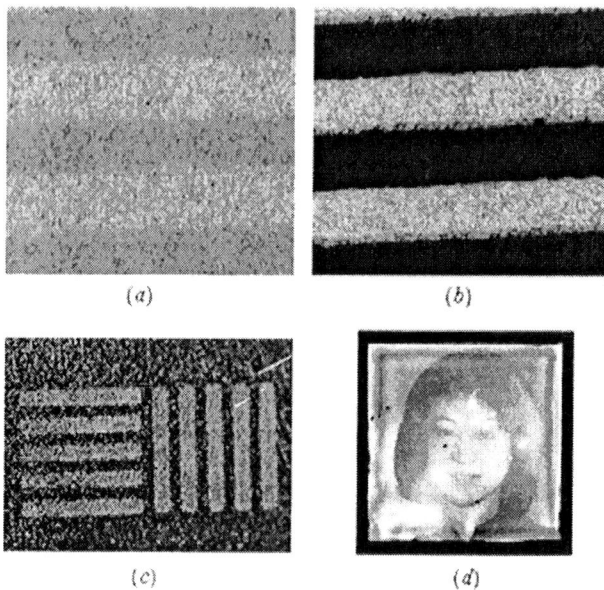

Figure 6. (a) Microscopic image (reflection mode) of the solid film of **2** obtained after irradiation for 10 s at 70 °C through a photomask with 200μm lines and spaces. The non-irradiated area was green, while irradiated area became darker. (b) Microscopic image (reflection mode) of the solid film of **2** obtained after irradiation for 30 s at 70 °C through a photomask with 200μm lines and spaces. The non-irradiated green area maintained the cholesteric structure. As the black area irradiated is a crystal, no reflection is observed (c) Microscopic image (transmission mode and cross polarized) of the solid films of **3** obtained after irradiation for 30 s at 90 °C through a photomask with 70μm lines and spaces. The black area is not-irradiated and has a homeotropic molecular alignment. Irradiated area is crystalline. (d) Displayed image prepared by annealing the cholesteric films of **2** to 73 °C after irradiation in glassy solid state (r.t.) through a photomask. The blue image is reproduced on the green background. (Reproduced with permission from reference 8. Copyright 2001 The Chemical Society of Japan.)

On the other hand, although the series of compound **1** show a monotropic CLC phase only during the cooling process, **2, 3** and **4** show an enantiotropic LC phase. Hence, we demonstrated the latent image recording based on the E/Z isomerization of the azobenzenes in the glassy solid states. Temperature is regarded as a "gate" for transmission of the information written by the photoisomerization. When the gate is opened by heating, the photochromic dopants direct the surrounding liquid crystalline molecules to realign. Upon irradiation (λ=366 nm) of the glassy solid films through a photomask at room temperature, no change occurs in the reflected color because E/Z isomerization of the azobenzene in the rigid glass can not affect the molecular alignment. However, the latent image is certainly drawn in the glassy solid by the E/Z isomerization of the azobenzene. When the films are then put on the hot stage controlled at the cholesteric temperature, the latent image is quickly displayed (within a few seconds). Furthermore, the displayed image is fixed in the glassy solid by subsequent rapid cooling. Figure 6(d) shows a photograph prepared by such a method. A similar method can be used for polymer CLCs, but the display speed is slower after annealing, because the realignment rate of the polymer molecules is slow (*1d*). For the image recording application, this method using **2, 3**, and **4** would have an advantage in saving energy at drawing and display of the image.

Conclusions

We have prepared a new series of photochromic LC compounds which formed glassy solids, maintaining the LC molecular alignment although they were non-polymeric materials. Their macroscopic optical properties based on molecular alignment were changed and fixed by the control of photochromic reaction and temperature. The controllable optical properties can be utilized for energy-saving information recordings.

References

(1) (a) Sackmann, E. *J. Am. Chem. Soc.* **1971**, *93*, 7088. (b) Yokoyama Y.; Sagisaka, T. *Chem. Lett.* **1997**, 687. (c) Vicentini, F.; Cho, J.; Chien, L.-C. *Liq. Cryst.* **1998**, *24*, 483. (d) Bobrovsky, A. Y.; Boiko, N. I. ; Shibaev, V. P. *Liq. Cryst.* **1998**, *25*, 679. (e) Brehmer, M.; Lub, J.; van de Witte, P. *Adv. Mater.* **1998**, *10*, 1438. (f) Bobrovsky, A. Y.; Boiko, N. I.; Shibaev, V. P. *Adv. Mater.* **1999**, *11*, 1025. (g) Ruslim C.; Ichimura, K. *J. Phys. Chem. B* **2000**, *104*, 6529. (h) Lee, H.-K.; Doi, K.; Harada, H.; Tsutsumi, O.; Kanazawa, A.; Shiono, T.; Ikeda, T. *J. Phys. Chem. B.* **2000**, *104*, 7023. (i) George, M.; Mallia, V. A.; Antharjanam, P. K. S.; Saminathan, M.; Das, S. *Mol. Cryst. Liq. Cryst.* **2000**, *350*, 125.

(2) (a) Tamaoki, N.; Song, S.; Moriyama, M.; Matsuda, H. *Adv. Mater.* **2000**, *12*, 94. (b)Moriyama, M.; Song, S.; Matsuda, H.; Tamaoki, N. *J. Mater. Chem.* **2001**, *11*, 1003.

(3) (a) Tamaoki, N.; Parfenov, A. V.; Masaki, A.; Matsuda, H. *Adv. Mater.* **1997**, *9*, 1102. (b) Tamaoki, N.; Kruk, G.; Matsuda, H. *J. Mater. Chem.*

1999, *9*, 2381. (c) Kruk, G.; Tamaoki, N.; Matsuda, H.; Kida, Y. *Liq. Cryst.* **1999**, *26*, 1687.
(4) (a) Murata, K.; Aoki, M.; Shinkai, S. *Chem. Lett.* **1992**, 739. (b) Murata, K.; Aoki, M.; Suzuki, T.; Harada, T.; Kawabata, H.; Komori, T.; Ohseto, F.; Ueda, K.; Shinkai, S. *J. Am. Chem. Soc.* **1994**, *116*, 6664. (c) Kawabata, H.; Murata, K.; Harada, T.; Shinkai, S. *Langmuir.* **1995**, *11*, 623.
(5) George, M.; Das, S. *Chem. Lett.* **1999**, 1081.
(6) van der Veen, J.; de Jeu, W. H.; Grobben, A. H.; Boven, J. *Mol. Cryst. Liq. Cryst.* **1972**, *17*, 291.
(7) Upon UV irradiation even at 254 nm, diyne compounds **2** and **3** did not topochemically polymerize in the crystalline state. Therefore, the photo-controllable and fixative optical properties of **2** and **3** are not attributed to the topochmical polymerization.
(8) Moriyama, M.; Tamaoki, N. *Chem. Lett.* **2001**, 1142.

Chapter 25

Photonic Papers: Colloidal Crystals with Tunable Optical Properties

Hiroshi Fudouzi, Yu Lu, and Younan Xia*

Department of Chemistry, University of Washington, Seattle, WA 98195
*Corresponding author: xia@chemistry.washington.edu (email)

This paper demonstrates the fabrication of colloidal crystals with tunable optical properties and their utilization as photonic papers for displaying colored letters and patterns. In a typical procedure, monodispersed spherical colloids were assembled into a three-dimensional crystal, followed by infiltration and curing of an elastomer such as poly(dimethylsiloxane). When an ink (i.e., any liquid capable of swelling the elastomer) was applied to the surface of this crystal, the lattice constant (and thus the color of Bragg-diffracted light) was changed. If the difference between the colors of the initial and final states is sufficiently large to be distinguishable by the naked eye, this system can be used to write and print colored patterns with an edge resolution as high as ~50 μm.

Colloidal crystals are long-range ordered lattices assembled from spherical colloids such as polymer latexes and silica beads *(1)*. The ability to organize colloidal particles into crystalline lattices has enabled us to obtain interesting (and often useful) functionality not only from the constituent material of the colloidal particles, but also from the periodic structure intrinsic to a crystalline lattice. The beautiful, iridescent color of an opal, for example, originates from the optical diffraction of a crystalline lattice made of silica colloids that display no color by themselves *(2)*. In recent years, colloidal crystals have been actively explored as photonic band gap (PBG) materials to control the propagation of electromagnetic waves in the three-dimensional space *(3)*. They have also been demonstrated for use as functional elements in fabricating diffractive optical devices. The wavelength of light diffracted from the surface of a colloidal crystal is determined by the Bragg equation *(4)*:

$$m \cdot \lambda = 2 \cdot d_{hkl} \cdot (n_a^2 - \sin^2\theta)^{1/2} \tag{1}$$

where m is the diffraction order (e.g., the 1st or 2nd order), λ is the wavelength of the diffracted light (or the so-called stop band), n_a is the mean refractive index of the crystalline lattice, d_{hkl} is the interplanar spacing along the [hkl] direction, and θ is the angle between the incident light and the normal to the (hkl) planes. The average refractive index is related to the filling ratio, f, of spherical colloids in the crystalline lattice by the following equation:

$$n_a = f \cdot n_1 + (1-f) \cdot n_0 \tag{2}$$

where n_1 and n_0 are the refractive indices of the colloids and the surrounding medium, respectively. Equation (1) suggests that the wavelength of light Bragg-diffracted from the surface of a colloidal crystal is dependent upon the angle (θ) between the incident light and the normal to the (hkl) planes, average refractive index of the crystal (n_a), and the lattice constants. In particular, any variation in the lattice constants will lead to an observable shift in the stop band position, and thus the color displayed by the crystal. In this regard, a colloidal crystal is able to serve as an optical sensor to monitor, measure, and display environmental variation in terms of change in color that can be easily and clearly visulized by the naked eye.

By embedding colloidal crystals in appropriate polymer hydrogels, Asher *et al.* have demonstrated the fabrication of temperature-, pH-, and ion-responsive optical sensors *(5)*. The hydrogel colloidal crystals developed by Hu *et al.* and Lyon *et al.* have enabled them to tune the color of diffracted light by varying temperature or by applying an electric field *(6)*. Stein *et al.* have explored the use of a ceramic inverse opal (fabricated by replica molding against the lattice of a colloidal crystal) in detecting various organic solvents due to the changes in mean refactive index *(7)*. Sato *et al.* and Caruso *et al.* recently illustrated that the reversible color tuning of a colloidal crystal could be adopted to detect the binding events of a biological species *(8)*. Ford *et al.*, Foulger *et al.*, and Tsutsui *et al.* demonstrated that colloidal crystals embedded in thin films of appropriate polymers could serve as mechanical sensors to monitor *in situ* strains caused by uniaxial stretching or compressing *(9)*. In all of these demonstrations, the lattice constants (and thus the color displayed by the colloidal crystal as a result of Bragg diffraction) changed in response to the environmental change(s). In some cases, the variation in color could be readily picked up by the naked eye. Here we describe another application based on the same mechanism *(10)*, in which colloidal crystals with tunable colors were exploited for use as photonic papers.

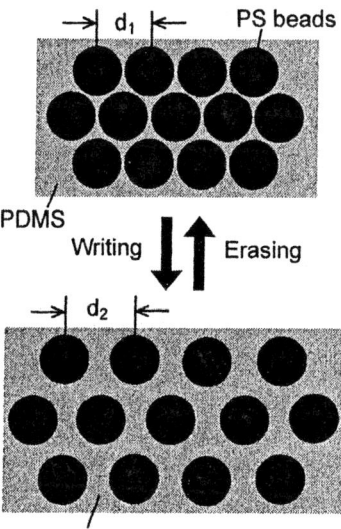

Figure 1. Schematic illustration of the mechanism by which the color diffracted from the crystalline lattice of a photonic paper is reversibly changed: the lattice constant (and thus the wavelength of diffracted light) is changed by swelling the PDMS matrix with an "ink". The PDMS matrix shrinks back to its original state once the ink molecules have completely evaporated.

Figure 1 illustrates how this new type of paper allows for color writing with a colorless ink. The paper is typically a crystalline lattice of polystyrene (PS) beads whose voids are completely filled with poly(dimethylsiloxane) (PDMS). The ink is a silicone fluid or any other liquid capable of swelling PDMS *(11)*. As the ink is applied to the surface of the paper, the position of stop band will be shifted too a new wavelength, and thus the color displayed by this crystalline lattice will change. If the colors of these two states are sufficiently different to be distinguishable by the naked eye, one can use their contrast to achieve color writing with materials that are colorless by themselves. As the ink evaporates, the PDMS will shrink back to its original state and the colored patterns will be automatically erased. The ability to write and erase colored patterns reversibly represents probably the most attractive feature associated with this new type paper. Our preliminary results indicate that such a reversible color change could be repeated for more than 10 times without observing any deterioration in the quality of color displayed by the colloidal crystal. If, necessary, the color could also be fixed by cross-linking the ink molecules to the PDMS network.

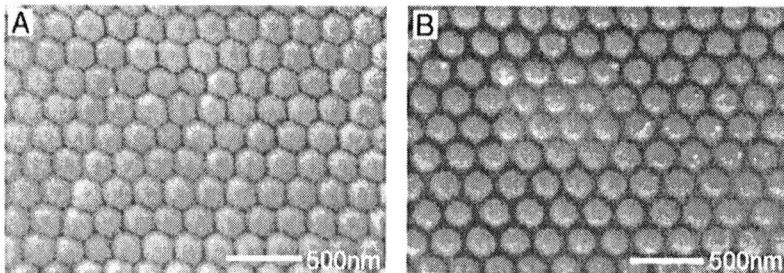

Figure 2. SEM images of a photonic paper (made of 202-nm PS beads) before and after its PDMS matrix had been swollen with a silicone fluid that contained vinyl-terminated siloxane oligomers. The swollen sample was fixed by thermally cross-linking the silicone oligomers with the PDMS network.

Figure 2A shows an SEM image of the (111) plane of a photonic paper, with 202-nm PS beads being arranged in an ABC stacking of the cubic-close-packed (ccp) lattice. The voids of this crystalline lattice (~26% by volume) had been infiltrated with PDMS. The (111) planes of this lattice were oriented parallel to the solid support, which was also consistent with optical diffraction studies. The PS beads were in physical contact within the (111) plane. Figure 2B shows an SEM image of the same colloidal crystal after it had been swollen with a vinyl-terminated silicone fluid that was subsequently cross-linked with the PDMS backbone. In this case, the PS beads were further separated from each other within the (111) plane, and the center-to-center distance between PS beads had been increased from 206 to 230 nm (or by 11.6%) as a result of the swelling of the PDMS matrix.

Plate 1 shows the transmission spectra of a photonic paper assembled from 175-nm PS beads, before (a) and after (b-e) it has been swollen with silicone fluids having different molecular weights (and viscosities). The incident light was aligned perpendicular to the (111) plane of this colloidal crystal for all measurements. The magnitude of swelling is mainly determined by the strength of interactions between the PDMS network and the silicone fluid, and in this case, by the molecular weight of silicone oligomers contained in the fluid *(12)*.

As the molecular weight of silicone oligomers was decreased, we observed two major changes for the stop band associated with the crystalline lattice: its midgap position was continuously shifted towards longer wavelengths; and its attenuation was gradually increased. For this particular paper made of 175-nm PS beads, its stop band could be changed to cover the spectral region from 450 to 580 nm. Accordingly, the color diffracted from the surface of this photonic paper could be easily tuned from blue through green to orange.

The ink could be applied to the surface of a photonic paper using a number of ways to generate colored patterns. For example, we could directly spot the ink on the surface of a paper using a Pilot pen. Plate 2A shows a photograph of two letters (in green color) that were written with spotted dots of octane. The colloidal crystal was assembled from 175-nm PS beads, followed by infiltration with PDMS elastomer. The color displayed by the pristine surface of this paper was violet at normal incidence, while the regions that were covered and then swollen by octane diffracted green light. As octane started to evaporate, these letters would disappear within a period of 5 min. In another set of experiments, we also generated letters and other test patterns on the surface of a paper by delivering the ink with an elastomeric PDMS stamp widely used in microcontact printing (13). In a typical procedure, the surface of a PDMS stamp was inked with a silicone fluid by wiping with a Q-tip and then brought into contact with the surface of a paper for a few seconds. Only silicone fluid on the raised regions of the stamp could be transferred onto the surface of the paper, just as one would have experienced with the microcontact printing process. Plate 2B shows a photograph of letters printed using a rubber stamp. The paper was crystallized from 202-nm PS beads, and it exhibited a green color when viewed at normal incidence. The regions swollen by the silicone fluid (T05) displayed a red color. It is worth mentioning that these pattern could have an edge resolution as high as ~50 μm, a feature that will hold the promise for color writing/printing at reasonably high resolution. In addition, the crystallization method described here was able to routinely generate photonic papers with a uniform color over areas as large as 25 cm^2, and it was also possible to fabricate the photonic papers on transparency films instead of rigid substrates.

In addtion to color tuning within the visible region (where the performance of writing strongly depends on the contrast between the two visible colors), it is also possible to fabricate photonic papers whose colors can be switched between an invisible state (e.g., with the stop band located in the region either below 400 nm or above 780 nm) and a visible one by selecting colloids with appropriate sizes and solvents with appropriate properties. For such systems, the photonic papers would appear colorless while the recorded patterns would be brightly colored, or vice versa. Figure 3A shows the transmission spectra of a photonic paper made of 155-nm PS beads. As shown by curve-i, this paper exhibited a stop band in the UV region with its stop band position located at ~380 nm. After

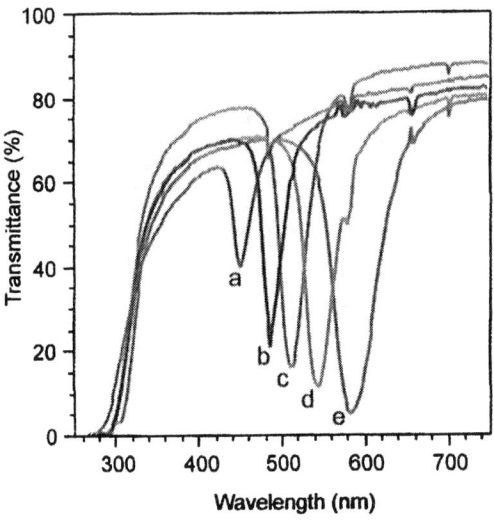

Plate 1. UV-Vis transmission spectra taken from a photonic paper (assembled from 175-nm PS beads) before (curve, a) and after (curves, b-e) it had been swollen with silicone liquids of different molecular weights (and viscosities): b) T12 ($M_w=2000$, 20 cSt), c) T11 ($M_w=1250$, 10 cSt), d) T05 ($M_w=770$, 5 cSt), and e) T00 ($M_w=162$, 0.65 cSt). (See Page 7 of color insert.)

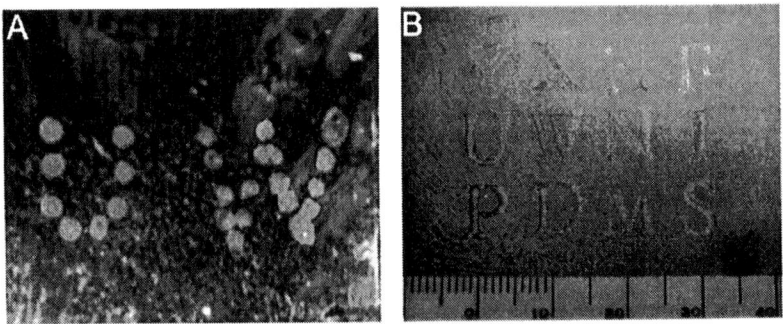

Plate 2. (A) A photograph of two dotted letters written on a photonic paper by delivering octane droplets to its surface using a Pilot pen. (B) A photograph of letters formed on the surface of a photonic paper by stamping with a silicone fluid (T11, $M_w=1250$). This paper was assembled from PS beads of 202 nm in diameter, and it exhibited a green color when viewed at normal incidence. Note that the pattern shown in (B) had an edge resolution better than 50 μm. (See Page 8 of color insert.)

this paper had been swollen with octane, the stop band was shifted to the visible region (~540 nm, green color, see curve-*ii*). Accordingly, the paper changed its appearance from colorless to green after it has been swollen with the octane ink. Figure 3B shows the transmission spectra of another photonic paper that was made of 202-nm PS beads embedded in PDMS matrix. This photonic paper, see curve-*iii*, exhibited a stop band in the visible region around 590 nm. The curve-*iv* shows the transmission spectrum of this paper after its surface had been coated with silicone fluid (DMS-T00, 0.65 cSt). As the PDMS matrix was swollen, the stop band of this photonic paper was relocated to 775 nm, which is essentially at the crossline between the visible and near IR regions. This paper also displayed a second-order Bragg diffraction peak at ~390 nm from its (111) planes, which was located in the UV region. As a result, the initial color of this photonic paper was red, and it would change to a transparent, colorless state after it had been swollen with the silicone fluid. In reality, these two new systems could lead to the writing and display of messages with high contrasts because either the pattern or the background will be colorless.

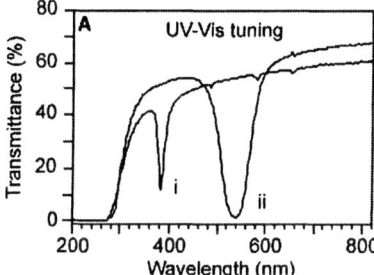 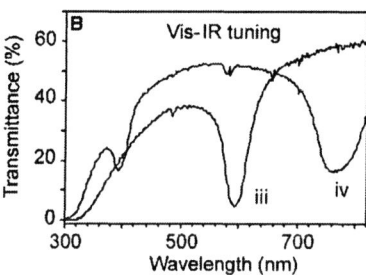

Figure 3. (A) UV-visible transmission spectra taken from a paper made of 155-nm PS beads before (curv-i) and after (curve-ii) it had been swollen with iso-propanol. The film changed color from colorless (in the UV region) to green. (B) UV-visible transmission spectra taken from a photonic paper before (curve-iii) and after (curve-iv) it had been swollen with iso-propanol. The film changed color from orange to colorless (within the near IR region). The paper was assembled from 202-nm PS beads, and its PDMS matrix had been swollen and grafted with the vinyl-terminated silicone fluid before it was swollen with the iso-propanol ink.

In summary, we have demonstrated a photonic paper system by embedding colloidal crystals of polymer beads in an elastomeric matrix and by judicially selecting liquids capable of swelling the matrix. This new paper system could be combined with conventional tools to generate colored letters and patterns by using colorless materials only. This system may provide an alternative route to the realization of reusable papers *(14)* or recording media where no pigment will be required for color displaying at relatively high edge resolutions. As we have demonstrated, colored patterns with an edge resolution of ~50 μm could be routinely generated by printing the ink with a rubber stamp.

Acknowledgment

This work was supported in part by an AFOSR-MURI grant awarded to the UW, and a Fellowship from the David and Lucile Packard Foundation. Y.X. is an Alfred P. Sloan Research Fellow (2000) and a Camille Dreyfus Teacher Scholar (2002). H.F. is grateful to the Ministry of Education, Culture, Sports, Science and Technology of the Japanese Government for the financial support to study at an abroad university or research institute. Y.L. thanks the Center for Nanotechnology at the UW for a fellowship award.

References

(1) Recent reviews on colloidal crystals: a) Gast, A. P.; Russel, W. B. *Physics Today* **1998**, *Dec.*, 24. b) *From Dynamics to Devices: Directed Self-Assembly of Colloidal Materials* Grier, D. G., Eds.; a special issue in *MRS Bulletin*, **1998**, *23*, 21. c) Xia, Y.; Gates, B.; Yin, Y.; Lu, Y. *Adv. Mater.* **2000**, *12*, 693. d) Dinsmore, A. D.; Crocker, J. C.; Yodh, A. G. *Curr. Opin. Colloid Interface* **1998**, *3*, 5.
(2) Sanders, J. V. *Acta Crystallographica* **1968**, *A24*, 427.
(3) See, for example, a) a special issue on photrnic crystals, *Adv. Mater.* **2001**, *13*, 369. b) Polman, A.; Wiltzius, P. *Materials Science Aspects of Photonic Crystals*, a special issue in *MRS Bull.* **2001**, *26*, 608. c) Velev, O. D.; Lenhoff, A. M. *Curr. Opin. Coll. Interf. Sci.* **2000**, *5*, 56. d) Stein, A.; Schroden, R. C. *Curr. Opin. Solid State Mat. Sci.* **2001**, *5*, 553. e) Blanco, A.; Chomski, E.; Grabtchak, S.; Ibisate, M.; John, S.; Leonard, S. W.; Lopez, C.; Meseguer, F.; Miguez, H.; Mondia, J. P.; Ozin, G. A.; Toader, O.; van Driel, H. M. *Nature* **2000**, *405*, 437. f) Vlasov, Y. A.; Bo, X. Z.; Sturm, J. C.; Norris, D. J. *Nature* **2001**, *414*, 289. g) Bertone, J. F.; Jiang, P.; Hwang, K. S.; Mittleman, D. M.; Colvin, V. L. *Phys. Rev. Lett.* **1999**, *83*, 300. h) Wang, D.; Caruso, R. A.; Caruso, F. *Chem. Mater.* **2001**, *13*, 364. i) Lee, W.; Prunzinski, S. A.; Braun, P. V. *Adv. Mater.* **2002**, *14*, 271. j) Ozin, G. A.; Yang, S. M. *Adv. Func. Mater.* **2001**, *11*, 95.
(4) a) Flaugh, P. L.; O'Donnell, S. E.; Asher, S. A. *Appl. Spectrosc.* **1984**, *38*, 847. b) Krieger, I. M.; O'Neill, F. M. *J. Am. Chem. Soc.* **1968**, *90*, 3114. c)

Hiltner, P. A.; Krieger, I. M. *J. Phys. Chem.* **1969**, *73*, 2386. d) Goodwin, J. W.; Ottewill, R. H.; Parentich, A. *J. Phys. Chem.* **1980**, *84*, 1580.
(5) a) Weissman, J. M.; Sunkara, H. B.; Tse, A. S.; Asher, S. A. *Science* **1996**, *274*, 959. b) Holtz, J. H.; Asher, S. A. *Nature* **1997**, *389*, 829. c) Reese, C. E.; Baltusavich, M. E.; Keim, J. P.; Asher, S. A. *Anal. Chem.* **2001**, *73*, 5038.
(6) a) Hu, Z. B.; Lu, X. H.; Gao, J. *Adv. Mater.* **2001**, *13*, 1708. b) Bebord, J. D.; Eustis, S.; Debord, S. B.; Lofye, M. T.; Lyon, L. A. *Adv. Mater.* **2002**, *14*, 658.
(7) Blanford, C. F.; Schroden, R. C.; Al-Daous, M.; Stein, A. *Adv. Mater.* **2001**, *13*, 26.
(8) a) Cassagneau, T.; Caruso, F. *Adv. Mater.* **2002**, *14*, 1629. b) Gu, Z.-Z.; Horie, R.; Kubo, S.; Yamada, Y.; Fijishima, A.; Sato, O. *Angew. Chem. Int. Ed.* **2002**, *41*, 1154.
(9) a) Jethmalani, J. M.; Ford, W. T. *Langmuir* **1997**, *13*, 3338. b) Foulger, S. H.; Jiang, P.; Lattam, A. C.; Smith, D. W.; Ballato, J. *Langmuir* **2001**, *17*, 6023. c) Foulger, S. H.; Jiang, P.; Ying, Y. R.; Lattam, A. C.; Smith, D. W.; Ballato, J. *Adv. Mater.* **2001**, *13*, 1898. d) Sumioka, K.; Kayashima, H.; Tsutsui, T. *Adv. Mater.* **2002**, *14*, 1284.
(10) Fudouzi, F.; Xia, Y. *Adv. Mater.* **2003**, *15*, 892.
(11) *Siloxane Polymers*; Clarson, S. J.; Semlyen, J. A., Eds.; PTR Prentice Hall: Englewood, NJ, 1993; p 64.
(12) a) Bueche, A. M. *J. Poly. Sci.* **1955**, *15*, 97. b) Mathison, D. E.; Yates, B.; Darby, M. I. *J. Mater. Sci.* **1991**, *26*, 6. c) Campbell, D. J.; Beckman, K. J.; Calderon, C. E.; Doolan, P. W.; Ottosen, R. M.; Ellis, A. B.; Lisensky, G. C. *J. Chem. Educ.* **1999**, *76*, 537.
(13) Xia, Y.; Whitesides, G. M. *Angew. Chem. Int. Ed.* **1998**, *37*, 550.
(14) Comiskey, B.; Albert, J. D.; Yoshizawa, H.; Jacobson, J. *Nature* **1998**, *394*, 253. b) Tamaoki, N. *Adv. Mater.* **2001**, *13*, 1135. c) Brehmer, M.; Lub, J.; van de Witte, P. *Adv. Mater.* **1998**, *10*, 1438.

Chromic Polymers for Chemical Sensors and Biosensors

Chapter 26

Synthesis and Tunable Chiroptical Properties of Amphiphilic Helical Polyacetylenes

Kevin K. L. Cheuk and Ben Zhong Tang*

Department of Chemistry, Open Laboratory of Chirotechnology, and Institute of Nano Materials and Technology, Hong Kong University of Science and Technology, Clear Water Bay, Kowloon, Hong Kong, China

A group of new amphiphilic macromolecules comprised of hydrophobic polyacetylene backbone and hydrophilic pendant groups of naturally occurring species such as amino acids, saccharides, and nucleosides are synthesized. The polymers exhibit solvatochromism. The macromolecular chains show helical conformations that depend on the molecular structures of the pendants, solvent, temperature, pH, and additives.

The native conformations of biopolymers such as proteins are determined by the folding information encoded in their amino acid sequences and are stabilized by a variety of noncovalent forces such as hydrogen bonding, solvation effect, hydrophobic stacking, and electrostatic interaction. The susceptibility of these noncovalent forces to external perturbations endows the proteins with structural flexibility, allowing them to change their conformations to adapt to or cope with the changes in their environmental surroundings. The conformational changes can also facilitate the execution of the biological functions of the proteins. The breakage of hydrogen bonds and salt bridges, for example, transfers hemoglobin from deoxy to oxy conformation, enhances its affinity for oxygen, and aids the oxygen binding to the protein (*1*).

Incorporation of the building blocks of the biopolymers into the molecular structures of synthetic polymers may generate biomimetic materials capable of undergoing conformational changes in response to environmental variations. If the polymers are conjugated, the conformational changes may cause changes in their electronic structures or photonic transitions, thus offering the opportunity of creating new polymeric materials with tunable optical properties. In this work, we attached naturally occurring building blocks as pendant groups to conjugated polyacetylene backbones. The polyacetylene chains are induced to take helical conformations, and the chiroptical properties of the conjugated helical polymers are tunable.

Design and Synthesis of Polymers

From the viewpoint of molecular design, the best building blocks for the construction of biomimetic macromolecules are naturally occurring species with molecular chirality and hydrogen-bonding capability. Such species are normally hydrophilic, and their incorporation into hydrophobic polymers would produce amphiphilic polymers (*2–6*). Polyacetylenes are hydrophobic, electroactive, and photoconductive polymers (*7–12*) and naturally occurring building blocks such as amino acids, saccharides, and nucleosides are hydrophilic, chiral, and capable of hydrogen bonding. It is known that the polyacetylene chains can be induced to rotate in a crew sense by optically active pendants (*13–22*) and it is envisioned that the naturally occurring species will endow polyacetylenes with added new properties such as amphiphilicity, proteomimetism, environmental adaptability, and compatibility to living cells (*23–29*).

We fused the naturally occurring building blocks with acetylene triple bond at the monomer stage in an effort to ensure the structural homogeneity of their polymers, that is, to make sure that every one of their monomer repeating units will precisely carry one pendant group of natural origin (*2–6, 23–29*). Most of the acetylene derivatives are synthesized by amidation and/or esterification of *p*-ethynylbenzoic acid with the naturally occurring species. The polymerizations of the acetylene monomers are carried out using organorhodium complexes as catalysts. Typical examples of the polymers synthesized in our laboratory are given in Charts 1 and 2.

We systematically studied the polymerization reactions of the monomers. Typical examples of the polymerization results are given in Table 1. In most cases, the polymerizations afford polymers with high molecular weights (M_w up to ~1.8×10^6 Da) in high yields (up to 100%). All the polymers are *Z*-rich (up to 99%) in IUPAC stereostructure terminology or trans-rich in conventional term (cf. Chart 3). Generally [Rh(nbd)Cl]$_2$ is a good catalyst for the polymerizations of the amino acid-containing monomers, while [Rh(cod)Cl]$_2$ works well for the polymerizations of the monosaccharide-containing monomers. The nucleoside monomers are difficult to polymerize, because the oligomers precipitate out from the mixtures at the early stages of the polymerization reactions. To mitigate this solubility problem, we copolymerized these monomers with other comonomers such as D-mannose.

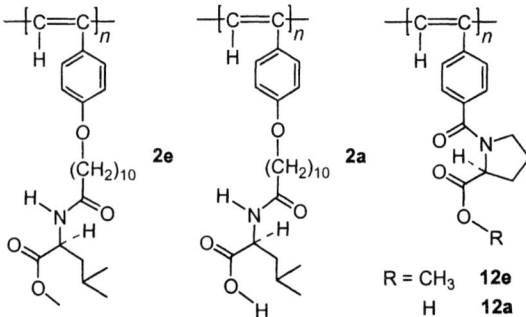

R =	CH$_2$CHMe$_2$	1e	1a	
	CHMe$_2$	3e	3a	
	CH(Me)CH$_2$Me	4e	4a	
	Me	5e	5a	
	CH$_2$Ph	6e	6a	6eo
	Ph	7e(D, L)^a	7a^b	7eo
	CH$_2$CO$_2$Me	8e		
	CH$_2$CO$_2$H		8a	
	(CH$_2$)$_2$CO$_2$Me	9e		
	(CH$_2$)$_2$CO$_2$H		9a	
	(CH$_2$)$_2$SMe	10e	10a	
	tryptophan	11e	11a	tryptophan

^{a} Polymers 7e(D) and 7e(L) bear amino acid pendants in D- and L- configurations, respectively.
^{b} Racemic mixture.

Chart 1. *Molecular structures of polyacetylenes containing amino acid moieties.*

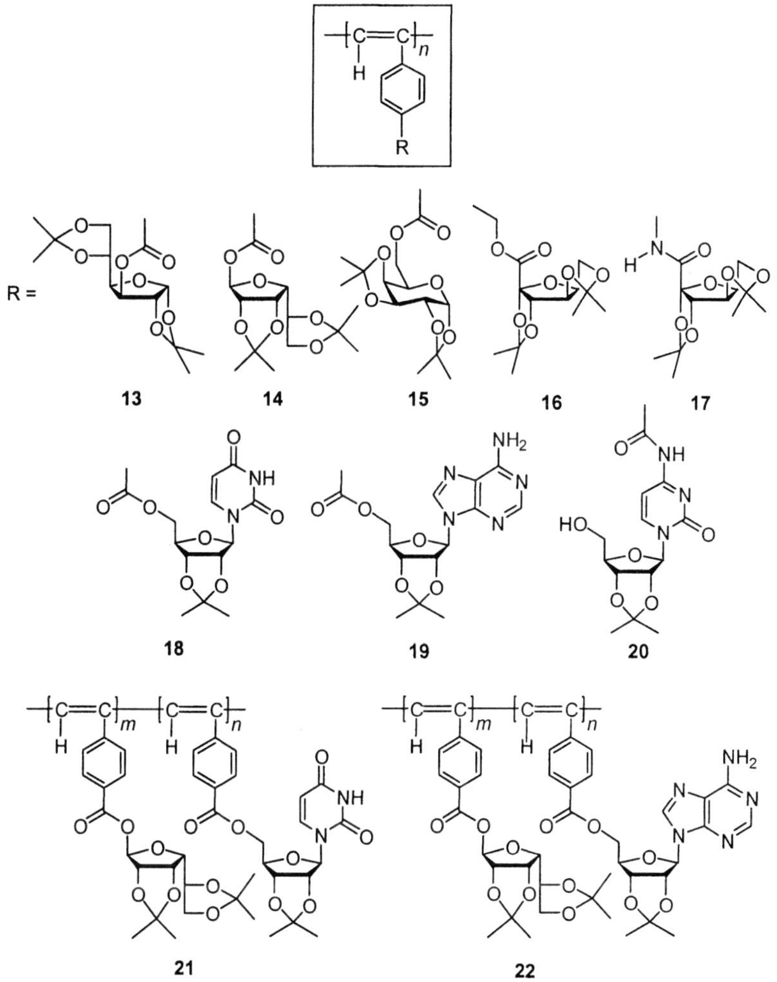

Chart 2. Molecular structures of polyacetylenes containing monosaccharide and nucleoside moieties.

The methyl protection group in the amino acid-containing polymers can be cleaved by base-catalyzed hydrolysis in a selective way: the "polyesters" **1e–12e** (with "e" standing for ester) are fully converted to their corresponding free-acid forms **1a–12a** ("a" for acid) without harming the amide functionality. Except **14**, all of the other monosaccharide-containing polymers (**13** and **15–17**) can also be selectively deprotected by ketal hydrolysis, although the resultant "polyols" with multiple hydroxyl groups are partially crystalline and difficult to dissolve. Polymer **14** is an exception, whose ester bonds were also hydrolyzed when we tried to remove the acetonide groups.

Table 1. Selected Examples of Polymerization Results[a]

no.	polymer	catalyst	solvent	yield (%)	$M_w{}^b$	$Z\%^c$
1	1e	[Rh(nbd)Cl]$_2$	THF	78.2	1 240 000	83.1
2	2e	[Rh(nbd)Cl]$_2$	THF/Et$_3$N	99.0	20 000	d
3	3e	[Rh(nbd)Cl]$_2$	THF	88.6	279 000	94.1
4	4e	Rh(cod)(NH$_3$)Cl	THF	96.9	77 000	92.5
5	5e	[Rh(nbd)Cl]$_2$	THF	72.0	1 201 000	89.5
6	6e	[Rh(nbd)Cl]$_2$	DMF/Et$_3$N	75.6	1 773 000	88.7
7	6eo	[Rh(nbd)Cl]$_2$	THF/Et$_3$N	97.1	1 072 000	92.9
8	7e(L)	[Rh(nbd)Cl]$_2$	dioxane	83.4	1 801 000	d
9	7e(D)	[Rh(nbd)Cl]$_2$	THF/ET$_3$N	93.6	43 000	97.6
10	7eo	[Rh(nbd)Cl]$_2$	CH$_2$Cl$_2$	97.5	176 000	97.0
11	8e	[Rh(nbd)Cl]$_2$	THF/Et$_3$N	85.3	428 000	83.6
12	8e	[Rh(nbd)Cl]$_2$	THF/Et$_3$N	95.1	441 000	87.2
13	10e	[Rh(nbd)Cl]$_2$	THF/Et$_3$N	96.3	d	
14	11e	[Rh(nbd)Cl]$_2$	THF/Et$_3$N	100.0	103 000	d
15	12e	Rh(cod)(tos)(H$_2$O)	THF/Et$_3$N	96.8	17 000	94.8
16	13	[Rh(nbd)Cl]$_2$	toluene	93.8	995 000	80.0
17	14	[Rh(cod)Cl]$_2$	THF/Et$_3$N	86.6	1 074 000	97.8
18	15	[Rh(cod)Cl]$_2$	CH$_2$Cl$_2$	71.7	1 236 000	96.2
19	16	[Rh(nbd)Cl]$_2$	THF/Et$_3$N	75.4	161 000	99.0
20	17	[Rh(cod)Cl]$_2$	CH$_2$Cl$_2$/Et$_3$N	43.6	224 000	88.5
21	18	Rh(cod)(NH$_3$)Cl	THF	42.2	5 000	d
22	19	Rh(cod)(NH$_3$)Cl	DMF	51.2	d	
23	20	[Rh(nbd)Cl]$_2$	CH$_2$Cl$_2$/Et$_3$N	91.3	3 000	d
24	21[e]	[Rh(nbd)Cl]$_2$	THF/Et$_3$N	76.9	373 000	d
25	22[f]	[Rh(nbd)Cl]$_2$	THF/Et$_3$N	59.9	48 000	d

[a] Carried out at room temperature under nitrogen for 24 h; $[M]_o = 0.1$ M, [cat.] = 5 mM.

[b] Estimated by GPC in THF on the basis of a polystyrene calibration.

[c] Determined by ^{1}H NMR analysis.

[d] Not determined.

[e] Molar ratio of uridine- and D-mannose-containing repeat units in the copolymer: 1:1.

[f] Molar ratio of adenosine- and D-mannose-containing repeat units in the copolymer: 1:4.

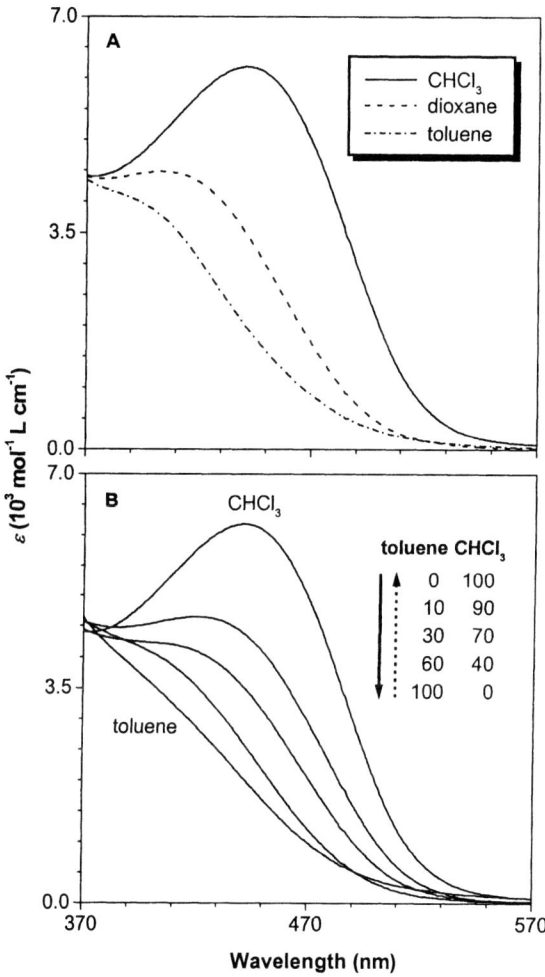

*Figure 1. (A) UV spectra of **13**, a polyacetylene bearing D-glucose pendants, in chloroform, dioxane and toluene. (B) Change of its absorption spectrum with composition of a toluene/chloroform mixture.*

Solvatochromism and Hydrogen Bonding

The amphiphilic polyacetylene chains may take different conformations in different environments (e.g., in different solvents) and the different conformers may undergo different electronic transitions. The polymers exhibit interesting solvatochromism: the absorption spectra of their solutions change with a change in the solvent.

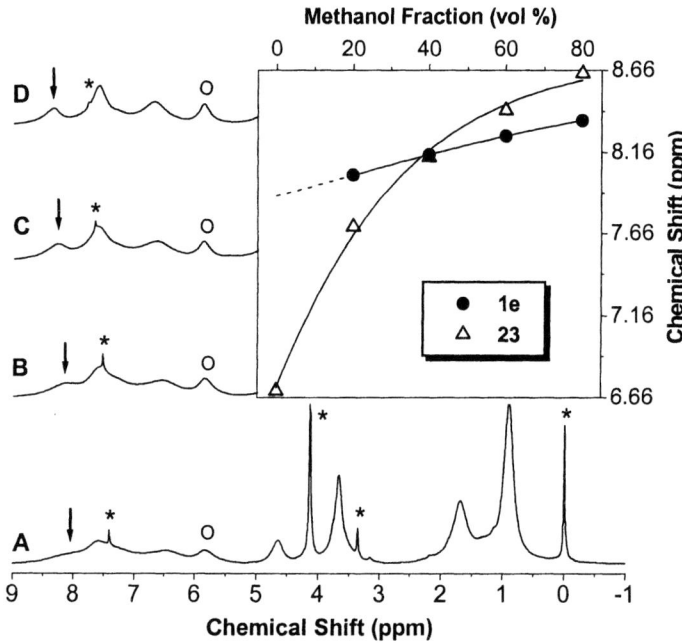

Figure 2. ^1H NMR spectra of methanol-d_4/chloroform-d solutions of an L–leucine-containing polyacetylene (**1e**; 50 mg/mL) with varying ratios of methanol-d_4 (vol %): (A) 20, (B) 40, (C) 60, and (D) 80. Insert: solvent dependence of the chemical shift of the amide proton resonance of **1e** and its monomer **23** (50 mg/mL) in methanol-d_4/chloroform-d mixtures. The resonance peaks of the amide (HNCO) and vinyl (HC=) protons are respectively marked with downward arrows (↓) and open circles (o), while those of the solvents are marked with asterisks (*).

Figure 1 shows the UV spectra of polymer **13** in different solvents. In chloroform, the backbone of the polymer absorbs strongly at ~440 nm. This backbone absorption weakens and blue shifts when the solvent is changed to dioxane. Further weakening and blue-shift are observed when the spectrum is measured in toluene. Changing the composition of a toluene/chloroform mixture changes the absorption spectrum in a continuous and reversible manner.

Chloroform is a good solvent to both the hydrophobic backbone and the hydrophilic pendants. The polymer chains may be well solvated by the solvent and take an extended planar conformation, in which the polyacetylene backbone is better conjugated. Toluene is, however, a poor solvent of the hydrophilic pendants. To minimize the exposure of their pendants to the unfavorable hydrophobic solvent environment, the polymer chains may take a coiled nonplanar conformation, in which the polyene backbone may be less conjugated. Since the conformational change is induced by the noncovalent solvent–polymer interaction, it is easy to understand why the solvatochromic change is continuous and reversible.

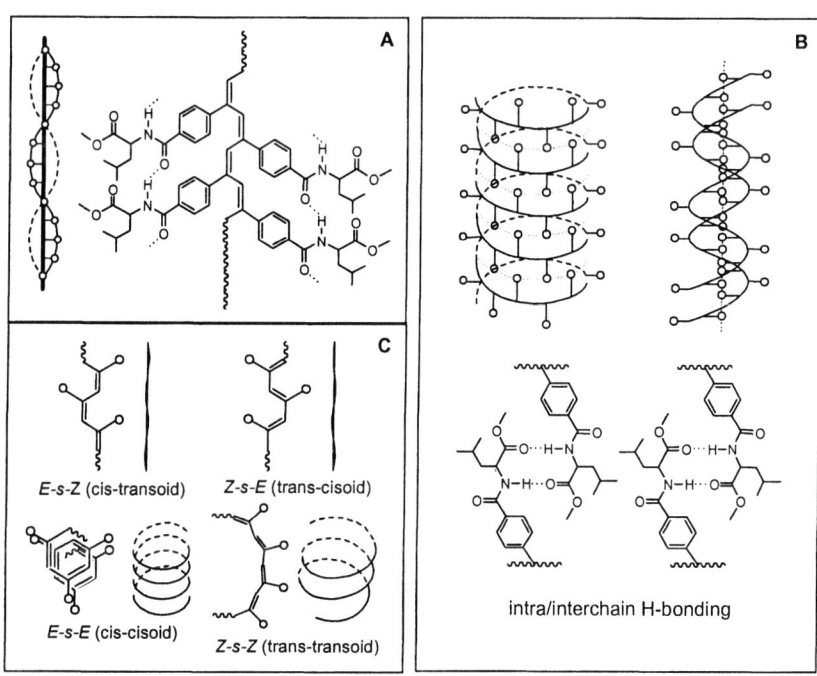

Chart 3. Diagrammatic illustrations of (A and B) single- and double-stranded helical chains of leucine-containing polyphenylacetylene **1e** stabilized by intra- and inter-chain hydrogen bonds and (C) theoretically possible conformations of chain segments of a substituted poly(phenylacetylene).

The UV and NMR spectra of the polymers changed with solvent. As shown in Figure 2 for the case of polymer **1e**, in a methanol-d_4/chloroform-d mixture with a methanol content of 20 vol %, the amide proton of the polymer resonates at δ ~8.1. This resonance peak progressively downfield-shifts to δ ~8.4, when the methanol content is increased to 80%. The shift in the resonance peak of the amide proton is obviously associated with the hydrogen bonding between the pendant group and the methanol solvent (*30*).

A similar progressive shift in the resonance peak of amide proton is also observed in the ^{1}H NMR spectra of **23**, the monomer of **1e**. The scale of the shift is, however, much bigger (from δ ~6.7 to ~8.7; inset of Figure 2). The smaller range of the chemical shift change and the broader span of the resonance peaks of the polymer suggest that there exist intra- and interchain hydrogen bonds between the pendant groups in the chain segments, as schematically illustrated in panels A and B of Chart 3. Such hydrogen bonds will downfield-shift the amide resonance peak of the polymer, and indeed, its δ value in the mixture of low methanol content is higher than that of its monomer. The steric effect of the polymer chains, on the other hand, may obstruct the methanol molecules from approaching the amide groups of the L-leucine moieties buried inside the coiled chains and thus impedes the formation of hydrogen bonds between the solvents and the pendants. This explains why the δ values of polymer are lower than those of its monomer in the mixture of high methanol contents.

Induced Helical Structures

The bulky pendants of neighboring repeat units may not be coplanar due to steric constraints. When the pendants are chiral, their cooperative twisting may generate an asymmetric force field to induce the segments of the polymer backbone to spiral in a helical fashion. To check whether the chain helicity is really induced in our polymers, we measured their circular dichroism (CD) spectra.

As can be seen from Figure 3, the polyacetylene bearing D-phenylglycine pendants [**7e**(D)] is clearly CD-active. In chloroform, it exhibits a CD band at ~372 nm with a high molar ellipticity ([θ] ~36430 deg cm^2 dmol^{-1}). Since its monomer is CD-*in*active at λ > 300 nm, the first Cotton effect of **7e**(D) at ~372 nm thus must be due to the absorption of its polyene backbone, unambiguously proving that the polymer chain takes a helical conformation with an excess of one handedness. The entropy penalty for the formation of the regular helical structures may be compensated by the stabilization effects of the intra- and interchain hydrogen bonds between the pendant groups. This balance may, however, be broken by external stimuli and/or internal perturbations, and the system may adapt to the new environments and reach new equilibrium states under the new sets of internal and external conditions. The chain helicity and its pitch can be changed via a single-bond rotation (Chart 4C); such atropisomerism [*31*] further supports the susceptibility of the chain helicity to perturbations. Indeed, when the solvent, a "simple" external condition, is changed, the CD spectrum of **7e**(D) varies accordingly (Figure 3, upper part). When its internal structure or pendant chirality is changed, the signs of the Cotton effects are reversed (Figure 3, lower part). The CD spectra of **7e**(L) are the mirror images of those of its counterpart **7e**(D), revealing that the helical preference of the polymer chain is determined by the molecular structure of the macromolecule, or more specifically, the chirality of the stereogenic center of the pendant.

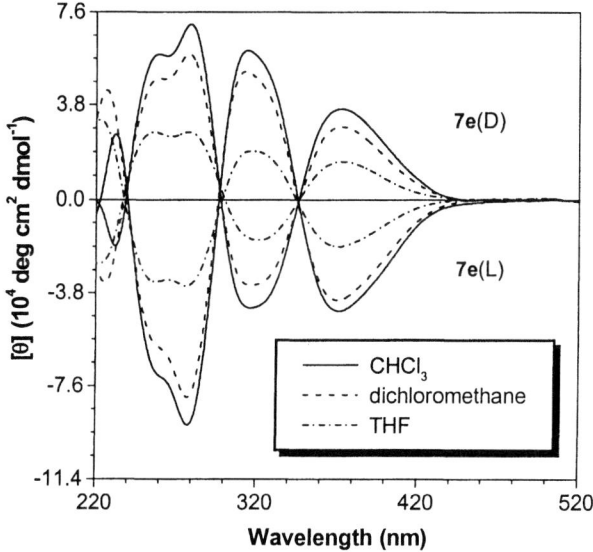

Figure 3. CD spectra of polyacetylenes bearing phenylglycine pendants (7e) of opposite configurations (D and L) in different solvents.

The helical conformation and its change with the pendant structure and solvent environment is a general feature for all the polymers carrying the chiral pendants. As summarized in Figures 4–6, all the polymers exhibit the first Cotton effects in the spectral region of backbone absorption; that is, the polyacetylene backbones are helically rotating in a one-handed screw sense. Their chain helicities change with solvents, and the patterns of the changes vary with the pendants, again confirming that the chain conformations are determined by both internal and external conditions. Different pendants possess different steric effects and hydrogen bonding ability, and the solvent molecules of varying solvating powers experience different interactions with the polymer chains. The interplay of the two factors arbitrates the direction of the helical rotation and the persistence length of the helical segment, resulting in the various changes of the CD data with solvent in the different polymer systems.

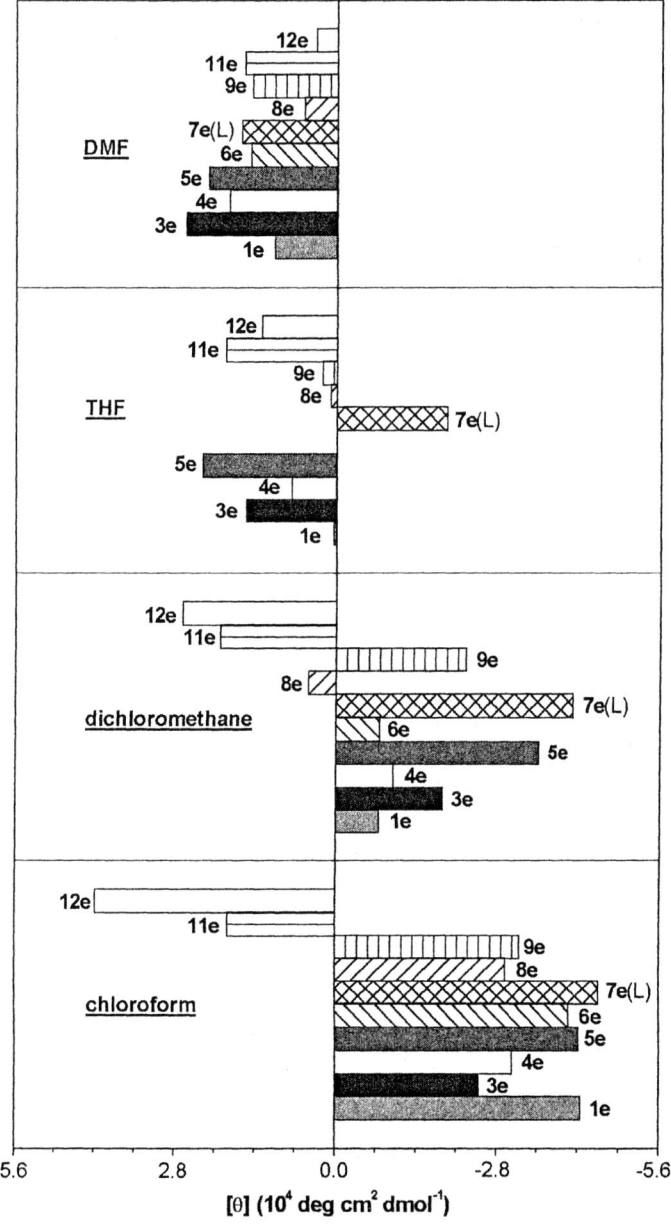

Figure 4. Solvent dependence of the first Cotton effect (λ_{max} ~375 nm) of the polyacetylenes bearing pendants of L-amino acid methyl esters.

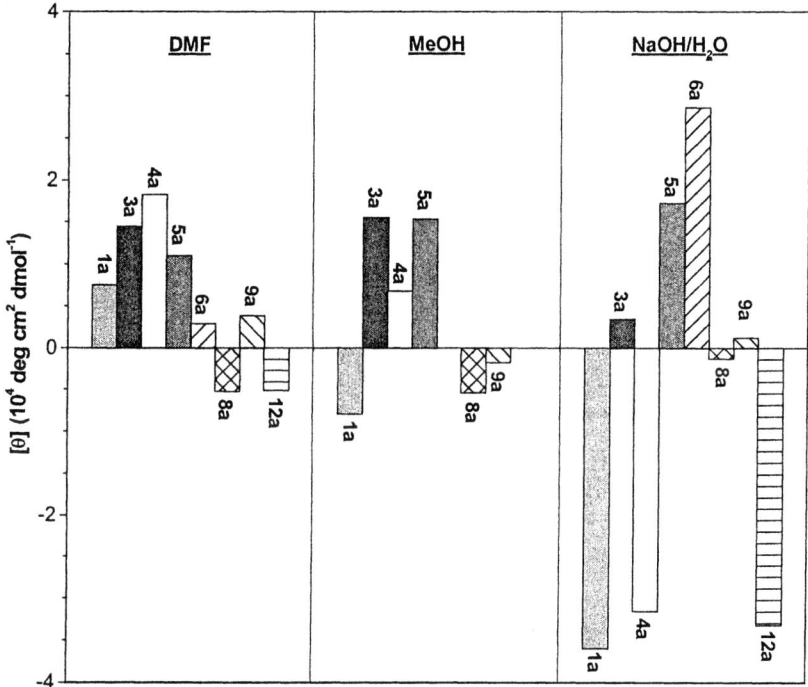

Figure 5. Environmental dependence of the first Cotton effect (λ_{max} ~375 nm) of the polyacetylenes bearing L-amino acid pendants. Concentration of sodium hydroxide in water ($NaOH/H_2O$): 0.2 M.

Several general trends can, however, be extracted from the CD data. The polymers bearing the pendants of amino acid methyl esters display the highest molar ellipticities in chloroform (Figure 4). The signs of their first Cotton effects in the chlorinated solvents of $CHCl_3$ and CH_2Cl_2 are the same (with one exception of **8e**). The Cotton effects in the polar solvents of THF and DMF are similar and are all positively signed except for **7e**(L). The Cotton effects of the polymers bearing the pendants of "free" amino acids are similar to those of their ester congeners in the similar solvents with a few exceptions of sign reversal. The changes in the CD data of the polymers in the $NaOH/H_2O$ mixture are complex because of the complications of involved pH change (Figure 5). Comparing to their CD data in methanol, the molar ellipticities of the polymers in $NaOH/H_2O$ are enhanced in intensity (**1a**) or reversed in sign (**4a** and **9a**) in some cases, while in other cases they remain almost the same (**5a**) or decrease in intensity (**3a** and **8a**).

The behaviors of the monosaccharide-containing polymers are, however, different from the amino acid-containing polymers discussed above. Although the change in the solvent causes some change in the intensity of the molar ellipticity, sign reversal is not observed in almost all the polymers expect **16** in THF (Figure 6). The main structural difference between the monosaccharide- and amino acid-containing polymers is the bulkiness and rigidity of the pendant groups. The monosaccharide groups are bulkier and more rigid, and their rotation around the polymer backbone is thus more sterically demanding. This restricted rotation of pendants may fix the "natural" chain conformations to a large extent, making the polymer chain less responsive to the variation in its environmental surrounding. This argument is further supported by the data of polymers **11e** and **12e**, which possess bulky and rigid tryptophan and proline pendants, respectively: the signs of their first Cotton effects do not change with the changes in their solvents (cf., Figure 4).

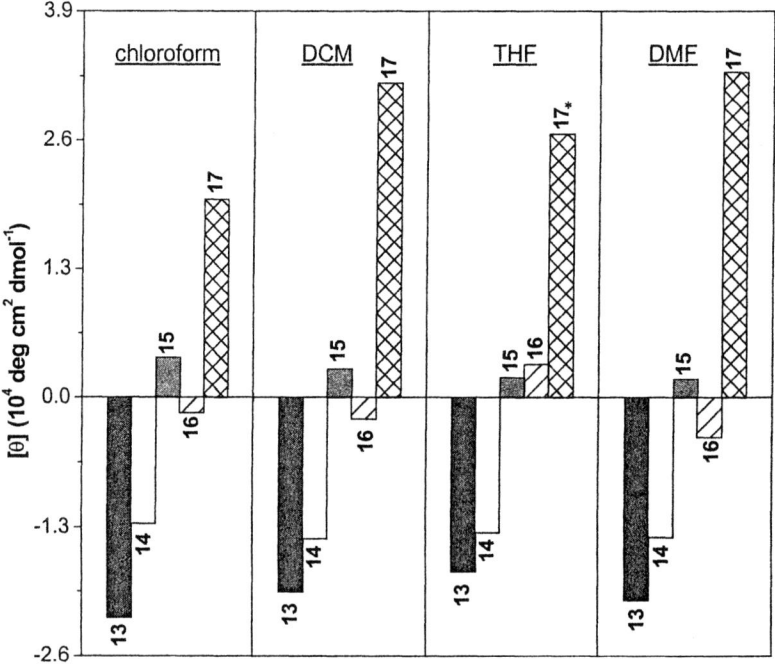

*Figure 6. Solvent dependence of the first Cotton effect (λ_{max} ~390 nm) of the polyacetylenes bearing pendants of D-monosaccharide acetonides. Data marked with * was measured in a mixture of THF/CHCl$_3$ (19:1 by volume).*

Tuning of Chain Helicity by External Stimuli

It is clear now that solvent environment can affect the chain conformations of the polyacetylenes carrying the pendants of naturally occurring building blocks. This suggests the possibility of tuning the chain helicity by external stimuli and we thus pursued the helical manipulation.

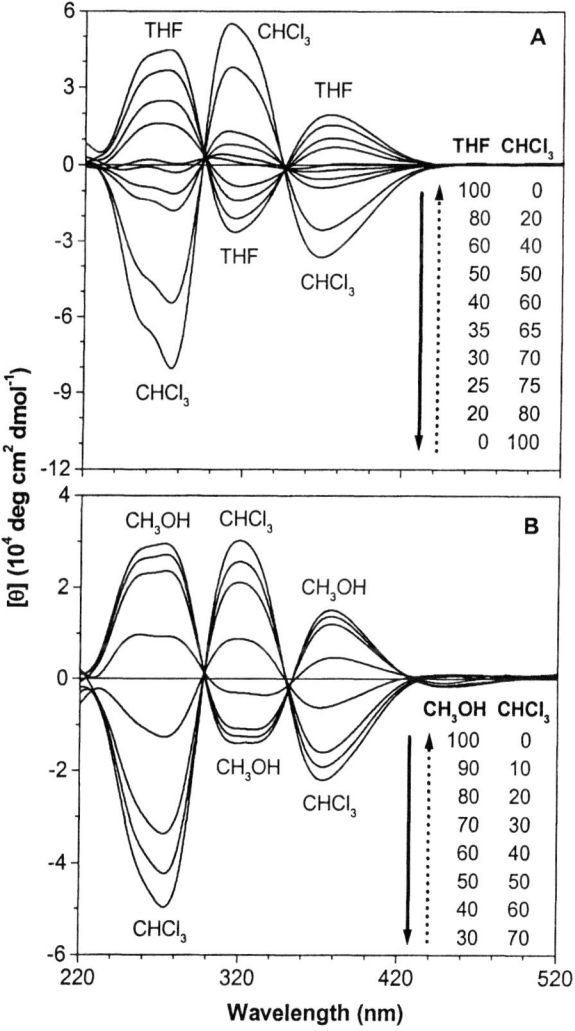

Figure 7. Changes of CD spectra of (A) 3e and (B) 3a with solvent compositions.

We already know that in some cases the handedness of a helical chain segment can be reversed by changing solvent from one to another. We are intrigued to know the detailed process of the helical reversal and explored the possibility of modulating the chain conformation in a continuous and reversible way by changing the composition of the mixture of two solvents. As shown in Figure 7A, with a progressive addition of chloroform into a THF solution of **3e** (a valine-containing polyacetylene), its molar ellipticity progressively changes; for example, its first Cotton effect moves downward. Clearly, chloroform induces a helical sense opposite to that in THF. The population of the chain segments with this opposite handedness is increased with an increase in the volume fraction of chloroform in the mixture and becomes dominant after a threshold point, as revealed by the sign reversion of the CD spectrum at the high fractions of chloroform. This spectral change is reversible, as shown by the dotted upward arrow in the figure: the CD spectrum changes in the opposite direction with an increase in the THF fraction in the solvent mixture. Similar continuous and reversible change of CD spectra with the composition of solvent mixture is observed in **3a**, the structural congener of **3e** (Figure 7B), as well as in other amino acid- and monosaccharide-containing polymers. This switchable phenomenon manifests that the "smart" polymer chains can "remember" their conformational "codes" under specific environmental conditions.

We further tried to tune the chain helicity by other external stimuli including temperature, pH, and additive. In above discussions, we have speculated that the helical conformations of the chain segments are stabilized by hydrogen bonding (cf., Chart 3). The noncovalent hydrogen bonds may be partially broken by thermal agitations, and the entropy-driven randomization of the relieved chain segments may lead to a decrease in the chain helicity. As can be seen from Figure 8A, the handedness of the helical chains of **3e** linearly decreases with an increase in temperature. Conversely, decreasing temperature restores original helical status of the chains, demonstrating that the hydrogen bonds can be reversibly reestablished at the low temperatures.

Increasing pH by adding an inorganic base into a polyacid solution can ionize the carboxylic acid with metal ions. The resultant polyelectrolyte chains may undergo different conformational changes, e.g., chain randomization by electrical repulsion (decreasing helicity) or chain braiding via salt bridge (increasing helicity), which exert opposite effects on the chain helicity. The CD activity of **3a** decreases, although nonlinearly, with the addition of KOH to its methanol solution (Figure 8B). The polymer chains are randomized by the K^+-induced ionization and the nonlinearity is due to the "zipping effect" of entropy-driven randomization coupled with the "polymer effect" of macromolecular complexation (*23, 32, 33*). Remarkably, however, the original CD spectrum is fully recovered when the alkaline solution is neutralized by the addition of an acid, indicating that the helicity tuning by pH is also reversible.

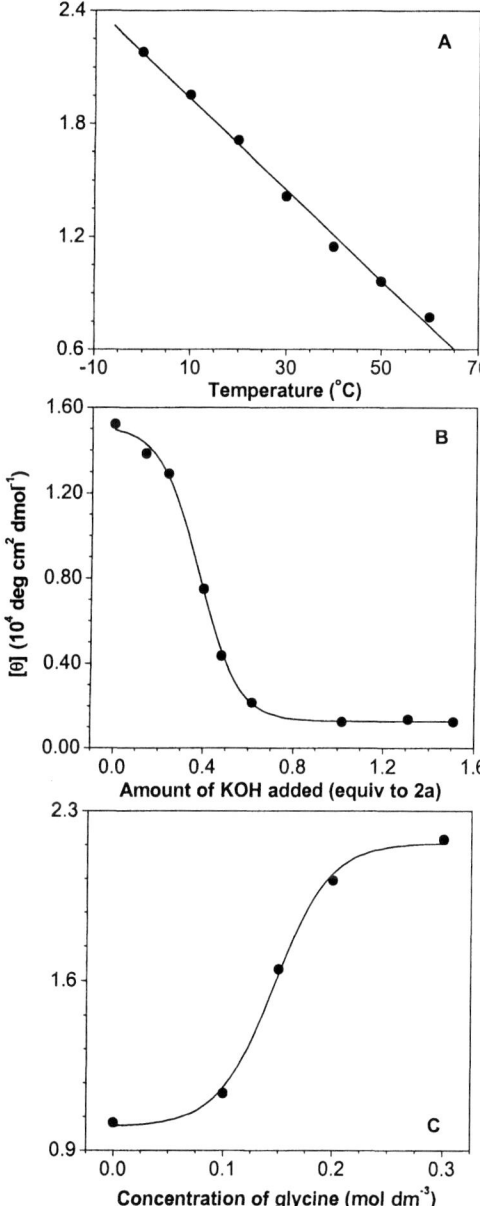

Figure 8. Effects of (A) temperature, (B) pH, and (C) additive (glycine) on the first Cotton effects at ~375 nm of (A) 3e in THF, (B) 3a in KOH/methanol, and (C) 3a in methanol/H_2O (1:1 by volume).

Interestingly, the chain helicity can even be tuned by achiral additives, an example of which is given in Figure 8C. The first Cotton effect of **3a** is intensified with the addition of glycine into its methanol/water solutions. This signal intensification may be related to the steric effect from the complexation of glycine with the valine pendants. Intriguingly, such helicity amplification is observed only in certain polymer systems (e.g., **1a**). This is probably associated with an enzyme-like "lock and key" binding process (*34, 35*).

Concluding Remarks

In this work, we synthesized a group of new biomimetic polymers, whose conjugated polyene backbones are appended with naturally occurring species of amino acids, monosaccharides, and nucleosides. The bathochromic shifts and hyperchromic effects in the electronic transitions of the polyacetylene solutions are induced by the solvent changes, leading to solvatochromism. The NMR and CD analyses reveal the existence of intra- and interchain hydrogen bonds. The segments of the polymer chains are induced by the chiral pendants to spiral in a screw sense, whose handedness, pitch, and persistence length are affected by both internal and external perturbations. The chiroptical properties of the polymers can be continuously and reversibly tuned by external stimuli.

Acknowledgements. This work was supported by the Research Grants Council (Projects Nos.: HKUST 6187/99P, 6121/01P, and 6085/02P) and the University Grants Committee (Project No.: AoE/P-10/01-1-A) of the Hong Kong Special Administrative Region, China.

Literature Cited

(1) Zubay, G. L. *Biochemistry*, 4th ed.; Wm. C. Brown: Boston, 1998.
(2) Cheuk, K. K. L.; Li, B.; Tang, B. Z. *Curr. Trends Polym. Sci.* **2002**, *7*, 41–55.
(3) Tang, B. Z. *Polym. News* **2001**, *26*, 262–272.
(4) Tang, B. Z.; Cheuk, K. K. L.; Salhi, F.; Li, B.; Lam, J. W. Y.; Cha, J. A. K.; Xiao, X. *ACS Symp. Ser.* **2001**, *812*, 133–148.
(5) Lam, J. W. Y.; Tang, B. Z. *J. Polym. Sci. Part A: Polym. Chem.* **2003**, *41*, in press.
(6) Cheuk, K. K. L.; Li, B. S.; Tang, B. Z. In *Encyclopedia of Nanoscience and Nanotechnology*; Nalwa, H. S., Ed.; American Scientific Publishers: CA, in press.
(7) Shirakawa, H. *Angew. Chem. Int. Ed.* **2001**, *40*, 2575–2580.

(8) MacDiarmid, A. G. *Angew. Chem. Int. Ed.* **2001**, *40*, 2581–2590.
(9) Heeger, A. J. *Angew. Chem. Int. Ed.* **2001**, *40*, 2591–2611.
(10) Mi, Y.; Tang, B. Z. *Polym. News* **2001**, *26*, 170–179.
(11) Tang, B. Z.; Chen, H. Z.; Xu, R. S.; Lam, J. W. Y.; Cheuk, K. K. L.; Wong, H. N. C.; Wang, M. *Chem. Mater.* **2000**, *12*, 213–221.
(12) Sun, J. Z.; Chen, H. Z.; Xu, R. S.; Wang, M.; Lam, J. W. Y.; Tang, B. Z. *Chem. Commun.* **2002**, 1222–1223.
(13) Tang, B. Z.; Kotera, N. *Macromolecules* **1989**, *22*, 4388–4390.
(14) Yamaguchi, M.; Omata, K.; Hirama, M. *Chem. Lett.* **1992**, 2261–2262.
(15) Aoki, T.; Kokai, M.; Shinohara, K.; Oikawa, E. *Chem. Lett.* **1993**, 2009–2012.
(16) Kishimoto, Y.; Itou, M.; Miyatake, T.; Ikariya, T.; Noyori, R. *Macromolecules* **1995**, *28*, 6662–6666.
(17) Yashima, E.; Maeda, Y.; Okamoto, Y. *Nature* **1999**, *399*, 449–451.
(18) Mitsuyama, M.; Kondo, K. *J. Polym. Sci. Part A: Polym. Chem.* **2001**, *39*, 913–917.
(19) Nomura, R.; Tabei, J.; Masuda, T. *J. Am. Chem. Soc.* **2001**, *123*, 8430–8431.
(20) Schenning, A. P. H. J.; Fransen, M.; Meijer, E. W. *Macromol. Rapid Commun.* **2002**, *23*, 266–270.
(21) Percec, V.; Obata, M.; Rudick, J. G.; De, B. B.; Glodde, M.; Bera, T. K.; Magonov, S. N.; Balagurusamy, V. S. K.; Heiney, P. A. *J. Polym. Sci. Part A: Polym. Chem.* **2002**, *40*, 3509–3533.
(22) Dong, Y.; Lam, J. W. Y.; Cheuk, K. K. L.; Tang, B. Z. *J. Polym. Mater.* **2003**, *20*, 189–193.
(23) Li, B. S.; Cheuk, K. K. L.; Salhi, F.; Lam, J. W. Y.; Cha, J. A. K.; Xiao, X.; Bai, C.; Tang, B. Z. *Nano. Lett.* **2001**, *1*, 323–328.
(24) Salhi, F.; Cheuk, K. K. L.; Sun, Q.; Lam, J. W. Y.; Cha, J. A. K.; Li, G.; Li, B.; Luo, J.; Chen, J.; Tang, B. Z. *J. Nanosci. Nanotechnol.* **2001**, *1*, 137–141.
(25) Li, B. S.; Cheuk, K. K. L.; Ling, L.; Chen, J.; Xiao, X.; Bai, C.; Tang, B. Z. *Macromolecules* **2003**, *36*, 77–85.
(26) Li, B. S.; Cheuk, K. K. L.; Chen, J.; Xiao, X.; Bai, C.; Tang, B. Z. In *Nano Science and Technology—Novel Structures and Phenomena*; Tang, Z. K., Sheng, P., Eds.; Taylor & Francis: London, 2003; Chapter 9, pp 98–104.
(27) Cheuk, K. K. L.; Li, B. S.; Lam, J. W. Y.; Chen, J.; Bai, C.; Tang, B. Z. *Nanotechnology* **2003**, in press.
(28) Li, B. S.; Cheuk, K. K. L.; Yang, D.; Lam, J. W. Y.; Wan, L.; Bai, C.; Tang, B. Z. *Macromolecules* **2003**, *36*, in press.
(29) Cheuk, K. K. L.; Lam, J. W. Y.; Chen, J.; Lai, L. M.; Tang, B. Z. *Macromolecules* **2003**, *36*, in press.

(30) Silverstein, R. M.; Bassler, G. C.; Morrill, T. C. *Spectrometric Identification of Organic Compounds*, 5th ed.; Wiley: New York, 1991.
(31) Green, M. M.; Cheon, K. S.; Yang, S. Y.; Park, J. W.; Swansburg, S.; Liu, W. H. *Acc. Chem. Res.* **2001**, *34*, 672–680.
(32) *Macromolecular Complexes*; Tsuchida, E., Ed.; VCH: New York, 1991.
(33) *Polymer Reaction Engineering*; Reichert, K.-H., Geiseler, W., Eds.; VCH: New York, 1989.
(34) Breslow, R. *Supramol. Chem.* **1995**, *6*, 41–47.
(35) *The Lock-and-Key Principle: the State of the Art–100 Years on*; Behr, J.-P., Ed.; Wiley: New York, 1994.

Chapter 27

DNA-Sensors Using a Water-Soluble, Cationic Poly(thiophene) Derivative

Hoang-Anh Ho[1], Maurice Boissinot[2], Michel G. Bergeron[2], Geneviève Corbeil[1], Kim Doré[1,3], Denis Boudreau[3], and Mario Leclerc[1,*]

[1]Canada Research Chair in Electroactive and Photoactive Polymers, CERSIM, Department of Chemistry, Université Laval, Québec City, Québec G1K 7P4, Canada
[2]Centre de Recherche en Infectiologie, Centre Hospitalier Universitaire de Québec, Université Laval, Québec City, Québec G1V 4G2, Canada
[3]Laboratory for Laser Spectrochemical Analysis, Department of Chemistry, Université Laval, Québec City, Québec G1K 7P4, Canada

> We designed a novel cationic, water-soluble polythiophene derivative which can easily transduce oligonucleotide hybridization into optical or electrical signal. This polymeric sensor proved to be specific enough to distinguish between single nucleotide mismatches. This simple methodology has the potential to be integrated into various applications involving detection and identification of oligonucleotides.

Sequence-specific methods are needed for the rapid detection of oligonucleotides for genotyping and diagnosis of infections and various genetic diseases. Most current detection methods use a fluorescent or electroactive tag bound to the analyte or to the probe (*1-5*). An assay that does not require any chemical manipulation (*6,7*) of nucleic acids or complex reaction mixtures would be advantageous. Here, we describe new cationic polythiophene biosensors for the detection of DNA. The present approach is based on previous studies on affinitychromic poly(3-alkoxy-4-methylthiophene)s (6). A new water-soluble cationic polythiophene was synthesized and this polymer is able to transduce oligonucleotide hybridization with a specific 20-mer capture probe into a clear optical (colorimetric or fluorometric) or electrical signal. This simple, rapid, and versatile methodology does not require any chemical modification of the probes or the analytes; it is based on different electrostatic interactions and conformational structures between electroactive and photoactive cationic poly(3-alkoxy-4-methylthiophene) and single-stranded oligonucleotides or double-stranded (hybridized) nucleic acids.

The new monomer 1 bearing the positive charge (imidazolium group) was obtained by using simple reactions (8-10) and the corresponding polymer was prepared from a chemical polymerization in chloroform by using $FeCl_3$ (11) as the oxidizing agent (Figure 1). As expected, the resulting polymer 1 is soluble in aqueous solutions.

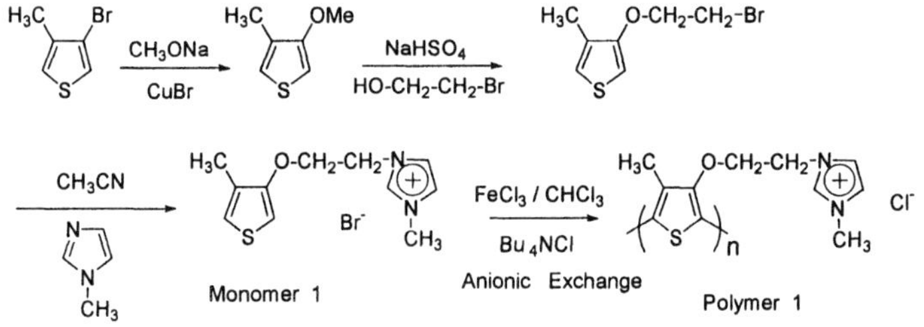

Figure 1. Synthesis of monomer 1 and polymer 1.

As shown in Plate 1, this polymer should be able to complex single-stranded and double-stranded oligonucleotides. In order to verify the specificity of these complexations, four types of negatively charged oligonucleotides (20-mers) were synthesized: a capture probe sequence (**X1**: 5'-CATGATTGAACCATCCACCA-3'), a perfect complementary target (**Y1**: 3'-GTACTAACTTGGTAGGTGGT-5'), a two-mismatch complementary target (**Y2**: 3'-GTACTAACTT<u>CG</u>AAGGTGGT-5') and an one-mismatch complementary target (**Y3**: 3'-GTACTAACTT<u>C</u>GTAGGTGGT-5').

Colorimetric Detection

At 55°C, aqueous solutions (0.1 M NaCl or TE buffer at pH=8 (10 mM Tris(hydroxymethyl)aminomethane, 1 mM EDTA) and 0.1M NaCl) of the cationic polymer 1 are yellow (λ_{max}= 397 nm) (Plate 2A,a and 2B,a). This absorption maximum at a relatively short wavelength is related to a random coil conformation of the polythiophene derivative, any twisting of the conjugated backbone leads to a decrease of the effective conjugation length (6). As with any water-soluble cationic polyelectrolyte, these polythiophene derivatives or another cationic conjugated polymers can make strong complexes with negatively-charged oligomers and polymers (11-13). Upon addition of 1.0 equivalent (on a monomeric unit basis) of an oligonucleotide X1 (20-mers), the mixture becomes red (λ_{max}= 527 nm) within 5 minutes (Plate 2A,b and 2B,b), due to the formation of a so-called duplex between the polythiophene and the oligonucleotide probe. Duplexes are stable up to 65 ^{0}C. After 5 minutes of mixing in the presence of 1.0 equivalent of the perfect complementary target Y1, the solution becomes yellow (λ_{max}= 421 nm); presumably due to the formation of a new complex termed triplex, formed by complexation of the polymer with the hybridized nucleic acids (Plate 2A,c and 2B,c).

To verify the detection specificity, two different oligonucleotides (20-mers differing by only 1 or 2 nucleotide mismatches) were investigated. A very distinct, stable (up to several hours), and reproducible UV-visible absorption spectrum is observed in the case of oligonucleotide target Y2 having two mismatches (Plate 2A,d and 2B,d) when compared to perfect hybridization (Plate 2A,c and 2B,c). It is even possible to distinguish only one mismatch (Plate 2A,e and 2B,e). In the specific case of one mismatch, the kinetics of complexation was monitored and it was observed that after 30-60 minutes at 55 ^{0}C, the color shifted back towards yellow. However, it is possible to stop the color shift after 5 minutes by cooling the solution to room temperature. Therefore, it is very easy to distinguish one or two mistmaches from the perfect complement in 5 minutes. The detection limit of this colorimetric method is about 1 x 10^{13} oligonucleotide strands (20-mers), in a total volume of 100 μL.

The duplex formed from polymer 1 and capture probe X1 therefore seems specific when tested with various target oligonucleotides. Figure 2 shows the UV-visible absorption spectrum of polymer 1 when using target oligonucleotides ranging from 0 to 5 mismatches (sequence Y1: 0 mismatch; Y3: 1 mismatch; Y2: 2 mismatches; Y4: 3 mismatches [5' TGGTGGATACATCAATCATG 3']; Y5: 5 mismatches [5' TGGTGGAAACAACAATCATG 3']. Figure 3 illustrates the UV-visible absorption results of polymer 1 employing target oligonucleotides with two mismatches but at different positions (sequence Y21 [5' TGGTAGATGCTTCAATCATG 3']; Y22 [5' TGGTGGTTGCTTCAATCATG 3']; Y23 [5' TGGTGGATGCTTTAATCATG 3']; Y24 [5' TGGTGGATGCTTCATTCATG 3']; Y25 [5' TGGTGGATGCTTCAATTATG 3']). These results show that the polymer can discriminate between perfectly matched and mismatched hybrids, independently of the nature of the mismatched nucleotide bases, and independently of the position or the length of the

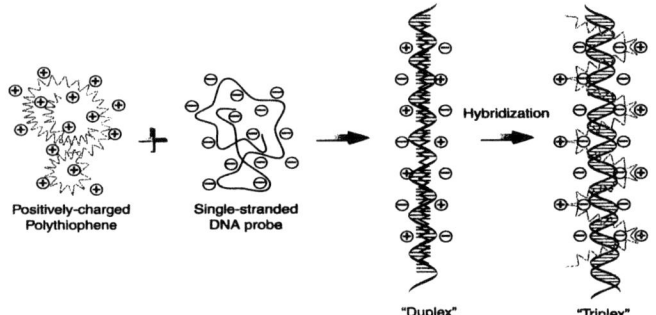

Plate 1. Schematic description of the formation of polythiophene/single stranded oligonucleotide duplex and polythiophene/hybridized oligonucleotide triplex. (See Page 8 of color insert.)

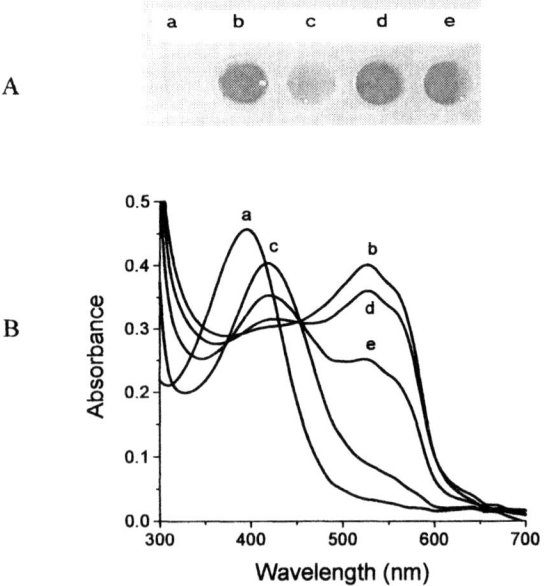

Plate 2. A) Photographs of 7.9 x 10^{-5} M (on a monomeric unit basis) solutions of a) polymer 1, b) polymer 1 / X1 duplex, c) polymer 1 / X1 / Y1 triplex, d) polymer 1 / X1 / Y2 mixture, and e) polymer 1 / X1 / Y3 mixture after 5 minutes of mixing at 55 C in 0.1 M NaCl/H_2O. B) UV-visible absorption spectrum corresponding to the different assays of photograph A. (See Page 8 of color insert.)

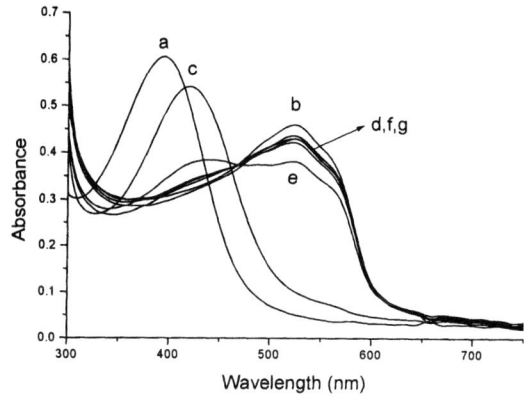

Figure 2. UV-visible spectroscopy spectrum of 7.9×10^{-5} M solutions in TE buffer, at $55°C$ of a) polymer 1; b) polymer 1/ X1 ; c) polymer 1/ X1/ Y1; d) polymer 1/ X1/ Y2 (two mismatches); e) polymer 1/ X1/ Y3 (one mismatch); f) polymer 1/ X1/ Y4 (three mismatches); g) polymer 1/ X1/ Y5 (five mismatches).

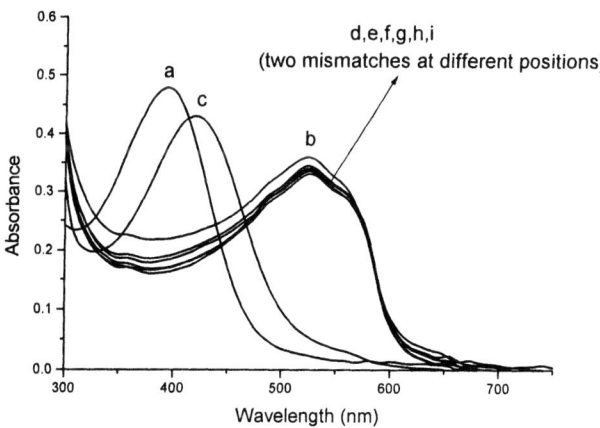

Figure 3. UV-visible spectroscopy spectrum of 7×10^{-5} M solutions in TE buffer, at $55°C$ of a) polymer 1; b) polymer 1/ X1; c) polymer 1/ X1/ Y1; d) polymer 1/ X1/ Y2; e) polymer 1/ X1/ Y21; f) polymer 1/ X1/ Y22; g) polymer 1/ X1/ Y23; h) polymer 1/ X1/ Y24; i) polymer 1/ X1/ Y25.

mismatches. Moreover, it is even possible to discriminate a single mismatch from multiple mismatches.

It is also important to note that the circular dichroism (CD) measurements of polymer 1 or duplex form (polymer 1/ X1) did not reveal any optical activity but a bisignate CD spectrum centered at 420 nm in the triplex (polymer 1/X1/Y1) (*14*), characteristic of a right-handed helical orientation of the polythiophene backbone (*15*). Such a right-handed helical structure is compatible with binding of the polymer to the negatively-charged phosphate backbone of DNA. The presence of CD signal only in the triplex form combined with the presence of a clear isosbestic point as a function of the amount of the complementary strand (not shown here) seem to indicate the coexistence of only two distinct conformational structures (triplex and duplex) for the polythiophene. In addition, these measurements have clearly indicated that triplexes are stable up to 80 ^{0}C. This difference in thermal stability between duplexes and triplexes could be extremely useful for the future design of solid-state sensors where the probes could be covalently attached onto different substrates.

Fluorometric Detection

A fluorometric detection of oligonucleotide hybridization is also possible since the fluorescence of poly(3-alkoxy-4-methylthiophene)s is quenched in the

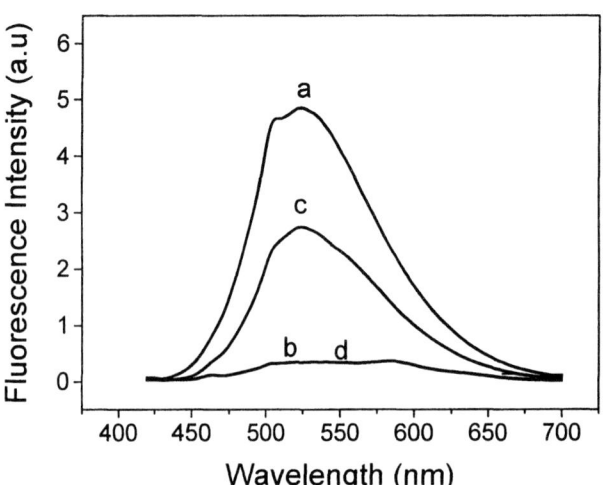

Figure 4. Fluorescence spectrum of a 2.0 x 10^{-7} M (on a monomeric unit basis) solution of a) polymer 1, b) polymer 1/ X1 duplex, c) polymer 1 / X1 / Y1 triplex, d) polymer 1 / X1 / Y2 (100 equivalents) mixture at 55 ^{0}C, in water containing 0.1M NaCl.

planar, aggregated form (*6*). In principle, this method should be much more sensitive than the colorimetric method. For instance, at 55 °C, the yellow form of polymer 1 is fluorescent (quantum yield of 0.03 with a maximum of emission at 530 nm (Figure 4, a) but upon addition of 1.0 equivalent of a negatively-charged capture oligonucleotide probe X1, the emission is quenched and slightly red-shifted (Figure 4, b). When hybridization with the complementary strand Y1 takes place, the formation of a polymeric triplex leads to a 8-fold rise in fluorescence intensity (Figure 4, c). Interestingly, upon addition of 1 or even 100 equivalents (Figure 4, d) of the target oligonucleotide Y2 with two mismatches, the fluorescence intensity is not significantly modified. By measuring the fluorescence intensity at 530 nm, it is possible to detect one mismatch nucleotide (*14*) and the presence of as few as 3×10^6 molecules of the perfect complementary oligonucleotide (20-mers) in a volume of 200 µL. Moreover, covalently attaching the oligonucleotide to the fluorescent conjugated polymer or using an optimized fluorescence detection scheme based on a high-intensity blue diode (as the excitation source) and a non-dispersive, interference filter-based system should yield even more sensitive and more specific detection capability. All these possibilities will be tested in the near future.

Electrochemical Detection

Interestingly, the electrochemical properties of these cationic polythiophene derivatives can also be exploited for the detection of DNA hybridization events in aqueous solutions. Using the layer-by-layer deposition technique (*16, 17*), a monolayer of oligonucleotide X1 was first deposited onto an ammonium-functionalized indium tin oxide (ITO) electrode (*18, 19*), and the presumed hybridization reaction was carried out by adding a solution of oligonucleotide Y1 (complementary target). After rinsing with pure water, the modified electrode was dipped into an aqueous solution (10^{-4} M on a monomeric unit basis) of polymer 1, for 5 minutes. As a control experiment, a solution of oligonucleotide X1 was added to X1-modified ITO electrode and then transfered into an aqueous solution of polymer 1. Therefore, as shown in figure 5, the anodic peak current is larger in the case of perfect hybridization (figure 5B, a), compared to the blank control (figure 5B, b) (the cyclic voltammogram of the control experiment is similar to that obtained with the polymer alone). In addition, a shift to a higher potential (*ca.* 40-50 mV) is observed when the specific nucleic acid hybridization occurs. The higher oxidation current could be explained by the stronger affinity of the polymers for hybridized double-stranded oligonucleotides whereas the positive shift of the oxidation potential could be caused by the formation of a less conjugated polymer structure when the triplex is formed, in agreement with previous optical measurements. An assay using smaller electrodes (S=10 mm²) has allowed the detection of 2 x 10¹¹ oligonucleotide molecules. Moreover, by decreasing the size of the electrodes and by increasing the size of the target molecules, much lower detection limits should be reachable. Future electrochemical investigations will also involve the covalent binding of the oligonucleotide probe onto the electrode to provide more stringent washing conditions.

A

B

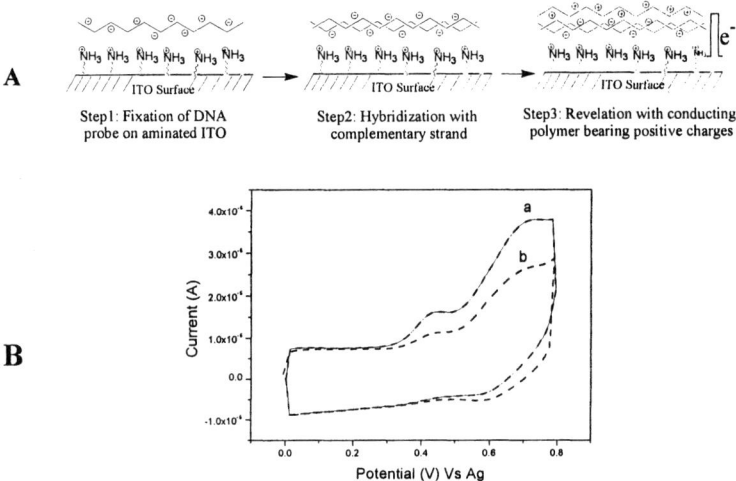

Figure 5. A)Schematic description of the preparation steps for electrochemical detection . B) Cyclic voltammogram of polymer 1 on a modified indium tin oxide (ITO) electrode (S = 50 mm²) in the case of X1/ Y1 / polymer 1 triplex (continuous line) and X1 / X1 / polymer 1 mixture (dotted line). Scan rate of 100 mV/sec vs Ag in 0.1 M NaCl/H_2O, at room temperature.

Acknowledgements

The authors would like to thank the Natural Sciences and Engineering Research Council of Canada, the Canada Research Chair Program, the Canadian Institutes of Health Research, and Infectio Diagnostic (IDI) Inc. for their financial support of this research.

References

(1) Fodor, S.P.A.; Read, J.L.; Pirrung, M.C.; Stryer, L.; Lu, A.T.; Solas, D. *Science* **1991**, *251*, 767-773.

(2) Livache, T.; Roget, A.; Dejean, E.; Barthet, C.; Bidan, G.; Teoul, R.. *Nucleic Acids Res.* **1994**, *22*, 2915-2921.
(3) Tyagi, S.; Kramer, F.R. *Nature Biotechnology* **1996**, *14*, 303-308.
(4) Mikkelsen, S.R. *Electroanalysis* **1996**, *8*, 15-19.
(5) Taton, T.A.; Mirkin, C.A.; Letsinger, R.L. *Science* **2000**, *289*, 1757-1760.
(6) Leclerc, M. *Adv. Mater.* **1999**, *11*, 1491-1498.
(7) Youssoufi, H.K.; Garnier, F.; Srivastava, P.; Godillot, P.; Yassar, A. *J. Am. Chem. Soc.* **1997**, *119*, 7388-7389.
(8) Faïd, K.; Leclerc M. *J. Chem. Soc. Chem. Commun.* **1996**, 2761-2762.
(9) Lucas, P.; Mehdi, N.E.; Ho, A.H.; Bélanger, D.; Breau, L. *Synthesis* **2000**, *9*, 1253-1258.
(10) Balanda, P.B.; M.B. Ramey,; Reynolds, J.R. *Macromolecules* **1999**, *32*, 3970-3978.
(11) Chayer, M.; Faïd, K.; Leclerc, M. *Chem. Mater.* **1997**, *9*, 2902-2905.
(12) B.S. Gaylord, A.J. Heeger, G.C. Bazan, *Proc. Natl. Acad. Sci. USA* . **2002**, 99, 10954-10957 .
(13) K.P.R. Nilsson, O. Inganäs, *Nature Mat.* **2003**, 2, 419-424.
(14) Ho, H.A.; Boissinot, M.; Bergeron, M.G.; Corbeil, G.; Doré, K.; Boudreau, D.; Leclerc, M. *Angew. Chem. Int. Ed.* **2002**, *41*, 1548-1551.
(15) Ewbank, P.C.; Nuding, G.; Suenaga, H.; McCullough, R.D.; Shinkai, S. *Tetrahedron Lett.* **2001**, *42*, 155-157.
(16) Lvov, Y.; Decher, G.; Sukhorukov, G. *Macromolecules* **1993**, *26*, 5396-5399.
(17) Lvov, Y.M.; Lu, Z.; Schenkman, J.B.; Zu, X.; Rusling, J.F. *J. Am. Chem. Soc.* **1998**, *120*, 4073-4080.
(18) Kumpumbu-Kalemba, L.; Leclerc, M. *Chem. Commun.* **2000**, 1847-1848.
(19) Faïd, K.; Leclerc, M. *J. Am. Chem. Soc.* **1998**, *120*, 5274-5278.

Chapter 28

Synthesis of Tunable Electrochromic and Fluorescent Polymers

John D. Tovar[1,3] and Timothy M. Swager[2,*]

[1]Department of Chemistry and [2]Center for Materials Science and Engineering, Massachusetts Institute of Technology, 77 Massachusetts Avenue, Cambridge, MA 02139
[3]Current address: Postdoctural Fellow, Department of Materials Science and Engineering, Center for Nanofabrication and Molecular Self-Assembly, Northwestern University, 2220 Campus Drive, Evanston, IL 60208-3108
*Corresponding author: tswager@mit.edu (email)

Abstract

Approaches to thiophene containing electroactive conjugated polymers are described. The synthetic methods involve a combination of metal catalyzed cross coupling reactions and electrochemically induced cyclizations and polymerizations. Materials with polycyclic aromatic structures were examined for their high stability which is required in demanding electrochromic applications. The large redox active moieties are demonstrated to be versatile building blocks that when inserted into a number of conjugated polymer frameworks can give stable electroactive materials. New highly crosslinked mateials wherein thiophene polymers are bridged by ortho-phenyl groups are also reported as viable electrochromics.

Taken in part from the thesis of John D. Tovar (PhD MIT, 2002.)

Introduction

The discovery of near-metallic conductivity in doped poly(acetylene) in 1977 has led to the development of several technologies that take advantage of the now large variety of conjugated organic polymers (1,2). New paradigms in the plastic electronics field have provided breakthroughs in flexible flat-panel displays and easily fabricated transistors (3). Conjugated polymers such as poly(thiophene)s have also been exploited for electrochromic applications that provide tuneable and transmissive coatings (4). Upon chemical or electrochemical doping, the charge carriers introduced into the conjugated backbone induce structural changes in the polymer that result the creation of mid-gap states and thus, lower-energy optical transitions (most commonly in the near IR). These changes manifest themselves in pronounced color changes as the material undergoes electrochemical cycling between oxidizing and reducing potentials. Furthermore, conjugated polymers have found use in sensory applications ranging from small analyte detection to biosensors (5).

Several groups have investigated new electrochromics and other tuneable systems that offer lower oxidation potentials, higher stability and higher contrast ratios between the differently colored states. Such efforts will ultimately complement advances in materials processing and device construction. The Reynolds group has looked at several modifications of dioxy-thiophene and –pyrrole during their search for stability and low-potential switching (6). The incorporation of oxygen atoms directly onto the heteroaromatic core significantly lowers the monomer (and polymer) oxidation potential, and such low band-gap materials become near-transparent in the oxidized state. Through synthetic alterations, they have obtained an arsenal of differently colored materials by controlling monomer electronics and inter-polymer interactions. The Bard group has studied larger aromatic platforms and ladder polymers in order to utilize the greater stability of the doped materials (7). From these large polycyclic aromatics, they have observed electrochromism and efficient electrogenerated chemiluminescence due to the high stability of the charged intermediates. The design of new synthetic methods for novel and functionalizable cores for incorporation into new polymers will play a key role in advancing this field.

Discussion

Our group has broad interests in new molecular systems of high stability that allow for inclusion into functional electrochromic and/or fluorescent conjugated polymers. Earlier work investigated segmented polymers where redox-active transition metals incorporated into monomers could influence the conjugation pathways of the resulting polymers, depending on the degree of oxidation or on metal binding events (8). Electrochromic measurements helped to verify the segmented conjugation pathways of these materials. Recently, we reported tandem cyclization-polymerizations of thienyl-based monomers where initial anodic activity resulted in a discrete monomer cyclization into a polycyclic aromatic system (9). Subsequently, this aromatized monomer underwent oxidative polymerization at more positive potentials. As generalized in Scheme 1, we studied α and β linked thienyl rings, where the α and β indicate the positions of covalent linkage onto the thienyl moieties. The synthesis of both materials started from common dibromide precursors: the β-linked system **1** via Suzuki coupling with 3-thiophene boronic acid and the α-linked system **2** via Stille coupling with 2-tributylstannyl thiophene (for structures, see Figure 1). Both routes proceeded in high yield and on multi-gram scales.

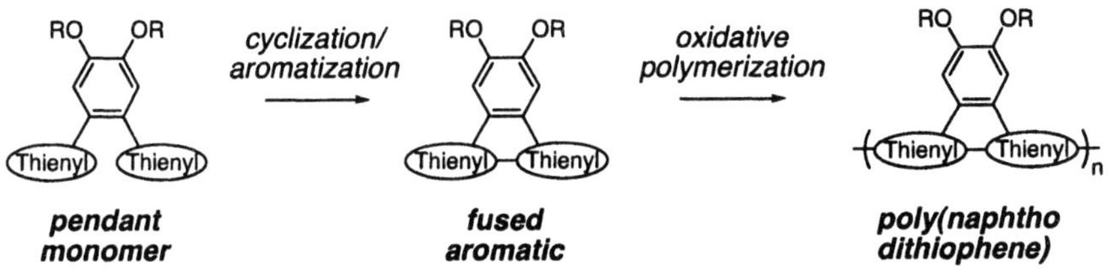

Scheme 1: *Generic cyclization-polymerization scheme. (Reprinted with permission from ref 9. Copyright 2001 Wiley-VCH Publishers.)*

Although a trivial synthetic alteration, the choice of thienyl linkages proved to have dramatic effects on the resulting energetics of the electrodeposited polymer. Anodic oxidation of both monomers resulted in similar cyclization-polymerization activity, although the deposited films of poly(**1**) appeared to have much faster electrochemical kinetics and broader electroactivity when compared to the isomeric poly(**2**) as illustrated in Figure 1. This indicates higher conductivity and a greater degree of delocalization available for poly(**1**). While poly(**1**) derived chemically had a bulk conductivity of 1.3 S/cm, a problematic chemical synthesis of poly(**2**) did not allow for a quantitative comparison. However, the naphthodithiophene moieties imparted extraordinary stability, allowing the electroactivity and electrochromicity of these films to persist after over 10,000 scans under ambient atmospheric conditions.

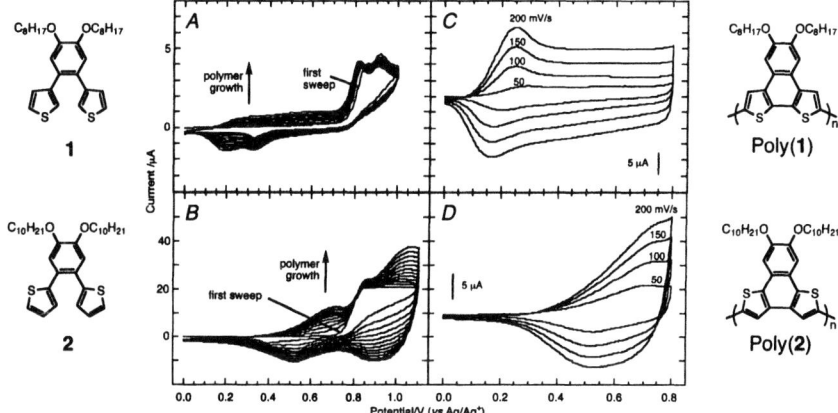

Figure 1: Monomer (A, B) and polymer (C, D) cyclic voltammetry: **1** (A; 0.27 mM), leading to poly(**1**) (C) and **2** (B; 1.72 mM) leading to poly(**2**) (D) at a 2 mm^2 Pt button electrode in 0.1 M n-Bu$_4$PF$_6$ (CH$_3$CN). Monomer CVs cycled at 100 mV/s; polymer film CVs obtained at 50-200 mV/s. $E_{1/2}(Fc/Fc^+) = +0.086$ V.

We can rationalize the greater delocalization available for poly(**1**) by considering the exaggerated, fully quinoidal resonance structures of poly(**1**) and poly(**2**) shown in Scheme 2. Within both isomers, the quinoidal resonance structures should play a major role in stabilizing the charge carriers (polarons and "bipolarons", etc.) (*10*). Poly(**1**) can maintain a quinoidal structure without significantly disrupting the aromaticity of the bridging benzo moieties fused to the dominant conjugation pathway. The quinoidal delocalization of poly(**1**) thus occurs most significantly through the poly(thiophene) portion of the backbone. In contrast, obtaining similar delocalization through poly(**2**) would require a formal disruption of aromaticity in the benzo fragments. One could expect that the less favorable energetics for such a disruption would minimize the extent of charge carrier delocalization.

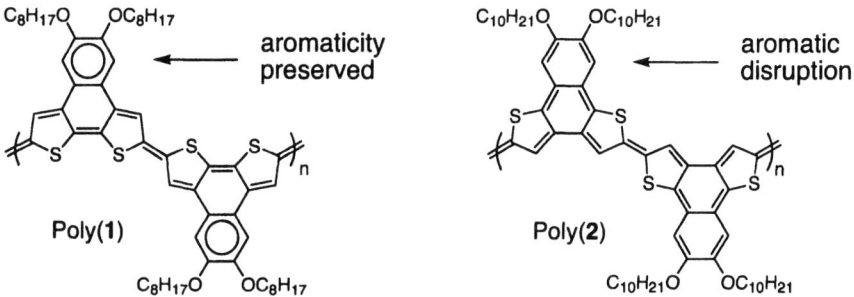

Scheme 2: Fully quinoidal resonance structures that may contribute to the structure of the doped polymers.

Spectroelectrochemical measurements of the electrochromicity of poly(**1**) and poly(**2**) supported these arguments. After depositing films of poly(**1**) or (**2**) (from **1** or **2**, respectively) onto transparent In-SnO$_2$ electrodes (ITO), we can observe the changes in absorbance of the polymer film as a function of applied potential through an optically transparent electrochemical cell. Figure 2 shows the superimposed spectra from these measurements as the applied potential through the cell increased at regular intervals (9). For poly(**1**) (top), a low energy absorption appeared centered at 620 nm along with a near-IR absorbance that increased with further oxidation. For poly(**2**), a low energy band centered at 550 nm appeared with increased oxidation along with a broad near-IR band that drifted to higher energy with further doping. The presence of the isosbestic point at 460 nm also indicated the presence of short or oligomeric conjugation pathways. The well-defined, higher energy transition coupled with the NIR drift in the bottom spectrum implies that poly(**2**) has a more localized electronic structure than that of poly(**1**), consistent with the resonance structures shown above.

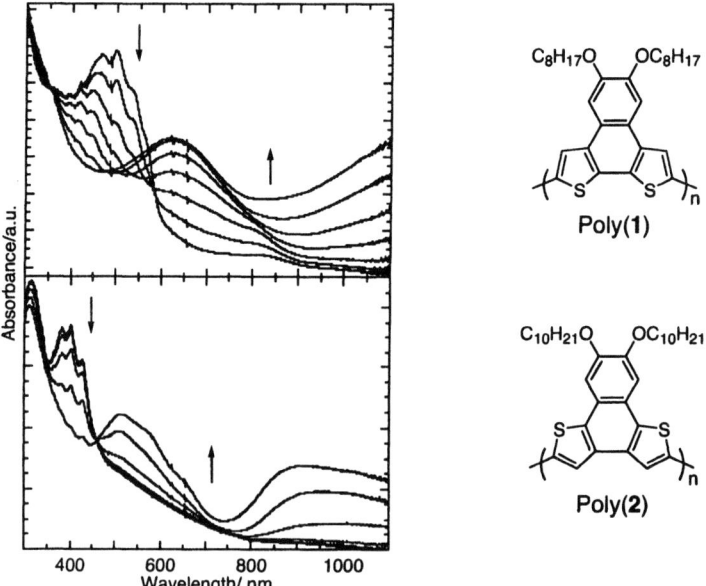

Figure 2: Changes in UV-vis absorption upon increased oxidation for films of poly(1) (top) and poly(2) (bottom) grown on ITO-coated electrodes and held at ca. 120 mV steps between 0.15-0.90 V. Other conditions as in Figure 1.

Other ongoing efforts in our group currently examine the effects that large aromatic cores impart to important photophysical parameters in relation to the design of highly fluorescent sensory polymers. From these studies, we have found that the inclusion of aromatic cores into the conjugation pathway of arylene-ethynylene (AE) polymers (*11*) leads to increased excited-state lifetimes by reducing the rate of radiative decay. Compared to standard phenylene-ethynylenes (PPEs), polymers containing more exotic triphenylene(*12a*) and dibenzochrysene (*12b*) scaffolds in the main chain provided greater degrees of exciton migration and excited-state lifetimes of up to 2.5 ns. Through optimization of such parameters, we may help to allow the excited-state to persist for a longer time in order to efficiently sample more receptors electronically coupled to a polymeric wire (*13*). As an example, Figure 3 shows two AEs that maintain similar conjugation pathways yet differ in the degree of chromophore rigidity. This synthetic alteration significantly changed the photophysics of **3** relative to chemically rigidified **4**, and the lifetime of **4** increased markedly (*12c*). Note that the emission profile of **3** resembled that of **4**, indicating that both materials share similar excited-state conjugation pathways despite the degree of molecular rigidity. We expect polymers such as **4** to offer greater environmental stability relative to the less aromatized co-monomer present in **3**.

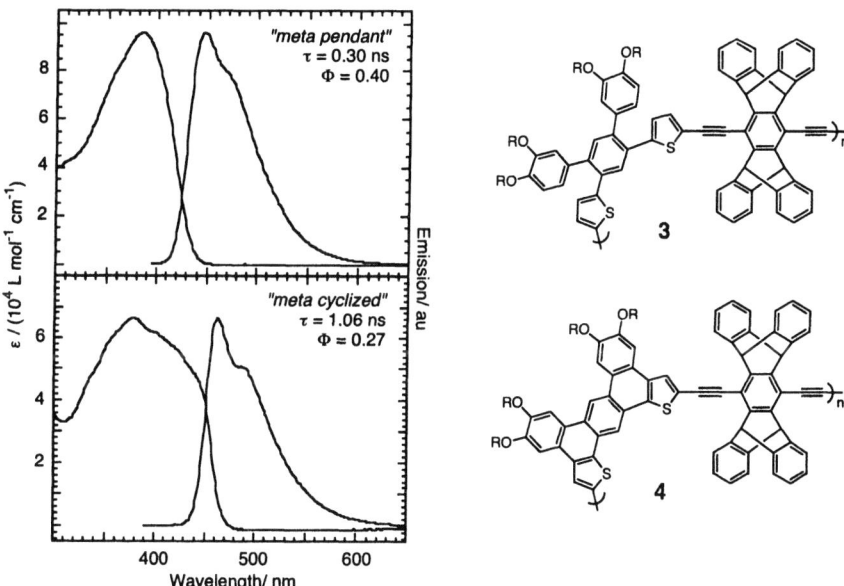

Figure 3: Absorption and emission spectra for **3** (top) and **4** (bottom) acquired at room temperature in CH_2Cl_2. R = 2-ethylhexyl. Quantum yield measurements are relative to quinine sulfate in 0.1 N H_2SO_4 (ϕ = 0.55); lifetime data were obtained by phase-modulation techniques.

The electrochemistry of the PPE family has not received the intense study as that of other conjugated polymers due to the difficulty in observing reversible anodic activity (*14*). However, we can readily study their electrochemistry and any associated spectroscopic changes by including large sulfur-containing aromatic cores. For example, Figure 4 shows the spectroelectrochemistry of an AE-based polymer (**5**). The polymer deposited on ITO during several CV scans performed in a solution of **5**. It would appear that the charged species remain localized on the fused thiophene moiety as opposed to delocalizing along the backbone on account of the high-energy peak (360 nm) that persists upon electrochemical oxidation. Zotti has reported similar phenomena within poly[bis(EDOT)-ethynylene]s where the charged species persist on the electron-rich thiophene moieties rather than delocalize through the alkyne linkers (*14b*).

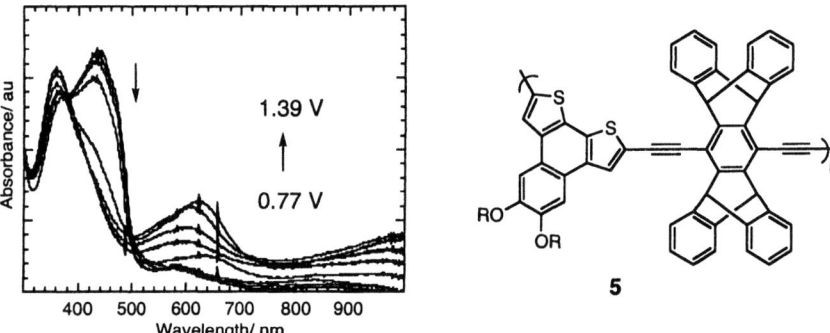

Figure 4: UV-Vis spectra of polymer **5** deposited on ITO and recorded at ca. 120 mV intervals; conditions as in Figure 1. R = 2-ethylhexyl. The pristine absorption profile persisted until application of 0.77 V vs. Ag/Ag$^+$.

We also have studied unique chromophores where the electroactive moieties lie physically constrained within van der Waals distances through covalent attachments to a common core. In this system, there exists a possibility to stabilize the intermediate charge carriers by way of π-dimers pre-existing within the polymer. Conducting polymer-based actuators can allow for counteranion and solvent swelling upon electrochemical doping as a mechanism for bulk spatial movement (*15*). This results from a collective expansion of a doped conducting polymer film as solvent and counteranions diffuse in to compensate charge. We felt that the cofacial orientation of molecular systems such as **6** would allow for unique optical signatures during such a process upon application of a mechanical stress. After electropolymerization from monomer **6**, the resulting cross-linked film displays optical characteristics similar to standard poly(thiophene)s (*16*). Routine Suzuki and Stille cross-couplings will allow for the synthesis and study of a variety of additional co-facially oriented chromophores.

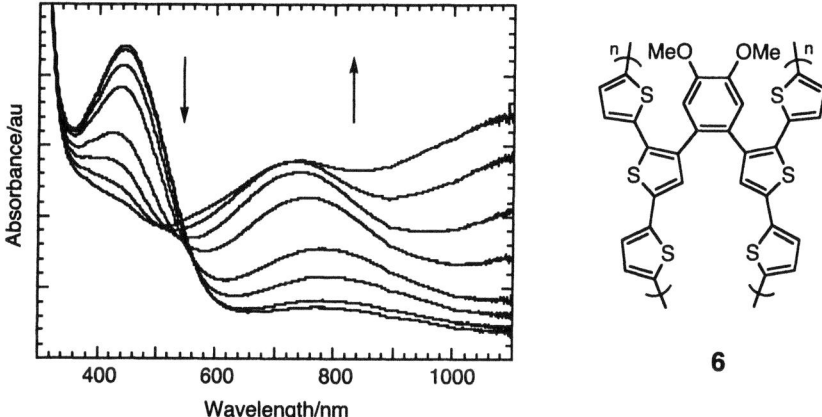

Figure 5: UV-Vis spectra of polymer **6** electrodeposited on ITO and recorded at ca. 120 mV intervals between 0.09-0.74 V; other conditions as in Figure 1.

Conclusions

We have presented some of our recent work focused on the design and synthesis of new tuneable optical polymers. By controlling the main chain connectivity through tandem cyclization-polymerizations, we could significantly alter the electrochromic properties of conductive thiophene-based polymers. Through incorporating large aromatic scaffolds, we could tune fluorescent polymer photophysics as well as incorporate redox-active units to provide electrochromism to AE-based systems. Finally, the design of co-facial redoxophores also holds promise for the synthesis of unique and responsive conducting materials.

Acknowledgment

This research was funded by the Tunable Optical Polymers MURI award from the Army Research office and made use of the Shared Experimental Facilities supported in part by the MRSEC Program of the National Science Foundation under award number DMR 02-13282.

References

(1) Chiang, C. K.; Fincher, C. R.; Park, Y. W.; Heeger, A. J.; Shirakawa, H.; Louis, E. J.; Gau, S. C.; MacDiarmid, A. G. *Phys. Rev. Lett.* **1977**, *39*, 1098-1101.
(2) *Handbook of Conducting Polymers*, 2nd ed; Skotheim, T. A.; Elsenbaumer, R. L.; Reynolds, J. R., Eds. Marcel Dekker: New York, 1998.
(3) (a) Horowitz, G.; *Adv. Mater.* **1998**, *10*, 365-377; (b) Dimitrakopoulos, C. D.; Mascaro, D. J. *IBM J. Res. Dev.* **2001**, *45*, 11-27; (c) Katz, H. E.; Bao, Z.; Gilat, S. L. *Acc. Chem. Res.* **2001**, *34*, 359-369; (d) Cornil, J.; Beljonne, D.; Calbert, J.-P.; Brédas, J.-L. *Adv. Mater.* **2001**, *13*, 1053-1067; (e) Dimitrakopoulos, C. D.; Malenfant, P. R. L. *Adv. Mater.* **2002**, *14*, 99-117.
(4) See Monk, P. M. S.; Mortimer, R. J.; Rosseinsky, D. R. *Electrochromism*. VCH: New York, 1995.
(5) McQuade, D. T.; Pullen, A. E.; Swager, T. M. *Chem. Rev.* **2000**, *100*, 2537-2574.
(6) (a) Kumar, A.; Welsh, D. M.; Morvant, M. C.; Piroux, F.; Abboud, K. A.; Reynolds, J. R. *Chem. Mater.* **1998**, *10*, 896-902; (b) Groenendaal, L.; Jonas, F.; Freitag, D.; Pielartzik, H.; Reynolds, J. R. *Adv. Mater.* **2000**, *12*, 481-494; (c) Schottland, P.; Zong, K.; Gaupp, C. L.; Thompson, B. C.; Thomas, C. A.; Giurgiu, I.; Hickman, R.; Abboud, K. A.; Reynolds, J. R. *Macromolecules* **2000**, *33*, 7051-7061.
(7) (a) Debad, J. D.; Bard, A. J. *J. Am. Chem. Soc.* **1998**, *120*, 2476-2477; (b) Debad, J. D.; Morris, J. C.; Lynch, V.; Magnus, P.; Bard, A. J. *J. Am. Chem. Soc.* **1996**, *118*, 2374-2379.
(8) (a) Kingsborough, R. P.; Swager, T. M. *J. Am. Chem. Soc.* **1999**, *121*, 8825-8834; (b) Shioya, T.; Swager, T. M. *J. Chem. Soc., Chem. Commun.* **2002**, *13*, 1364-1365.
(9) Tovar, J. D.; Swager, T. M. *Adv. Mater.* **2001**, *13*, 1775-1780.
(10) (a) Brédas, J. L. *J. Chem. Phys.* **1985**, *82*, 3808-3811; (b) Patil, A. O.; Heeger, A. J.; Wudl, F. *Chem. Rev.* **1988**, *88*, 183-200; (c) Hill, M. G.; Mann, K. R.; Miller, L. L.; Penneau, J.-F. *J. Am. Chem. Soc.* **1992**, *114*, 2728-2730; (d) Roncali, J. *Chem. Rev.* **1997**, *97*, 173-205.
(11) Bunz, U. H. F. *Chem. Rev.* **2000**, *100*, 1605-1644.
(12) (a) Rose, A.; Lugmair, C. G.; Swager, T. M. *J. Am. Chem. Soc.* **2001**, *123*, 11298-11299; (b) Yamaguchi, S.; Swager, T. M. *J. Am. Chem. Soc.* **2001**, *123*, 12087-12088; (c) Tovar, J. D.; Rose, A.; Swager, T. M. *J. Am. Chem. Soc.* **2002**, *124*, 7762-7769.
(13) Swager, T. M. *Acc. Chem. Res.* **1998**, *31*, 201-207.
(14) (a) Ofer, D.; Swager, T. M.; Wrighton, M. S. *Chem. Mater.* **1995**, *7*, 418-425; (b) Zotti, G.; Schiavon, G.; Zecchin, S.; Berlin, A. *Synth. Met.* **1998**, *97*, 245-254; (c) Evans, U.; Soyemi, O.; Doescher, M. S.; Bunz, U. H. F.; Kloppenburg, L.; Myrick, M. L. *Analyst* **2001**, *126*, 508-512.
(15) Otero, T. F. In *Polymer Sensors and Actuators*, Osada, Y.; DeRossi, D. E., Eds. Springer: New York, 2000. pps 295-323.
(16) Tovar, J. D.; Swager, T. M. *J. Polym. Sci. A, Polym. Chem.*, submitted.

Chapter 29

Enzymatically Synthesized Electronic and Photoactive Materials

Jayant Kumar[1], Wei Liu[1], Soo-Hyuong Lee[1], Suizhou Yang[1], Sukant Tripathy[1], and Lynne Samuelson[2]

[1]Center for Advanced Materials, Department of Chemistry and Physics, University of Massachusetts at Lowell, Lowell, MA 01854
[2]Materials Science Team, U.S. Army Soldier and Biological Chemical Command, Soldier Systems Center, Natick, MA 01760

Electronic and photo-active polymers as a class of advanced materials have attracted a lot of interest during last two decades. Typically these materials are synthesized chemically or electrochemically under harsh conditions such as extremely low pH, toxic catalysts and byproducts. New enzymatic approaches have been developed for the synthesis of electro and photo active polymers such as polyanilines, polyazophenols, and polypyrene derivatives. These enzymatically synthesized materials show interesting optical and electronic properties.

Introduction

Peroxidase-catalyzed oxidation of phenols and anilines has been extensively investigated due to its importance in the synthesis of electronic and photoactive polymer in an environmentally benign way (*1-3*). In the presence of H_2O_2, peroxidase, such as horseradish peroxidase (HRP) can oxidize phenols and anilines to generate corresponding radicals. These radicals may couple together through radical coupling and radical transfer to form oligomers and polymers. In a suitable solvent system, polymers such as polyanilines and polyphenols with modest molecular weight may be realized.

Typically, the enzymatically synthesized polyanilines (in both organic and aqueous solutions) are not electrically active due to the low molecular weight and presence of branched structure (4). Recently a new enzymatic approach was developed to synthesize water-soluble conducting polyaniline under mild conditions using sulfonated polystyrene (SPS) as a template (3). The properties of this polyaniline/SPS complex are comparable to previously reported chemically synthesized and self-doped sulfonated polyaniline. The use of the anionic polyelectrolytes as templates in peroxidase-catalyzed polymerization of aniline demonstrates the first enzymatic synthesis of conducting polyaniline, and opens the door for the synthesis of electrical active conducting polymers through biological approach.

Peroxidase-catalyzed synthesis involves a reaction mechanism that results in a direct ring-to-ring coupling of phenol and aniline monomers. The resulting polymers may have an aromatic backbone structure, with interesting electrical and optical properties. By using chromophore functionalized phenols or anilines as substrates, phenol-based (or aniline-based) macromolecular dyes with interesting optical or electrical properties (depending on the chromophore) may be synthesized. This approach offers the possibility to build in substantial chromophore density in the polymers. For example, by using azo-functionalized phenol and aniline as monomers, photo-active polyaniline and polyphenol have been enzymatically synthesized by our group (5). These biologically derived azopolymers have almost 100% dye content. Since both ortho and meta couplings occur through the phenol ring, a high articulated nanostructured macromolecular dye is realized, leading to large free volume and poor packing of the azo chromophores.

In this paper, the enzymatic synthesis of conducting polyaniline, and macromolecular dyes such as polyazophenols and polypyrenes is discussed.

Experimental

Materials. Horseradish peroxidase (HRP) (EC 1.11.1.7) (200 unit/mg) was purchased from Sigma with RZ > 2.2. A stock solution of 10 mg/ml in 0.1M phosphate buffer (PH = 6.0) was prepared. The para substituted azophenol monomers were synthesized following typical procedure. All other chemicals and solvents used were commercially available, of analytical grade or better and used as received.

Enzymatic polymerization of aniline. The enzymatic polymerization of aniline was typically carried out at room temperature in a 30 ml, 0.1 M sodium phosphate buffer solution of pH 4.3 which contained a 1:1 molar ratio of SPS to aniline, (6 mM) SPS (based on the monomer repeat unit) and 6mM aniline. SPS was added first to the buffered solution, followed by addition of the aniline with constant stirring. To the solution, 0.2 ml of HRP stock solution (10 mg/ml) was then added. The reaction was initiated by the addition of a stoichiometric amount of H_2O_2 under vigorous stirring. To avoid the inhibition of HRP due to excess H_2O_2, diluted H_2O_2 (0.02 M) was added dropwise, incrementally, over 1.5 hours. After the addition of H_2O_2, the reaction was left stirring for at least one hour and then the final solution was dialyzed (cutoff molecular weight of 2000D) against pH 4.3 deionized water overnight to remove any unreacted monomer, oligomers and phosphate salts.

Enzymatic polymerization of azophenols. Enzymatic polymerization of 4-phenylazophenol was carried out at room temperature in a 100 ml, 50% acetone and 50% 0.01 M sodium phosphate buffer mixture, which contained 2.0 g of 4-phenylazophenol. To this solution, 2.0 ml of HRP stock solution was added. The reaction was initiated by the addition of H_2O_2. To avoid the inhibition of HRP due to excess H_2O_2, a diluted stoichiometric amount of H_2O_2 (0.2 M) was added incrementally under vigorous stirring over a three hour time period. After the addition of H_2O_2, the reaction was left stirring for one more hour. The yellow precipitates formed during the reaction were then collected with a Buchner funnel, washed thoroughly with the mixed solvent of 20% acetone and 80% water (v/v) to remove any residual enzyme, phosphate salts and unreacted monomers and then vacuum dried for 24 hours. Similar experimental conditions were used for CH_3O, NO_2 and CN substituted azophenols.

Polymerization of the sulfonated and carboxylic substituted monomers was carried out in a mixture of 80% phosphate buffer and 20% acetone. The other conditions are similar to that described previously. In these reactions, no precipitates were formed and the resulting polymers were solubilized in the reaction media. The polymer solutions were dialyzed against deionized water for 24 hours using a dialysis bag (SPECTRUM®) with a molecular weight cut off of 3000 to remove unreacted monomer and phosphate salts. The resulting dialyzed solution was then condensed and dried in a vacuum oven at 50 °C.

Enzymatic polymerization of 4-hydroxypyrene. The enzymatic polymerization of 4-hydroxypyrene was performed as a similar procedure as described above for the azophenols in the solution of 50% ethanol and water mixture. The synthesized polymers were formed as precipitates, and purified by centrifuging and washing.

Optical quality polymer film preparation. The H, CN, NO_2, and CH_3O substituted azophenol polymers were soluble in most polar organic solvents.

These polymers were dissolved in spectroscopic grade dioxane, and then filtered through a 0.45 μm membrane. The solutions were then spin-coated onto glass slides. The film thickness was controlled to 0.2-2.0 μm by adjusting the solution concentration and spin speed. The spin-coated films were then dried under vacuum for 24 h at 40-50 °C and stored in a desiccator until further studies. Since the sulfonated and carboxylic substituted polyazophenols were water-soluble, deionized water was used as the solvent to dissolve these polymers (pH 11 water was used for the carboxylic substituted polyazophenol). The solutions were also filtered through a 0.45 μm membrane and the films were fabricated on glass substrates at temperature of 70 °C. The thickness of all spin-coated films was measured by using a Dektak IIA surface profilometer.

Surface relief grating formation. Surface relief gratings (SRG) were holographically recorded by a simple two-beam interference apparatus at 488 nm from an argon ion laser under ambient conditions with a typical laser intensity of 300 mW/cm^2. The formation process of the grating was probed by monitoring the first order diffraction of a lower power He/Ne laser beam at 633 nm, at which the absorption is negligible. After the holographic gratings were recorded, the surface relief structures of the gratings on the polymer films were imaged by atomic force microscopy (AFM, Autoprobe Cp, Park Scientific Instruments) under ambient conditions. A 100 μm scanner in the contact mode under a scan rate of 1 Hz was used in these measurements.

Results and Discussion

Enzymatically synthesized conducting polyaniline. The enzymatic polymerization of aniline in micelle solution was performed as shown schematically in Scheme 1. This approach is based on preferential electrostatic alignment of aniline monomer onto an anionic template to minimize branching and promote a linear polyaniline chain growth. Since aniline has a pK$_b$ of 4.63 (6), it is primarily positively charged at pH 4.3. Conversely, the sulfonate groups on the SPS are negatively charged (SPS is a strong polyelectrolyte that will totally dissociate in almost the entire pH range). Therefore it is believed that the aniline monomer interacts with the SPS electrostatically and preferentially complexes with the template prior to and during the reaction. This approach inherently minimizes the parasitic branching and promotes a more para-directed, head to tail polymerization of aniline and produces a water-soluble conducting polyaniline and SPS complex.

Scheme 1

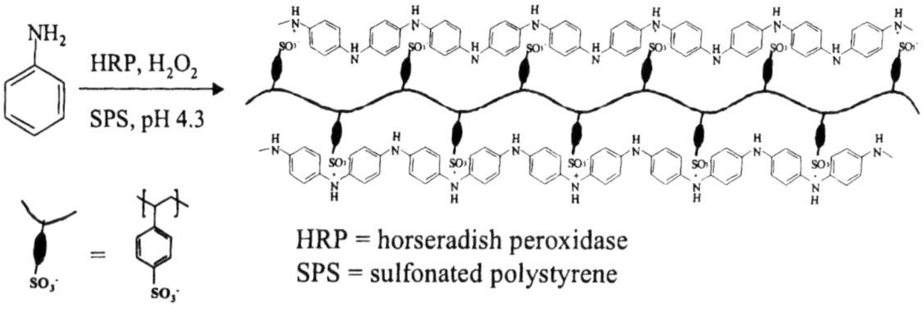

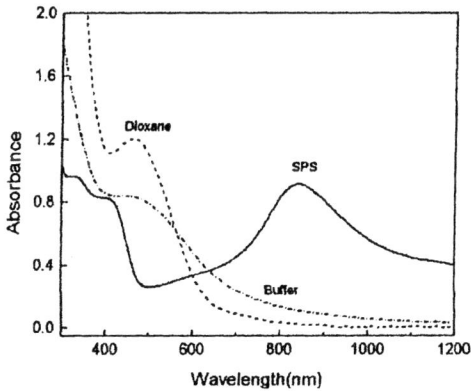

HRP = horseradish peroxidase
SPS = sulfonated polystyrene

To determine the role of the SPS template during the enzymatic polymerization, a series of control experiments were investigated. The polymerization was carried out in an 85% dioxane/15% water mixture with no SPS; an aqueous pH 4.3 buffered solution with no SPS and an aqueous pH 4.3 buffered solution with 1mM SPS. The absorption spectra of the three solutions prior to precipitation were measured and are given in Figure 1. The solutions which contained no SPS showed an absorption band at approximately 460 nm, indicating the presence of multiple branched structures in the polymer. In contrast, the polyaniline formed in the presence of SPS exhibits a significantly different absorption spectrum. In this case, three absorption bands are observed

Figure 1. UV-Vvis spectra of the polymer obtained by the polymerization of 1mM aniline in (...) phosphate buffer, (---) mixture of 85% dioxane and 15% buffer and (—) 1 mM SPS buffer solution at pH 4.3. Reproduced from reference 3. Copyright 1999 American Chemical Society.

which are consistent with the emeraldine salt form of PANI. One is due to a π-π* transition of the benzenoid ring at 325 nm and two absorption peaks at 414 and 843 nm are due to polaron band transitions (7). These peaks indicate that a conducting form of the PANI, which is spectroscopically similar to that presently obtained through either chemical or electrochemical methods, may now be synthesized enzymatically. These results also demonstrate that the role of the template is critical to this process. The SPS in this case promotes a less parasitic and more para directed polymerization, provides the necessary counterions for doping and maintains the water solubility of the polyaniline.

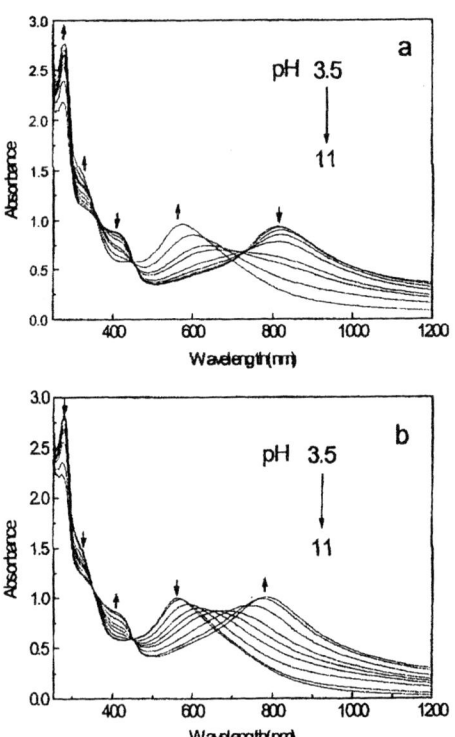

Figure 2. UV-Vis spectra change of PANI/SPS complex during titration by 1N NaOH and 1N HCl. The pH ranged from (a) 3.5 to 11 and (b) 11 to 3.5. The pH values were monitored by a pH meter during the titration. Reproduced from reference 3. Copyright 1999 American Chemical Society

The reversible reduction/oxidation behavior of the PANI/SPS complex was determined by monitoring the absorption spectra change in the pH range from 3.5 to 11. Figure 2a gives the shift in absorption spectra of the complex with increasing pH from 3.5 to 11 by titrating with 1N NaOH. At pH 3.5, the PANI in the complex is in the doped state as reflected by the presence of the polaron band transition at about 420 nm and 823 nm, as well as the π-π* transition of the benzenoid rings at 310-320 nm. As the pH of the complex is increased, the polaron bands at 420 and 823 nm gradually disappear and a strong absorption due to exciton transition of the quinoid rings at 560-600 nm begins to emerge. At the same time bands at 257 and 320 nm, which are due to π-π* transitions of the benzenoid rings in the SPS and PANI molecules, respectively, increase with a pH increase. At a pH of 11, a blue solution of PANI/SPS complex is formed, indicating that the PANI has been fully dedoped to the emeraldine base form. The dedoped PANI can be redoped by titrating with 1N HCl. A reversible color change is observed and the spectra are given in Figure 2b. This pH induced redox reversibility confirms the presence of the electroactive form of polyaniline in the PANI/SPS complex.

Enzymatically synthesized polyazophenols. Peroxidase-catalyzed polymerization of azophenols is schematically shown in Scheme 1. FTIR, FT-Raman and NMR (^{1}H, ^{13}C) spectroscopy show that the coupling reaction occurs

Scheme 2

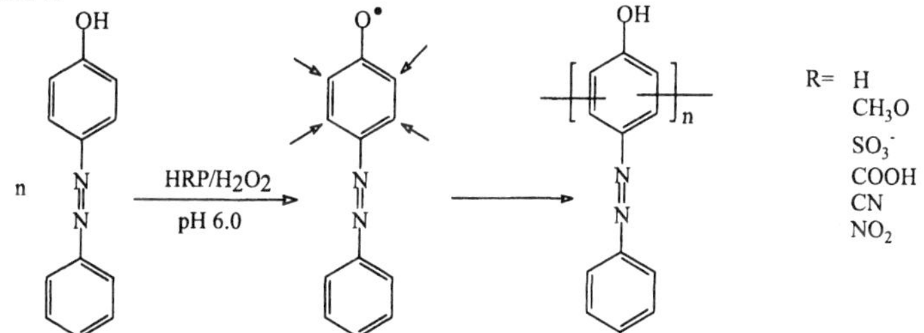

R= H
CH$_3$O
SO$_3^-$
COOH
CN
NO$_2$

primarily at the ortho positions with some coupling at the meta positions of the phenol ring of the monomer as well (data not shown in this paper). This results in the formation of a branched polyphenylene backbone with pendant azo functionalities on every repeat unit of the macromolecules. The enzymatic polymerization of 4-phenylazophenol was studied with UV-Vis spectroscopy. Figure 3 shows the absorption spectra of solutions of the monomer and polymer in dioxane. The monomer spectrum of 4-phenylazophenol is similar to that known for other azo benzene derivatives where a maximum absorption at 355 nm, characteristic of trans 4-phenylazophenol and a weak broad peak at about

440 nm due to cis 4-phenylazophenol (8). Comparison of these solution spectra shows that a significant absorption change occurs as a result of polymerization. The trans absorption at 355 nm in the monomer, blue shifts to 345 nm in the polymer, and the cis absorption at 440 nm becomes stronger in the polymer.

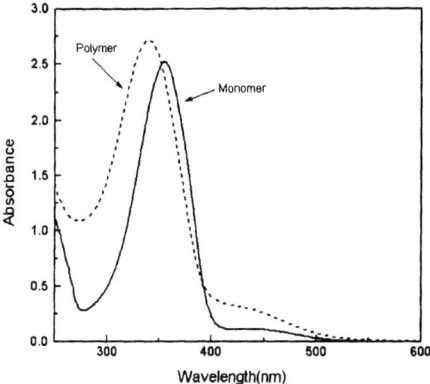

Figure 3. UV-Vis spectra for the polymer and monomer of 4-phenylazophenol in dixoane. Reproduced form reference 5. Copyright 2000 American Chemical Society.

One possible explanation for this trans to cis isomerization during polymerization is that incorporation of the phenol ring into the backbone of the polymer causes strong steric hindrance to the trans form. The backbone of the growing polymer has significant conformational constraint, and a blue shift of

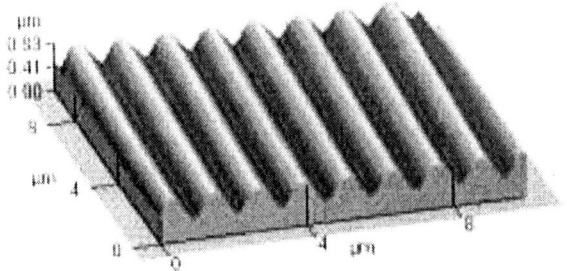

Figure 4. AFM image of SRG formed on enzymatically synthesized polyazophenol thin film.

the trans absorption is observed. This blue shift is accompanied by a partial isomerization to the cis form to help accommodate the structural constraints of the growing macromolecule.

One of the important applications of these enzymatically synthesized macromolecular dyes is photoinduced fabrication of SRG's. Taking advantage of the good solubility of these polymers, optical quality films may be fabricated by spin coating the polymer solution onto glass slides. Surface relief gratings were optically inscribed on these polymer films at room temperature. This photofabrication process is a simple, one step process that doesn't require any subsequent post processing. As an example, a three-dimensional view of the SRG written on the methoxy substituted polyazophenol film is shown in Figure 4. A SRG with surface modulation around 0.4 µm was formed on this film.

Enzymatically synthesized poly(1-hydroxypyrene). 1-Hydroxypyrene was enzymatically oxidized in a mixture of 50% ethanol and 50% buffer (0.01M phosphate) at pH 6.0 at room temperature as shown in Scheme 3.

Scheme 3

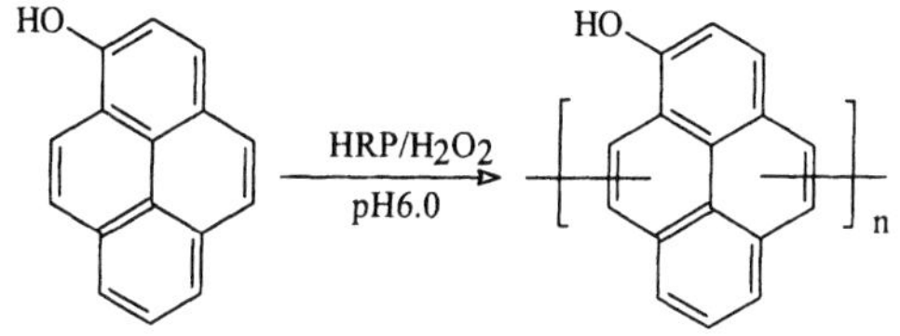

The enzymatically synthesized poly(1-hydroxypyrene) doesn't show any fluorescence in the mixture of ethanol and water during the reaction. However, the synthesized products do exhibit strong fluorescence in anhydrous solvents, such as ethanol, dioxane, and DMF. The fluorescence spectra of monomer and polymer of 1-hydroxypyrene in dioxane are shown in Figure 5. The monomer of 1-hydroxypyrene shows strong fluorescence peak at 394 nm in anhydrous dioxane with a shoulder peak at 416 nm. However, the peaks of the polymer fluorescence spectrum shift significantly to longer wavelength with maximum emitting peak at 482 nm and a shoulder peak at 509 nm. A red shift of ~70 nm for the major peak of the polymer was observed compared to that of the monomer. The pyrene derivatives aggregate at high concentration to form excimers, which usually causes red shift of the emission. In this case, the significant red shift of the emission of the synthesized products is not due to the formation of the excimer, since the low concentrations of monomer and polymer have been used in these measurements. The observed dramatic red shift of emission compared to the monomer may be due to an increase of the conjugation length.

It has been reported previously that the structures of enzymatically synthesized polyphenols are very complicated due to the presence of the coupling of both C-C (two carbons on the different aromatic ring coupled together, usually at ortho position) and C-O-C (the oxygen from hydroxy group on one aromatic ring coupled with the carbon on another aromatic ring) in the reaction. The main chain of the synthesized polyphenols is usually a mixture of phenylene and oxyphenylene units. The structure of the synthesized product was characterized by ^{1}H NMR and FTIR spectroscopy in the present work (data not shown). The primary results show a similar structural feature as that observed in the enzymatically synthesized polyphenols with both C-C and C-O-C coupling involved in the reaction. As one can see from the molecular structure of 1-hydroxypyrene, several radical resonance structures may be formed. Thus, more positions on the pyrene ring are available for coupling compared to that of the phenol. The structure of the enzymatically oxidized 1-hydroxypyrene may be more complicated compared to polyphenols. The details on the coupling positions and the final structure of the synthesized products are still under investigation.

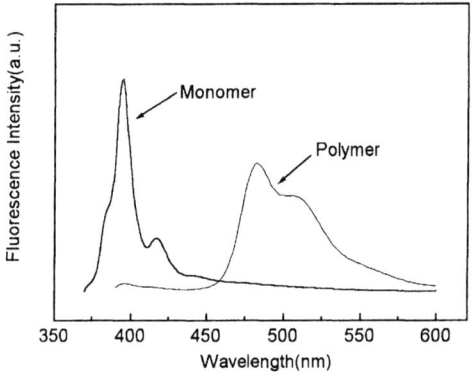

Figure 5. Fluorescence spectra of polymer and monomer of 1-hydroxypyrene.

Conclusions

Electronic and photo active polymers such as polyaniline, polyazophenols and poly(1-hydroxypyrene) were synthesized by peroxidase-catalyzed polymerization. These enzymatically synthesized polymers showed interesting electronic and photonic properties. Optical devices such as SRG can be fabricated on the enzymatically synthesized polymer thin film. The enzymatic synthesis of electronic and photoactive polymers provide new opportunities regarding the design and synthesis of processable macromolecular systems.

Acknowledgement. We thank Drs. Shaoping Bian and Lian Li for their helps in the fabrication of SRG. This work was finacially surported by US Army Natick Labs and the Office of Naval Research.

References

1. Premachandran, R.; Banerjee, S.; John, V. T.; McPherson, G. L.; Akkara, J. A.; Kaplan, D. L. *Chem. Mater.* **1997**, *9*, 1342.
2. Premachandran, R. S.; Banerjee, S.; Wu, X.-K.; John, V. T.; McPherson, G. L.; Akkara, J. A.; Ayyagari, M.; Kaplan, D. L. *Macromolecules* **1996**, *29*, 6452.
3. Liu, W.; Kumar, J.; Tripathy, S. K.; Senecal, K. J.; Samuelson, L. A. *J. Am. Chem. Soc.* **1999**, *121*, 71.
4. Kaplan, D.L; Dordick, J.; Gross, R.A.; Swift, G. Enzymes in Polymer Science: An Introduction. In: Gross, R.A., Kaplan, D.L. and Swift, G. Enzymes in polymer synthesis. *ACS symposium series* **1998**, 684, pp.4.
5. Liu, W; Bian, S; Li, L.; Samuelson, L.; Kumar, J.; Tripathy, S.K. *Chem. Mater.* **2000**, *10*, 1577-1584.
6. Lide, D. R. In *Handbook of Chemistry and Physics*; 68th ed.; CRC Press: Boca, Raton, FL, 1993; pp D159-161.
7. (a) Stafstrom, S.; Bredas, J. L.; Epstein, A. J.; Woo, H. S.; Tanner, D. B.; Huang, W. S.; MacDiarmid, A. G. *Phys. ReV. Lett.* **1987**, *59*, 1464. (b) Ginder, J. M.; Epstein, A. J. *Phys. ReV. B* **1990**, *41*, 10674. (c) Wudl, F.; Angus, R. O.; Lu, F. L.; Allemand, P. M.; Vachon, D. J.; Nowak, M.; Liu, Z. X.; Heeger, A. J. *J. Am. Chem. Soc.* **1987**, *109*, 3677.
8. Kumar, G. S.; Neckers, D. C. *Chem. Rev.* **1989**, *89*, 1915.

Chapter 30

Explosive Detection by Fluorescent Electrospun Polymer Membrane Sensor

Xianyan Wang[1], Christopher Drew[1], Soo-Hyoung Lee[1], Kris J. Senecal[2], Jayant Kumar[1,*], and Lynne A. Samuelson[2,*]

[1]Center for Advanced Materials, Departments of Chemistry and Physics, University of Massachusetts at Lowell, Lowell, MA 01854
[2]Natick Soldier Center, U.S. Army Soldier and Biological Chemical Command, Natick, MA 01760

Introduction

Chemical sensors for explosive detection have attracted increasing attention recently due to heightened awareness of terrorist and criminal activities. Numerous methods of direct explosive detection, including ion mobility spectrometry, neutron analysis, X-ray backscattering, and electron capture detection have been developed. However, there remains urgent need for new approaches that not only complement existing methods, but improve on them in terms of lower cost and greater instrumental simplicity (*1*).

Fluorescent optical chemical sensors are of particular interest due to their inherent sensitivity and simplicity (*2*). These types of sensors have many other advantages that optical sensors, in general, offer. One of the most attractive features is that they do not require a separate reference sensor, as a potentiometric chemical sensor does. In addition, they are not affected by electrical interference, sample flow rate, and stir speed which can be serious problems with electrochemical sensors. Fluorescent optical chemical sensors have been widely used for quantitative measurements of various analytes in environmental, industrial, clinical, medical, and biological applications (*2*).

Over the last decade, polymeric materials have attracted tremendous interest in optical sensor applications due to their unique attributes and advantages for sensor technologies (3). They are relatively low-cost materials. Their processing and fabrication techniques are quite simple, i.e. there is no need for special clean-room or high-temperature processes. They can be deposited on various types of substrates. Polymers posses a wide variety of molecular structures expanded by the possibility to build in various side chains to endow the films with desirable physical and chemical properties. However, conventional polymer materials alone are generally not active sensing materials in that their optical parameters cannot be significantly affected by the environment. Therefore, suitable fluorescence indicators are used as molecular recognition materials in polymeric sensors (4). These indicators must exhibit changes in fluorescence intensity in the presence of the desired analyte to be detected. The procedures for immobilization, the materials used, and the morphology of the sensing films have strong effects on the performance of the sensor in terms of stability and sensitivity.

In many cases, fluorescent dyes are immobilized by physical or chemical procedures onto the polymeric materials for fabrication. The physical procedures used for immobilization include adsorption (5), dissolution (6), entrapment in a porous network (7) and ion exchange (8). These methods are simple but suffer from the problem of insolubility of dyes in the polymeric support, which results in the dyes leaching-out. The chemical procedure to immobilize the dye entails the formation of covalent bonds between the dyes and support materials. Sensors with covalently immobilized dyes have the advantage of not suffering from dye loss over time. However, attaching the fluorescent group to the polymer is not always a trivial task in that the reaction can be quite complicated, involving multiple steps and difficult reactant purification (9).

It is well established that the sensitivity of the sensing film in a sensor is proportional to the surface area of the film to volume ratio. Thin films with very large surface areas can be easily fabricated by electrospinning, wherein a polymer solution is exposed to a high static voltage creating sub-micron or nanometer scale fibers collected as a non-woven membrane (10). Electrospun nanofibrous membranes can have a surface area approximately one to two orders of the magnitude higher than those found in continuous thin films. It has been demonstrated that this high surface area has the potential to provide unusually high sensitivity and fast response time for sensing applications (11).

In this chapter, the on-going research into the effects of the polymeric system on the performance of the fluorescent electrospun polymer nanofibrous membrane sensors for the detection of 2,4-dinitrotoluene are discussed.

Techniques

Fluorescence quenching

The fluorescence-quenching phenomenon has been used in the detection of explosive molecules such as 2,4,6-trinitrotoluene (TNT) and 2,4-dinitrotoluene (DNT) (*1*). Since TNT and DNT contain electron-withdrawing nitro groups on the aromatic benzene ring, they have relatively strong interaction with electron-rich species, such as many fluorescent dyes. These interactions result in quenching of the fluorescence of the dye. The degree of quenching is dependent on the amount of the quencher, TNT or DNT, present. In a homogeneous medium, such as a solution, the quantitative measure of fluorescence quenching is described by the Stern-Volmer constant, K_{sv} in the equation $I_0/I = 1 + K_{sv}[Q]$ (*2*).

In the above equation, I_0 and I are the intensities of fluorescence in the absence and in the presence of the quencher respectively. The equation shows that I_0/I increases in direct proportion to the concentration of the quencher. When all other variables are held constant, the higher the K_{sv}, the lower the concentration of quencher required to quench the fluorescence. In many microheterogeneous mediums, such as in polymer films, a negative deviation from the linear Stern-Volmer equation occurs at high quencher concentration (*12,13*). Recent studies (*11,14*) show that in some solid systems, the experimental data also fit Stern-Volmer equation well within certain concentration ranges.

The constant K_{sv} defines the quenching efficiency and is given by $K_{sv} = k_2\tau_1$. Where τ_1 is the luminescence decay time of the fluorophore in the absence of the quencher (=$1/k_1$), and k_2 is the bimolecular quenching rate constant. This equation implies two important practical consequences (*2*). First, the sensitivity of the quenching process is enhanced by employing fluorophores with long luminescence decay times (τ_1). Second, the sensitivity of the process can be tailored by controlling the quencher diffusion rate to fluorophores via the microstructural properties of the sensing film.

Electrospinning

Electrospinning has recently gained much attention as a unique technique to fabricate high surface area and highly responsive nanofibrous structures (*15, 16,*

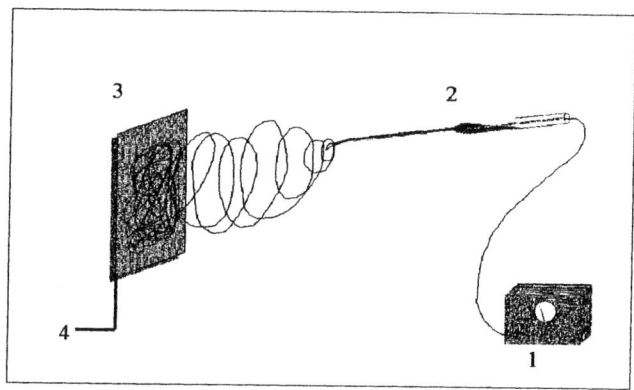

Figure 1. Electrospinning set-up (1: power source; 2: polymer spin-dope solutions; 3: collection target; 4: ground) (Reproduced from reference 11. Copyright 2002 American Chemical Society.)

17, 18). In electrospinning, a high voltage is applied to a viscous polymer solution creating electrically charged jets. These jets dry to form very fine polymer fibers, which are collected on a target as a non-woven membrane. The diameters of electrospun fibers typically range from several micrometers to less than 50 nanometers depending on the polymer and experimental conditions. As a result, electrospun nanofibrous membranes can have a surface area per unit volume of up to two orders of magnitude higher than that of continuous thin films. This high surface area has the potential to provide high sensitivity and fast response times for sensing applications.

The apparatus commonly used in the electrospinning process is shown in Figure 1. It consists of a glass capillary tube with a 1.5 mm inside diameter tip mounted on an adjustable, electrically insulating stand. The capillary tube was filled with a polymer solution or melt, into which a metal electrode was inserted. The polymer solution or melt held by its surface tension at the end of a capillary tube is subjected to an electric field. Charge is induced on the liquid surface by the electrode. Mutual charge repulsion causes a force counter to the surface tension. As the intensity of the electric field is increased, the hemispherical surface of the solution at the tip of the capillary tube elongates to form a conical shape known as the Taylor cone (*19*). When the electric forces reach a critical threshold at which the repulsive electric forces overcome the surface tension and viscosity forces, a charged jet of the solution is ejected from the tip of the Taylor cone. As the jet travels in air, the solvent evaporates, leaving behind a charged polymer fiber. Continuous fibers are collected in the form of a nonwoven membrane (*18*), which have high porosity but generally very small pore size. For fibers spun from polymer solutions, the presence of residual solvent in the electrospun fibers facilitates bonding of intersecting fibers, creating a strong cohesive porous structure. A variety of polymeric materials or blends have been electrospun successfully (*20, 21, 22, 23, 24*). One implication of these membrane morphologies is that electrospun nanofibrous membranes should provide broad enhancements in sensitivities in sensor applications.

Synthesis of the fluorescent polymers

Among the many known fluorescent dyes, pyrene methanol was chosen as a fluorescent indicator because of its large Stokes' shift, high quantum yield, high absorbance, excellent photostability, long lifetime, and nontoxicity (*25*). Covalent attachment of the pyrene methanol fluorophore to a polymeric molecule with an appropriate spacer serves to both minimize self-quenching and improve the processability of the material. Both water-insoluble and water-soluble fluorescent polymers poly (methylmethacrylate-pyrenemethanol) and poly (acrylic acid-pyrene methanol) were synthesized for the study.

Poly (methylmethacrylate-pyrenemethanol) (PMMA-PM)

Poly (methylmethacrylate-pyrenemethanol) was synthesized as shown in Figure 2. (*26*) The monomer was prepared by dissolving 1-pyrene butanol in tetrahydrofuran (THF) with pyridine and a trace amount of 2,6-di-tert-butyl-4-methylphenol. Methacryloyl chloride was added to the solution at 0 °C under an inert atmosphere. After 24 h, the solution was poured into water and the monomer was collected by filtering. The resulting pyrene functionalized monomer was mixed with methylmethacrylate (MMA) monomer and polymerized by free-radical methods. The obtained solid polymer was washed with water and ethanol and dried under vacuum. The solution was thoroughly degassed and heated in a sealed ampoule for two days. The solution was cooled and slowly poured into methanol to precipitate the polymer. After filtering, the polymer was washed with methanol and dried in a vacuum oven. Poly methylmethacrylate-pyrenemethanol has a maximum fluorescence emission wavelength of 381 nm when excited at 336 nm.

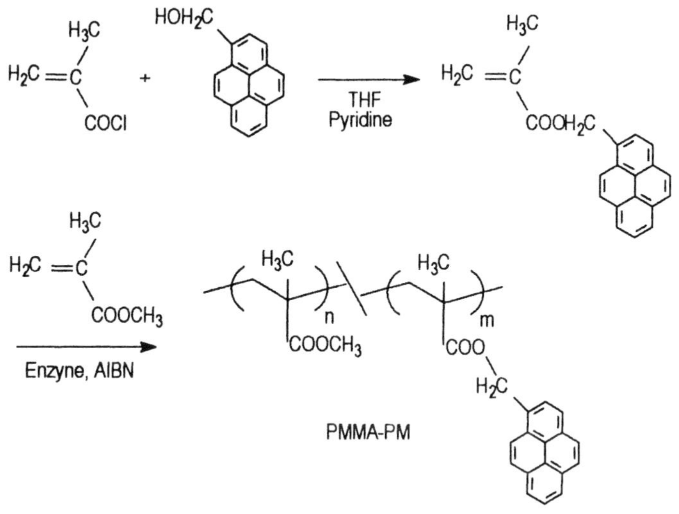

Figure 2. Synthesis of PMMA-PM.

Poly (acrylic acid-pyrene methanol) (PAA-PM)

Poly (acrylic acid-pyrene methanol) was synthesized as shown in Figure 3 (*11*). Catalysts 1,1'-carbonyldiimidazole (CDI) and 1,8-diazabicyclo[5.4.0]undec-7-ene (DBU) dissolved in N, N'-dimethylformamide was added to a solution of polyacrylic acid. After stirring the solution at elevated temperature until the evolution of carbon dioxide subsided, a solution of pyrene methanol solution was added and the solution was stirred for one day. The solution was then poured into ethyl ether to precipitate the polymer. After filtration, the obtained polymer was extensively washed with ether and acetone and dried in a vacuum oven. This method resulted in a degree of functionalization of about seven percent by NMR measurement. That is, about seven percent of the acid groups on the polymer had reacted with the pyrene methanol. Earlier investigations showed that there is a strong fluorescence self-quenching when a higher degree (>10%) of functionalization was used (*28*). Therefore, the polyacrylic acid can be functionalized with appropriately low pyrene methanol levels to minimize the self-quenching effect. The synthesized poly (acrylic acid-pyrene methanol) polymer shows the characteristic UV-Vis absorption and emission spectra of the pyrene methanol indicator with maximum UV-Vis absorbance and fluorescence emission at 348 nm and 453 nm respectively.

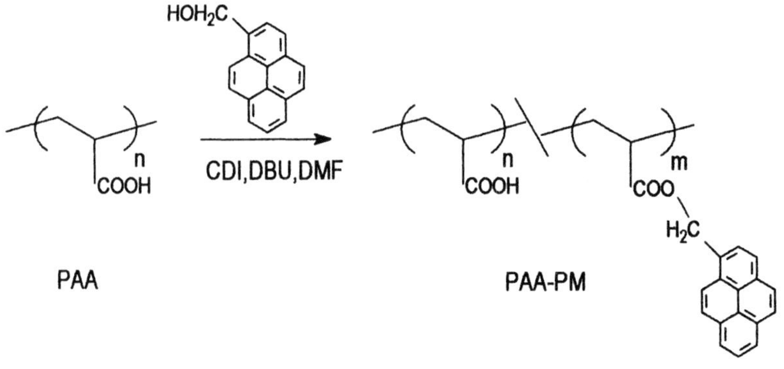

Figure 3. Synthesis of PAA-PM (From Ref. 11).

Sensor Fabrication and Performance

For poly (methylmethacrylate-pyrenemethanol), the spin-dope solution, which consisted of a 26%, by weight, solution of PMMA-PM, was dissolved in propylene glycol methyl ether acetate. Very fine fibers were collected on a glass slide coated with tin oxide. The applied electrospinning voltages ranged from 15 - 20 kv. The working distance between the tip of the pipette and the glass slide ranged from 15 to 20 cm. The collection time was between 30 to 45 seconds.

For poly (acrylic acid-pyrene methanol), the spin-dope solution, consisted of an 18.6% by weight solution of the copolymer and 36.5% of crosslinkable polyurethane, dissolved in DMF. Fibers were again collected on a glass slide coated with tin oxide. The applied electrospinning voltages, working distances, and collection times were the same as previously described. The electrospun membranes were dried in a vacuum oven at 80°C for one day. The polyurethane contained a melamine cross-linker that required a cure time of one to two minutes at about 250°C. The fluorescent polymer was immobilized with the cross-linked polyurethane to form an interpenetrating network structure and a water insoluble sensing membrane.

The scanning electron microscope (SEM) images of electrospun membranes of the PMMA-PM and cross-linked PAA-PM are shown in Figure 4. It was observed that the membranes have three-dimensional structures with a random fiber orientation that is uniformly distributed on the substrate. The diameters of the fibers were approximately 400 to 1000 nm for PMMA-PM and 100 to 400 nm for PAA-PM. The porous structure of electrospun membrane provides 1 to 2 orders of magnitude higher surface area to volume ratio than that known for continuous thin films.

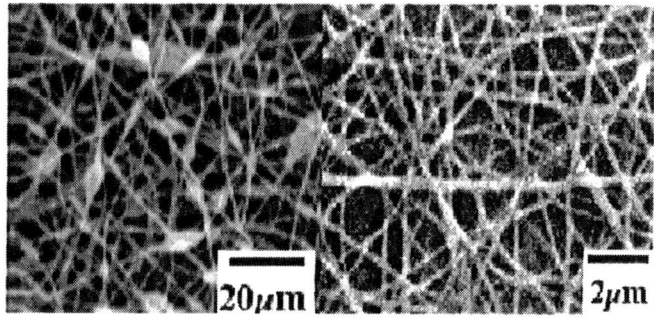

Figure 4. SEM images of electrospun membranes: PMMA-PA (left) and PAA-PM (right).

Fluorescence spectra as a function of different concentrations of 2,4-dinitro toluene (DNT) were measured from the electrospun membranes of PMMA-PM. The fluorescence intensity of an electrospun membrane decreased with DNT concentration and the degree of quenching depended on the amount of DNT, as shown in Figure 5.

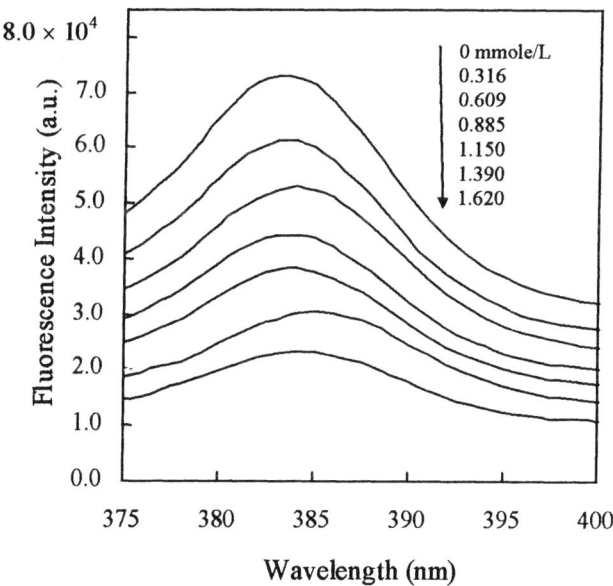

Figure 5. Fluorescence emission spectra of a PMMA-PM electrospun membrane with varying DNT concentration.

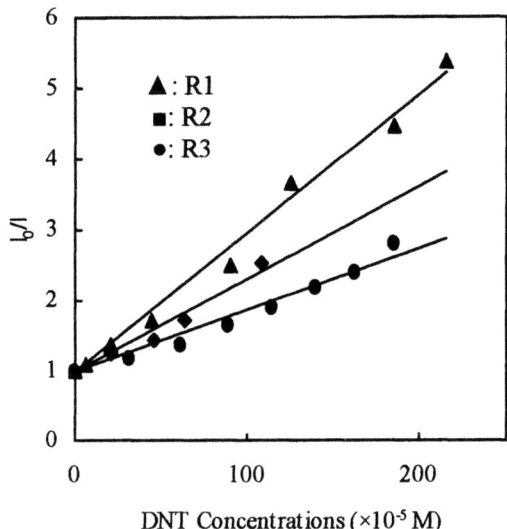

Figure 6. Stern-Volmer plots of electrospun membranes as a function of quencher concentration.

The ratio of fluorophore monomer to structural monomer was varied at 0.5%, 2 % and 4 % denoted as R1, R2 and R3. Similar behavior was observed in all the polymers. Linear relationships between concentration of quencher (DNT) and I_0/I were obtained. Stern-Volmer constants (K_{sv}) were calculated from the slopes of each plot and are 1.96×10^3, 1.30×10^3 and 8.70×10^2 for R1, R2 and R3 respectively. These values are one order of magnitude higher than those obtained from continuous films of the similar polymer system by electrostatic layer-by-layer self-assembly technique, which were previously reported from our group (28). It is interesting to note that a higher value of K_{sv} was observed in the polymer R1 which contained a lower concentration of fluorophore. This may be explained in that the change of the fluorescence intensity due to small amount of quencher is more significant if the baseline fluorescence intensity (I_0) is lower. Therefore, a lower concentration of fluorophore should be advantageous in increasing the sensitivity of the device.

Similar studies with an electrospun membrane of poly acrylic acid-pyrene methanol showing the change in fluorescence spectra as a function of different concentrations of DNT were measured and are shown in Figure 7. It was found that the fluorescence intensities decreased with increasing DNT concentration. This fluorescence intensity decrease is expected and believed to be due to the quenching of the pyrene-based indicator by electron poor species, DNT.

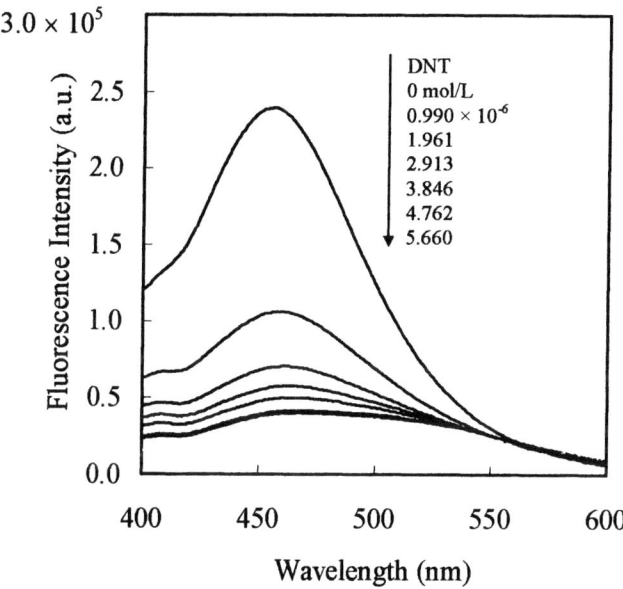

Figure 7. Fluorescence emission spectra of a PAA-PM electrospun membrane with varying DNT concentration.

For quencher concentrations in the range of 10^{-7} to 10^{-6} mol/L, a linear relationship between quencher concentration and I_0/I is obtained showing a Stern-Volmer relationship, where the Stern-Volmer constant is (K_{sv}) 9.8 × 10^5 M^{-1}. This value is roughly two to three orders of magnitude greater than that obtained previously from the thin film sensors (28). The sensitivities of these devices are in the range of tens parts per billion. The fact that PAA-PM system is more sensitive than PMMA-PM is presumably due to the smaller size of electrospun fibers for PAA-PM and the solvent swelling effect of the cross-linked network structure of the polymer in analyte solution. Thus the cross-linked, PAA-PM system has a greater effective surface area than the PMMA-PM system.

Conclusions

We have successfully demonstrated nanofibrous optical chemical sensors for DNT detection using electrospun fluorescent polymers. The device sensitivity is 1 to 2 orders of magnitude higher than that observed for continuous thin films. Further efforts will focus on exploring new sensing materials including fluorescent conjugated polymers, controlling the structural properties of the electrospun films and optimizing the sensitivities for the detection of explosives both from liquid phase and vapor phase.

Acknowledgements

Financial support from the U.S. Army is acknowledged. The authors are grateful to Dr. J. He, Dr. S. Balasubramanian, and Dr. L. Li for discussions of optical sensing. The authors also recognize Professor C. Sung and Ms. B. Kang for the SEM images, Mr. Bon-Cheol Ku for assistance with polymer synthesis and Dr. H. Schreuder-Gibson for advice and assistance with electrospinning. Thanks are also expressed to Soluol Chemical Co., INC for providing the crosslinkable polyurethane and helpful discussions with Dr. John Reisch. This work is dedicated to Professor S. K. Tripathy.

Reference

(1). Yang, J. S.; Swager, T. M. *J. Am. Chem. Soc.* **1998**, *120*, 11864-11873.
(2). Diamond, R. W. *Principle of Chemical and Biological Sensors;* John Wiley & Sons, INC, New York, NY, **1998**, 206-208.
(3). Harsanyi, G. *Mat. Chem. & Phys.* **1996**, *43*, 199-203.
(4). Choi, M. M. F.; Xiao, D. *Anal. Chim. Acta* **1999**,*387*, 197-205.
(5). Demas, J. N.; DeGraff, B. A.; Xu, W. *Anal. Chem.* **1995**, *67*, 1377-1380.
(6). Mills, A.; Lepre, A.; Theobald, B. R.; Slade, E.; Murrer, B. A. *Anal. Chem.* **1997**, *69*, 2842-2847.
(7). McDonagh, C.; MacCraith, B. D.; McEvoy, A. K. *Anal. Chem.* **1998**, *70*, 45-50.
(8). Zhujun, Z.; Seitz, W. R. *Anal. Chim. Acta* **1984**, *160*, 47-52.
(9). Lobnik, A.; Oehme, I.; Murkovic, I.; Wolfbeis, O.s. *Anal. Chim. Acta* **1998**, *367*, 159.
(10). Reneker, D. H.; Chun, I. *Nanotechnology* **1996**, *7*, 216-223.
(11). Wang, X. Y.; Drew, C.; Lee, S-H; Senecal, K. J.; Kumar, J.; Samuelson, L. A. *Nano Lett.* **2002**, *2*, 1273-1275.
(12). Bacon, J. R.; Demas, J. N. *Anal Chem* **1987**, *59*, 2780-2785.
(13). Carraway, E. R.; Demas, J. N.; DeGraff, B. A.; Bacon, J. R. *Anal. Chem.* **1991**, *63*, 337-342.
(14). Li, X. M.; Wong, K. Y. *Analytica Chimica Acta* **1992**, *262*, 27-32.

(15). Drew, C.; Wang, X. Y.;Senecal, K.; Schreuder-Gibson. H.; He, J.; Tripathy, S.; Samuelson, L. *J. Macromol. Sci.- Pure Appl. Chem.* **2002**, *39*, 1085-1094.

(16). Norris, I. D.; Shaker, M. M.; Ko, F. K.; MacDiarmid, A. G. *Synth. Met.* **2000**, *114*, 109-114.

(17). Megelski, S.; Stephens, J. S.; Chase, D. B.; Rabolt, J. F. *Macromolecules* **2002**, *35*, 8456-8466.

(18). Gibson, P.; Schreuder-Gibson, H.; Rivin, D. *Colloids & Surfaces A* **2001**, *187-188*, 469-481.

(19). Taylor, G. I. *Proc. Roy. Soc. Lond. A.* **1969**, *313*, 453-475.

(20). Kim, J-S and Lee, D-S. *S. Korea. Polym. J.* **2000**, *32*, 616-618.

(21). Demir, M. M., Yilgor, I., Yilgor, E. and Erman, B. *Polymer,* **2002**, *43*, 3303-3309.

(22). Zarkoob, S., Reneker, K. H., Eby, R. K., Hudson, S. D., Ertley, D. and Adams, W. W. *Polym. Prepr. (ACS), POLY* **1998**, *39*, 244-245.

(23). Bognitzki, M., Czado, W., Frese, T., Schaper, A., Hellwig, M., Steinhart, M., Greiner, A. and Wendorff, J. H. *Adv. Mater.* **2001**, *13*, 70-72.

(24). Fang, X. and Reneker, D. H. *J. Macromol. Sci. Phys., B*, **1997**, *36*, 169-173.

(25). Sharma, A. and Schulman, S. G. *Introduction to Fluorescence Spectroscopy*, John Wiley& Sons, New York, **1999**, 123-157.

(26). Wang, X. Y., Lee, S. H., Ku, B. C., Samueslon, L. A. and Kumar, J. *J. Macromol. Sci.- Pure Appl. Chem.* **2002**, *39*, 1241-1249.

(27). Wang, X. Y., Drew, C., Lee, S. H., Senecal, K. J., Kumar, J. and Samuelson, L. A. *J. Macromol. Sci.- Pure Appl. Chem.* **2002**, *39*, 1251-1258.

(28). Lee, S-H; Kumar, J. and Tripathy, S. K. *Langmuir* **2000**, *16*, 10482-10489.

Indexes

Author Index

Alam, Maksudul M., 188
Andrasik, Stephen, 122
Aziz, Hany, 220
Baek, N. S., 247
Bangcuyo, Carlito, 147
Bard, Allen J., 34
Belfield, Kevin D., 122
Bender, Jessica L., 233
Bergeron, Michel G., 359
Birur, G. C., 66
Boissinot, Maurice, 359
Bolognesi, A., 211
Botta, C., 211
Boudreau, Denis, 359
Bunz, Uwe H. F., 147
Burns, Alan R., 82
Carpick, Robert W., 82
Chandrasekhar, P., 66
Chang, Shun-Chi, 201
Chen, Philip H. M., 290
Chen, Shaw H., 290
Chen, Wen-Chang, 307
Cheng, Quan, 96
Cheuk, Kevin K. L., 340
Choi, Jai-Pil, 34
Corbeil, Geneviève, 359
DeLongchamp, Dean M., 18
Desjardins, Pierre, 51
Dinnocenzo, Joseph P., 135
Doré, Kim, 359
Douglas, D., 66
Drew, Christopher, 388
Eriksson, M. A., 82
Facchinetti, D., 211
Farid, Samir, 135
Fraser, Cassandra L., 233
Fudouzi, Hiroshi, 329
Fungo, Fernando, 34
Gillmore, Jason G., 135
Hammond, Paula T., 18
Hikmet, Rifat A. M., 278
Ho, Hoang-Anh, 359
Ho, S. W., 264
Holdcroft, Steven, 220
Huang, W. Y., 264
Huang, Z., 161
Imae, Ichiro, 173
Jandke, M., 211
Jenekhe, Samson A., 2, 34, 188
Katsis, Dimitris, 290
Kauder, L., 66
Kim, H. K., 247
Kim, O.-K., 161
Kirkpatrick, S., 161
Kiserow, Douglas, 2
Kreger, K., 211
Kriwanek, Jörg, 110
Kumar, Jayant, 377, 388
Kwei, T. K., 264
Leclerc, Mario, 359
Lee, Chia-Hua, 307
Lee, J. H., 247
Lee, Soo-Hyoung, 377, 388
Lee, Y., 247
Li, Yuning, 220
Liu, Wei, 377
Lötzsch, Detlef, 110
Lu, Yu, 329

403

Marcus, Matthew S., 82
Mastrangelo, John C., 290
McQueeney, T., 66
Mercogliano, C., 211
Moriyama, Masaya, 320
Nakano, Hideyuki, 173
Ohsedo, Yutaka, 173
Okamoto, Y., 264
Paik, K. L., 247
Patzak, André, 110
Pei, Qibing, 201
Peterman, E., 161
Popovic, Zoran D., 220
Pyo, S., 201
Relini, A., 211
Robello, Douglas R., 135
Rolandi, R., 211
Ross, D., 66
Samuelson, Lynne A, 377, 388
Sasaki, Darryl Y., 82
Schafer, Katherine J., 122
Seeboth, Arno, 110
Senecal, Kris J., 388
Shirota, Yasuhiko, 173
Song, Jie, 96

Stevens, Raymond C., 96
Strohriegl, P., 211
Sung, C. S. P., 161
Swager, Timothy M., 368
Swanson, T., 66
Takahashi, Toru, 173
Tamaoki, Nobuyuki, 320
Tang, Ben Zhong, 340
Thomas III, Samuel W., 135
Tonzola, Christopher J., 188
Tovar, John D., 368
Tripathy, Sukant, 377
Ujike, Toshiki, 173
Utsumi, Hisayuki, 173
Vamvounis, George, 220
Wang, Xianyan, 388
Wang, Zhi Yuan, 51
Wei, Ming-Hsin, 307
Wilson, James N., 147
Xia, Younan, 329
Yang, Suizhou, 377
Yang, Yang, 201
Yasuda, Yoshiaki, 173
Zay, B. J., 66
Zhu, Yan, 188

Subject Index

A

Absorbance switching profiles, poly(aniline) layer-by-layer films, 30*f*
Absorption spectra
  polarized, of double-layer polymer film, 215*f*
  poly(dialkylfluoryleneethynylene)s (PFEs), 153*f*
  thin films of PFE and doped PFE, 154*f*
  See also Electronic absorption spectral change
Acrylates. *See* Chiral gels
Additives, tuning chain helicity of polyacetylenes, 355*f*, 356
Alcohol initiators
  polymerizations, 237
  synthesis, 236*f*
Alternating copolymers. *See* Silicon-based alternating copolymers
Amino acid moieties
  polyacetylenes, 342
  See also Amphiphilic helical polyacetylenes
Amino acid termination
  amphiphilic polydiacetylenes, 98–102
  bolaamphiphilic diacetylene lipid, 102–103
Amorphous molecular glasses
  categories, 291
  description, 290–291
Amphiphilic helical polyacetylenes
  additive effects, 355*f*, 356
  behaviors of monosaccharide-containing polymers, 352
  CD (circular dichroism) activity for, bearing D-phenylglycine pendants, 348–349
  CD spectra of, bearing phenylglycine pendants, 349*f*
  changes of CD spectra with solvent compositions, 353*f*
  Cotton effects, 349, 351
  design and synthesis, 341, 343
  environmental dependence of Cotton effect of, bearing L-amino acid pendants, 351*f*
  examples of polymerization results, 344*t*
  ^{1}H NMR spectra of solutions of L-leucine-containing polyacetylene, 346*f*
  illustrations of single- and double-stranded helical chains of leucine-containing polyphenylacetylene, 347
  incorporation of biopolymer building blocks into, 340–341
  induced helical structures, 348–352
  methyl protection group in amino acid-containing, 343
  molecular structures of, containing amino acid moieties, 342
  molecular structures of, containing monosaccharide and nucleoside moieties, 343
  pH effects, 354, 355*f*
  solvatochromism and hydrogen bonding, 346–348
  solvent dependence of Cotton effect of, bearing pendants of D-monosaccharide acetonides, 352*f*
  solvent dependence of Cotton effect of, bearing pendants of L-amino acid methyl esters, 350*f*
  temperature effects, 354, 355*f*
  tuning of chain helicity by external stimuli, 353–356

405

UV spectra of polyacetylene bearing D-glucose pendants, 345f
valine-containing polyacetylene, 353f, 354
Amphiphilic polydiacetylenes
colorimetric properties, 100–102
colorimetric response vs. solution pH, 101f
molecular structures of amino acid terminated diacetylene lipids, 98f
monomer synthesis, microstructure formation, and characterization, 98–100
synthesis of amino acid terminated diacetylene lipids, 98f
transmission electron microscopy images, 99f
See also Poly(diacetylene)s (PDAs)
Amplified imaging system. See Quantum amplified isomerization (QAI)
Anilines
enzymatic polymerization, 378–379
peroxidase-catalyzed oxidation of, 377
Aqueous polymer systems
calorimetric measurements, 112–113
incorporation of salts, 114
Arylene-ethynylene (AE) polymer
conjugation pathway, 373
UV-visible spectra, 374f
Assembly
film method, 22
ordered block copolymers with labile metal crosslinks, 243–244
poly(aniline)-containing films, 23–24
See also Poly(aniline) (PANI) films
Atomic force microscope (AFM)
poly(benzobisimidazobenzophenanthroline) (BBB), ladder BBB (BBL), and poly(2,2'-[10-methyl-3,7-phenothiazylene]-6,6'-bis[4-phenylquinoline]) (PPTZPQ) films, 38, 39, 41
film structure of poly(diacetylene)s (PDAs), 86–87
friction anisotropy of PDAs, 90–91
intermittent contact, 84, 85–86
mechanochromism of PDAs, 87–89
poly(methyl methacrylate) hybrid material, 313f
surface relief gratings (SRG) on polyazophenol thin film, 384f
Azobenzene-based photochromic materials
electronic absorption spectral change with irradiation, 176f
formation of surface relief gratings, 176
glass-forming properties, 174
photochromic behavior, 175
structures, 175
synthesis, 174
Azophenols
enzymatic polymerization, 379
optical quality polymer film preparation, 379–380

B

Benzene derivatives (Dewar). See Quantum amplified isomerization (QAI)
Bilayer polymer films. See Electroluminescence
Biopolymers
incorporation into synthetic polymers, 340–341
thermotropic polymer gel networks, 111–112
See also Amphiphilic helical polyacetylenes
Biosensors. See Cationic polythiophene biosensors

Bipyridine (bpy) ligand
  Adduct-forming component, 235
  macroligand toolkit and polarity, 241
  See also Lanthanide complexes
Blended polymer films. *See* Poly(quinoline)s
Block copolymers
  bipyridine (bpy) macroligand toolkit, 241
  labile metal crosslinks, 243–244
  lanthanide complexes, 240–243
  microphase separation, 235*f*
  modular approach to Ln tris and adduct complexes, 241*f*
Blue light emitting polymers and devices
  blue light-emitting devices (LEDs), 202
  charge confinement at poly(paraphenylene) with triarylamine side groups (TA-PPP)/2-(4-*t*-butylphenyl)-5-biphenyloxadiazole (*t*-PBD) interfaces, 208
  differential scanning calorimetry (DSC) of TA-PPP, 206*f*
  DO-PF (poly(9,9-dioctylfluorene)), 203, 204*f*
  light emitting diode (LED) fabrication, 207
  photoluminescence (PL) spectra of DO-PF, 204*f*
  PL spectrum of DO-PF film after thermal treatment, 205*f*
  PL spectrum of TA-PPP film after thermal treatment, 205*f*
  poly(fluorenes) (PF), 202
  poly(paraphenylene) (PPP), 202
  poly(paraphenylene vinylene) (PPV), 202–203
  poly(pyridine), 265
  spiro-PPP, 207
  stability of PL, 205, 206*f*
  TA-PPP, 203
  thermogravimetric analysis of TA-PPP and DO-PF, 206*f*
  thin films of blend of TA-PPP with *t*-PBD, 207–208
  UV-vis absorption and PL spectra for TA-PPP, 204*f*
  See also Poly(paraphenylene) (PPP)
Bolaamphiphilic polydiacetylenes
  amino acid-terminated, 102–103
  lipid doping-induced structural transformation, 105–107
  morphology and surface packing arrangement, 103–104
  pH-induced optical and structural transformation, 104–105
  pH-triggered morphological transformation, 105*f*
  ribbon-to-vesicle microstructural transformation, 106–107
  transmission electron micrographs of doped L-Glu-Bis-3, 107*f*
  transmission electron microscopy (TEM), 103*f*
  See also Poly(diacetylene)s (PDAs)
Bragg equation, light diffraction, 330
Bromothymol Blue
  temperature dependence of transparency, 118*f*
  temperature dependence of UV/vis absorption spectra, 117*f*
  See also Chromogenic gel networks

C

Calorimetry, aqueous polymer gel networks, 112–113
ε-Caprolactone polymerization
  dibenzoylmethane (dbm) macroligands, 237–238
  kinetic plot, 237*f*
Carbazole unit, silicon-based copolymers, 251
Cationic polythiophene biosensors

circular dichroism (CD), 364
colorimetric detection, 360–361, 364
cyclic voltammogram, 366f
detection of oligonucleotides, 359
detection specificity, 361
electrochemical detection, 365–366
fluorescence spectra of polymer solutions, 364f
fluorometric detection, 364–365
formation of polythiophene/single-stranded oligonucleotide duplex and polythiophene/hybridized oligonucleotide triplex, 362
photographs of polymer solutions, 362
preparation steps for electrochemical detection, 366f
synthesis, 360
synthesis of monomer and polymer, 360f
testing with various target oligonucleotides, 361, 364
UV-vis absorption spectra of polymer solutions, 362
UV-vis spectra of polymers using target oligonucleotides, 363f
Chain helicity, tuning, of polyacetylenes, 353–356
Charge density, poly(aniline) layer-by-layer films, 27f
Charge transfer, modes, 163f
Charge transfer complex, formation in silicon-based copolymer, 256f, 261f
Chemical amplification process, 136
See also Quantum amplified isomerization (QAI)
Chemical sensors, explosive detection, 388
Chiral gels
applications, 279–280
cholesteric liquid crystal phase, 279
critical voltage, 283
effect of excited state quencher concentration, 287
effect of UV intensity, 287
functions of diacrylate molecules, 282
Helfrich's deformation, 283
materials, 280
optical effects, 279
phase separation upon polymerization, 285–286
photographs of cell at room temperature, 285
position of reflection band vs. temperature, 284f
reflection band vs. temperature, 288f
reflection band vs. time, 286f
reflection band vs. voltage, 288f
schematic of cholesteric phase, 279f
schematic of gel formation, 281f
shifting of reflection band to lower wavelengths, 282–283
structures of acrylates, 280f
switching by electric field, 281
temperature dependence of reflection band, 287, 289
temperature dependent reflection, 284
threshold voltage, 282–283, 284
transmission as function of voltage, 282f, 283f
width of reflection band vs. time, 286f, 287f
Chiroptical properties. See Amphiphilic helical polyacetylenes
Cholesteric gels
liquid crystal phase, 279
See also Chiral gels
Cholesteric liquid crystal films
optical properties, 294–295
phototunable, 300f
reflective coloration, 300f
selective reflection, 294f

tunable reflective coloration, 299–300
See also Glassy liquid crystals (GLCs)
Cholesteric liquid crystals
  materials for color information recordings, 321
  molecular alignment, 320–321
  photochromic reactions, 320–321
  See also Photochromic liquid crystalline compounds
Cholesteric order
  high-temperature glassy cholesterics, 299*f*
  molecular self-organization, 292*f*
  optical spectra of unpolarized beam through cholesteric film, 298*f*
  synthesis of glassy cholesterics, 297
  See also Glassy liquid crystals (GLCs)
Cholesteric reflection, photochromic liquid crystalline compounds, 324*t*
Chromatic transition, poly(diacetylene)s, 97
Chromicity
  conjugated polymer examples, 147
  definition, 147–148
  interplay between, and rotational conformation, 151*f*
  poly(heteroaryleneethynylene)s, 156–157
  See also Poly(aryleneethynylene)s (PAEs)
Chromism, tuning, 52
Chromogenic effects in polymers
  electrochromic (EC) polymers and devices, 4–5
  mechanochromic polymers, 8–9
  photochromic polymers, 3–4
  photoluminescence, 5, 6*f*
  photonic band gap composites, 12
  piezochromic polymers, 9–12
  reversible change of optical properties, 2

reversible piezochromism, 11–12
solvatochromism, 2–3
tunability of optical properties, 2–3
tunable emission and electroluminescence (EL), 5–6, 8
voltage-tunable multicolor EL emission from bilayer polymer LEDs, 6–8
phenomena, 2*f*
Chromogenic gel networks
  reversible transparency and color control with temperature, 117–120
  temperature dependence of transparency of Bromothymol Blue in, 118*f*
  temperature dependence of transparency of Phenol Red in, 119*f*
  UV/vis absorption spectra of Bromothymol Blue in, 117*f*
  UV/vis absorption spectra of Phenol Red in, 118*f*
Chromophores. *See* Conjugated chromophores
Chronoamperometry, double potential step, poly(aniline) layer-by-layer films, 26–27
Circular dichroism (CD)
  cationic polythiophene derivative, 364
  changes of CD spectra with solvent compositions, 353*f*
  polyacetylene bearing D-phenylglycine pendants, 348–349
  See also Amphiphilic helical polyacetylenes
Colloidal crystals
  application of ink to photonic paper surface, 333, 334
  average refractive index, 330
  Bragg equation, 330
  description, 330

embedding, 330–331
fabrication of photonic papers with switching colors, 333, 335
hydrogels containing, 330
mechanism for color writing with colorless ink, 331–332
photonic band gap (PBG) materials, 330
photonic paper system by embedding elastomeric matrix, 335–336
scanning electron microscopy (SEM) images of photonic paper before and after matrix swelling, 332f
thin films containing, 330–331
transmission spectra of photonic paper before and after silicone swelling, 332, 334
UV-visible transmission spectra of paper before and after isopropanol swelling, 335f
wavelength of light diffracted from surface, 330
See also Photonic papers
Coloration, modes, 299
Coloration, reflective
glassy cholesteric films, 300f
photoresponsive glassy liquid crystals, 301–302
reversible tunability, 301, 302f
See also Glassy liquid crystals (GLCs)
Coloration efficiency, ruthenium complex polymers, 62–63
Color control
chromogenic gel networks, 117–120
thermochromic gel networks, 115–117
Colorimetric properties, amphiphilic polydiacetylene microstructures, 100–102
Color writing, colorless ink, 331–332
Columnar order

molecular self-organization, 292f
See also Glassy liquid crystals (GLCs)
Composite films. See Poly(aniline) (PANI) films
Composites, polymer-based photonic band gap, 12
Conducting polymer-based electrochromics
controller, 74
electrochemical control and reflectance, emittance, and solar absorptance measurements, 69
electrochromic and thermal control technologies, 76–77
experimental, 68–70
features of technology, 68, 70–71
four-device panel, 72f
infrared (IR) diffuse reflectance, 72f
IR specular reflectance, 72f
materials and device assembly, 68
military applications, 75–76
military camouflage data, 77f, 78f
military requirement, 68
NASA's ST5 mission, 74–75
need for visible-to-far-IR region electrochromics, 67–68
non-rad-hard controller, 74f
photos of terrestrial device, 70f
principle of operation, 70–71
radiation, solar wind exposure, outgassing and space durability tests, 70
satellite constellation orbit, 76f
schematics of electrochromic device, 69f
spacecraft device, 72f
spacecraft requirement, 67
space-use controller, 74
spectral, thermal, space durability and switching data, 71, 73
ST5 microsatellite, 76f
terrestrial IR, 70

thermal cycling, thermal vacuum
and calorimetric measurements,
69
thermal vacuum and calorimetric
tests, 73*f*
typical emittance data, 71*t*
UV-vis-NIR reflectance, 72*f*
Conductivity, poly(ethylene oxide) gel
electrolytes, 44, 45*f*
Conjugated chromophores
absorption and emission spectra of
dithienothiophene (DTT)-based
chromophores, 166*f*
degree of charge-transfer (CT), 162
DTT, 162
donor and/or acceptor
chromophores and CT modes,
163*f*
fluorine-based oligomeric
chromophores, 164–165
molecular concept and synthetic
strategy, 162–164
optical properties of fluorine-based
chromophores, 167*t*
optical properties of
oligothiophene-based
chromophores, 166*t*
π-center role in DTT, 164
role of π-centers, 163–164
spectroscopic properties, 165–167
structures, 165*f*
synthesis, 164–165
TPA (two-photon absorption), 161–162
TPA cross-section of various DTT-based, 168*t*
TPA cross-section of various
fluorine-based, 169*t*
TPA properties, 167–170
Conjugated polymers
absorption and emission spectra of
arylene-ethynylene (AE)
polymers, 373*f*
AE polymers, 373

changes in UV-vis absorption upon
increased oxidation for polymer
films, 372*f*
chromicity, 147–148
cyclic voltammetry of thienyl-
based monomers and polymers,
371*f*
device applications, 188–189
effects of aromatic cores, 372–373
electrochromic applications, 369
electrochromic measurements, 370
generic cyclization-polymerization
scheme, 370
lanthanide chromophores, 234
poly(thiophene) derivatives, 374,
375*f*
quinoidal resonance structures of,
371
sensing devices, 96
spectroelectrochemical
measurements, 372
sulfur-containing aromatic cores,
374
UV-vis spectra of AE-based
polymer, 374*f*
See also Conjugated
chromophores;
Electroluminescence
Copolymers. *See* Silicon-based
alternating copolymers
Copolymers with methacrylate,
pendant oligo(ethylene oxide),
182–184
Cotton effects
polyacetylene backbones, 349,
351–352
See also Amphiphilic helical
polyacetylenes
Coupling chemistry, poly(3-bromo-4-
hexylthiophene) transformation,
222, 224
Current density, poly(aniline) layer-
by-layer films, 27*f*
Cyclic voltammetry

poly(benzobisimidazobenzophenanthroline) (BBB), ladder BBB (BBL), and poly(2,2'-[10-methyl-3,7-phenothiazylene]-6,6'-bis[4-phenylquinoline]) (PPTZPQ) films, 37–38, 40f, 42f
cationic polythiophene biosensors, 366f
poly(aniline) layer-by-layer films, 24–26
thienyl-based monomers and polymers, 371f

D

Deoxyribonucleic acid (DNA) sensors. *See* Cationic polythiophene biosensors
Dewar benzene derivatives. *See* Quantum amplified isomerization (QAI)
Dialkyl-poly(p-phenyleneethynylene)s (dialkyl-PPEs). *See* Poly(aryleneethynylene)s (PAEs)
Diarylene, glassy liquid crystals (GLCs) containing, 301–302
Dibenzoylmethane (dbm)
 initiators, 236–237
 macroligands, 237–238
 See also Lanthanide complexes
Dicarbonylhydrazine (DCH). *See* Ruthenium complex polymers
Dichroism, irradiation of amorphous film, 178–179
9,10-Dicyanoanthracene sensitizer. *See* Quantum amplified isomerization (QAI)
Differential scanning calorimetry (DSC)
 aqueous polymer gel networks, 112–113
 liquid crystals, 292f
 photochromic liquid crystalline compounds, 322f, 323f

poly(methyl methacrylate) hybrid material, 314f
poly[3-(6-methoxyhexyl)thiophene], 213
triarylamine-poly(p-phenylene) and poly(9,9-dioctylfluorene), 206f
Diketonate ligands, polymeric metal complexes, 235
2,4-Dinitrotoluene (DNT)
 fluorescence of poly(methyl methacrylate)–pyrene methanol (PMMA–PM) with DNT concentration, 395–397
 See also Fluorescent optical chemical sensors
Dithienothiophene (DTT)
 absorption and emission spectra of DTT-based chromophores, 166f
 charge transfer, 162
 optical properties of chromophores, 167t
 π-center role in two-photon absorption, 164
 spectroscopic properties, 165–167
 two-photon absorption cross-section of, chromophores, 168t
 See also Conjugated chromophores
Dithienylethene-based photochromic materials
 application for dual image formation, 178–179
 electronic absorption spectral change with irradiation, 178f
 glass-forming properties, 177
 photochromic behavior, 177–178
 structures, 177
 synthesis, 177
Donor-acceptor arrangements, intramolecular charge transfer, 36
Doping/de-doping, conducting polymers, 67
Double layer light-emitting diodes (LEDs). *See* Polarized electroluminescence

Double potential step chronoamperometry, poly(aniline) layer-by-layer films, 26–27
Dual image formation, 1,2-dithienylethene-based photochromic materials, 178–179
Dynamic tunability, optical properties, 2

E

Electrically controlled light filters, thermotropic hydrogels, 114
Electrically switchable reflectors. *See* Chiral gels
Electrochemical detection, cationic polythiophene biosensors, 365–366
Electrochemistry
  conducting polymer-based electrochromics, 69, 71–73
  poly(benzobisimidazobenzophenanthroline) (BBB), ladder BBB (BBL), and poly(2,2'-[10-methyl-3,7-phenothiazylene]-6,6'-bis[4-phenylquinoline]) (PPTZPQ) films, 37–38, 40f, 42f
  ruthenium complex polymers, 58–59
Electrochromic device (ECD)
  assembly of conducting polymer-based, 68, 69f
  conductivity of gel electrolyte, 44, 45f
  construction and characterization, 43–48
  desired characteristics, 44
  diagrams, 35f
  electrochromic switching, 46f
  electronic absorption spectra of ladder poly(benzobisimidazobenzophenanthroline) (BBL) and poly(benzobisimidazobenzophenanthroline) (BBB)/PMMA/V$_2$O$_5$, 47f
  low band gap polymers, 36
  polymeric materials, 35–36
  poly(2,2'-[10-methyl-3,7-phenothiazylene]-6,6'-bis[4-phenylquinoline]) (PPTZPQ), BBL, and BBB ECD using V$_2$O$_5$ as counter electrode, 44–47
  potential uses, 35
  PPTZPQ-BBL plastic ECD, 47–48
  rear view mirrors, 44
  reversible electrochromic material, 43–44
  secondary electrode, 43–44
  self-erasing processes, 44
  solid, 44
  two-electrode voltammograms, 48f
  types of materials, 35–36
  visible spectra in transmittance mode of PPTZPQ and BBL, 48f
Electrochromic films. *See* Poly(aniline) (PANI) films
Electrochromicity
  conjugated polymers, 369–370
  thienyl-based polymers, 372
Electrochromic organic materials
  applications, 173
  copolymers with methacrylate containing oligo(ethylene oxide) moiety, 182–184
  methacrylate polymers containing pendant oligothiophenes, 182–184
  molecular gels, 179–180
  novel classes of organic compounds for gel formation, 180
  polymers containing pendant oligothiophenes, 181–184
  response time for electrochemical doping and de-doping of methacrylate homo- and copolymers, 184t

vinyl polymers containing pendant oligothiophenes, 181, 182
Electrochromic polymers
  atomic force microscopy images of poly(benzobisimidazobenzophe nanthroline) (BBB) and poly(2,2'-[10-methyl-3,7-phenothiazylene]-6,6'-bis[4-phenylquinoline]) (PPTZPQ) films, 38, 39
  cyclic voltammetry of PPTZPQ- and BBB-coated electrodes, 40f, 42f
  description, 4–5
  electrochemical characterization, 37–38
  electronic absorption spectra of PPTZPQ films on electrode, 43f
  examples of conjugated polymer EC materials, 5f
  film characterization, 37–43
  morphological and structural characterization, 38, 41
  pendant oligothiophenes, 181–184
  spectroelectrochemical characteristics, 41, 43
  structures, 36f
  See also Conducting polymer-based electrochromics; Ruthenium complex polymers
Electrochromism
  characterization, 67
  color change process, 35
  poly(aniline) layer-by-layer films, 30–31
  polycyclic aromatics, 369
  visible and strong near-infrared (NIR), 52
Electroluminescence
  absorption spectra of binary blend and components, 193, 194f
  atomic force microscopy (AFM) phase images of blend and components, 193
  color-tunable, 189

color-tunable multicolor, 189
current-voltage and luminance-voltage curves of blend light-emitting diodes (LEDs), 196, 198f, 199
devices based on polymeric thin layers, 248
electroluminescence (EL) micrographs of blend, 195
EL micrographs of bilayer polymer, 197
EL spectra of bilayer polymer, 197f
experimental, 189–192
fabrication and characterization of LEDs, 191–192
materials, 189–190
morphology and photophysics of binary blends of conjugated polymers, 192–193
optical absorption and photoluminescence (PL) spectra of polyquinolines, 192f, 193
performance of color-tunable bilayer polymer, 199
photophysics and surface morphology, 191
PL emission spectrum of blend, 193, 195f
PL of conjugated polymers, 5, 6f
poly(thiophene) polymers, 229–230
polyquinolines, 190
preparation of blends and thin films, 190
schematic of single-layer and bilayer polymer LEDs, 191f
silicon-based alternating copolymers, 258f, 259f
single-layer LEDs from polyquinoline, 196
tunable emission and, 5–8
voltage-tunable multicolor polymer LEDs, 6–8, 194–196, 199
See also Polarized electroluminescence; Poly(thiophene)s

Electronic absorption
  ladder
    poly(benzobisimidazobenzophenanthroline) (BBB) and BBB/PMMA/V_2O_5 devices, 46, 47f
  poly(2,2'-[10-methyl-3,7-phenothiazylene]-6,6'-bis[4-phenylquinoline]) (PPTZPQ) film as function of potential, 41, 43f
Electronic absorption spectral change
  azobenzene-based photochromic film, 176f
  dithienylethene-based photochromic film, 178f
  See also Absorption spectra
Electronic and photo active materials
  atomic force microscopy (AFM) image of surface relief grating (SRG) on polyazophenol thin film, 384f
  enzymatically synthesized conducting polyaniline (PANI), 380–383
  enzymatically synthesized poly(1-hydroxypyrene), 385–386
  enzymatically synthesized polyazophenols, 383–385
  enzymatic polymerization of 4-hydroxypyrene, 379
  enzymatic polymerization of aniline, 378–379
  enzymatic polymerization of azophenols, 379
  experimental, 378–380
  fluorescence spectra of polymer and 1-hydroxypyrene, 386f
  materials, 378
  optical quality polymer film preparation, 379–380
  peroxidase-catalyzed synthesis, 378
  reversible reduction/oxidation behavior of PANI/sulfonated polystyrene (SPS) complex, 383
  role of SPS template, 381
  SRG formation, 380
  UV-vis spectra change of PANI/SPS complex, 382f
  UV-vis spectra for polymer and monomer of 4-phenylazophenol, 384f
  UV-vis spectra of polyaniline, 381f
Electrospinning, 390–392
Emission spectra
  Ru(II)-chelated copolymers, 254f
  silicon-based alternating copolymers, 252f, 253f
Emittance, conducting polymer-based electrochromics, 69, 73
Environmental dependence, polyacetylenes bearing L-amino acid pendants, 351f
Enzymatic synthesis
  conducting poly(aniline), 380–383
  poly(azophenol)s, 383–385
  poly(1-hydroxypyrene), 385–386
  See also Electronic and photo active materials
N-(2-Ethanol)-10,12-pentacosadiynamide (PCEA). See Poly(diacetylene)s (PDAs)
Ethoxylated poly(ethylene imine) (ePEI)
  chemical structure, 21f
  polyanion, 20
  See also Poly(aniline) (PANI) films
Europium complexes
  applications, 234
  emission spectra, 242f
  general structure, 235f
  homopolymers, 239–240
  luminescence lifetimes for tris complexes, 240t
  See also Lanthanide complexes
Excited state quencher, reflection band width for chiral gels, 287
Explosives, detection by chemical sensors, 388

Extinction coefficients, ruthenium complex polymers, 61

F

Film. *See* Poly(aniline) (PANI) films; Poly(diacetylene)s (PDAs); Trialkoxysilane-capped poly(methyl methacrylate)-silica
Film structure, poly(diacetylene)s (PDAs), 86–87
First-order transition, liquid crystals, 292–293
Fluorene
   changes in fluorescence emission spectra, 126, 127f
   film exposure via Air Force targets, 130
   near-IR fs laser irradiation, 125
   oligomeric chromophores, 164–165
   photoacid generator (PAG), 125
   three-dimensional two-photon fluorescence imaging, 128–129
   total integrated fluorescence intensity vs. pump power, 128f
   TPA (absorption, two-photon), 125
   TPA cross-section of, chromophores, 169t
   two-photon fluorescent images of films, 128–129
   two-photon upconverted fluorescence emission, 127, 128f
   UV-vis absorption spectra of irradiation of, and PAG, 126f
   See also Conjugated chromophores; Photosensitive polymers; Poly(fluorene)s (PFs)
Fluorescence emission spectra
   fluorene, 126, 127f
   monomer and polymer of 1-hydroxypyrene, 386f
   poly(acrylic acid)–pyrene methanol (PAA–PM) with varying 2,4-dinitrotoluene (DNT) content, 397f
   poly(methyl methacrylate)–pyrene methanol (PMMA–PM) with DNT concentration, 395f
   poly(styrene-*co*-maleic anhydride) modified polymer, 132f
Fluorescence imaging, 3D two-photon, multilayer structures, 128–129
Fluorescence quenching, technique, 390
Fluorescent optical chemical sensors
   advantages, 388
   electrospinning, 390–392
   electrospinning set-up, 391f
   fluorescence emission spectra of poly(acrylic acid–pyrene methanol) (PAA–PM) electrospun membrane vs. 2,4-dinitrotoluene (DNT) concentration, 396–397
   fluorescence quenching, 390
   fluorescence spectra with varying DNT, 395f
   immobilization of dyes onto polymeric materials, 389
   PAA–PM, 393, 394f
   poly(methyl methacrylate–pyrenemethanol) (PMMA–PM), 392, 393f
   polymeric materials, 389
   scanning electron microscopy (SEM) images of electrospun membranes of PMMA–PM and PAA–PM, 394, 395f
   sensitivities, 397
   sensitivity of sensing film, 389
   sensor fabrication and performance, 394–397
   spin-dope solution of PAA–PM, 394
   spin-dope solution of PMMA–PM, 394

Stern–Volmer plots of electrospun membranes vs. quencher concentration, 396
techniques, 390–392
Fluorometric detection, oligonucleotide hybridization, 364–365
Free radicals, photopolymerization, 136
Friction anisotropy, poly(diacetylene)s (PDAs), 90–91
Functionalization. *See* Poly(thiophene)s

G

Gel networks
chromogenic, 117–120
outlook, 120
thermochromic, 115–117
thermotropic, 111–114
Gels. *See* Chiral gels
Glass-forming properties
azobenzene-based photochromic materials, 174
dithienylethene-based photochromic materials, 177
Glass transition temperature
definition, 290
glassy liquid crystals, 295, 298–299
Glassy films, characterization, 291
Glassy liquid crystals (GLCs)
amorphous molecular glasses, 290–291
differential scanning calorimetry (DSC) of liquid crystals, 292f
elevated phase transition temperatures and superior morphological stability, 295, 298–299
first synthesis attempts, 293
high-temperature glassy cholesterics, 299f
high-temperature glassy nematics, 297f
liquid crystalline order via molecular self-organization, 292f
novel class of photoresponsive GLCs, 301–302
novel glassy materials containing photoresponsive diarylethenes, 302f
optical properties of cholesteric LC films, 294–295
optical spectra of unpolarized beam, 298f
phototunable glassy cholesteric film, 300f
reflective coloration by glassy cholesteric films, 300f
representative GLCs reported previously, 293f
representative morphologically stable, 296f
reversible tunability of reflective coloration, 301, 302f
selective reflection by left–handed cholesteric film, 294f
supercooled liquid crystals, 291–294
synthesis of glassy cholesterics, 297
tunable reflective coloration by glassy cholesteric films, 299–300
D-Glucose pendants, UV spectra of polyacetylenes with, 345f

H

Helfrich's deformation, 283
Helical structures
polyacetylenes bearing amino acid pendants, 348–352
See also Amphiphilic helical polyacetylenes

Helicity, chain, tuning, of polyacetylenes, 353–356
Hertzian model, phase shifts, 93
Hexamethyl Dewar benzene (HBDB) photoinitiated conversion, 136–137
See also Quantum amplified isomerization (QAI)
Horseradish peroxidase. *See* Peroxidase
Hybrid materials
optical waveguide applications, 308
See also Trialkoxysilane-capped poly(methyl methacrylate)-silica
Hybrid solar light filters, thermotropic hydrogels, 114
Hydrogels, embedding colloidal crystals, 330
Hydrogen bonding, amphiphilic helical polyacetylenes, 346–348
4-Hydroxypyrene, enzymatic polymerization, 379

I

Initiators, dibenzoylmethane (dbm), 236–237
In-plane anisotropy, poly(diacetylene)s (PDAs), 91–93
Intermittent contact atomic force microscope (IC–AFM)
contrast, 84, 85–86
imaging in-plane anisotropy, 91–93
phase shifts, 91–92
Intramolecular charge transfer, alternating donor-acceptor arrangements, 36
Irradiation, photochromic liquid crystalline compounds, 325–327
Isomerization. *See* Quantum amplified isomerization (QAI)

L

Ladder poly(benzobisimidazobenzophenanthroline) (BBL)
electrochemical characterization, 37–38, 40f, 42f
morphological and structural characterization, 38, 39, 41
spectroelectrochemical characteristics, 41, 43
structure, 36f
V_2O_5 film as counter electrode, 44–47
visible spectra in transmittance mode, 48f
See also Electrochromic devices (ECD)
Lanthanide complexes
applications, 234
bipyridine (bpy) macroligand toolkit, 241f
block copolymer assemblies with labile metal crosslinks, 243–244
block copolymer microphase separation, 235f
block copolymers and other adducts, 240–243
Claisen condensation of ester and ketone to afford dibenzoylmethane (dbm), 236f
conjugated polymers, 234
dbm initiators, 236–237
dbm macroligands, 237–238
emission spectra of Eu tris dbm poly(lactic acid) and bpy poly(caprolactone)$_2$ adduct, 242f
general Eu dbm complex structure, 235f
heteroarm stars, 235
homopolymers, 239–240
kinetics plot of ε-caprolactone (CL) polymerizations, 237f
luminescence lifetimes for tris complexes, 240t

modular approach to Ln tris and adduct complexes, 241f
morphological changes with thermal treatment, 243–244
polymeric ligands, 234–235
routes to three dbm macroligands, 238f
synthesis of alcohol initiators, 236f
Latent image recording, photochromic liquid crystalline compounds, 326–327
Layer-by-layer assembly
film method, 22
films containing poly(aniline), 20
processing, 19
See also Poly(aniline) (PANI) films
Leucine-containing polyphenylacetylene, single- and double-stranded helical chains, 347
Leucoemeraldine, poly(aniline) (PANI), 20
Light-emitting diodes (LEDs)
band diagrams for materials in blue LEDs, 208f
blue, using poly(pyridine), 265
development of blue and red, 248
fabrication with triaryl-poly(paraphenylene) (TA-PPP), 207
See also Electroluminescence; Polarized electroluminescence; Poly(paraphenylene) (PPP); Silicon-based alternating copolymers
Light filters, thermotropic hydrogels, 114
Lipid doping, bolamphiphilic polydiacetylene (PDA), 105–107
Liquid crystals
photochromic reactions, 320–321
See also Chiral gels; Glassy liquid crystals (GLCs)
Low band gap, polymers, 36
Luminescence. *See* Electroluminescence; Lanthanide complexes; Photoluminescence (PL); Polarized electroluminescence

M

Macroligands. *See* Lanthanide complexes
Mechanism, color writing with colorless ink on photonic papers, 331–332
Mechanochromism
description, 8–9
poly(diacetylene)s (PDAs), 9, 87–89
stress for conversion, 84
Metal crosslinks, labile, and temperature, 243–244
Metallization, complex polymers, 52–53
Methacrylate polymers, pendant oligothiophenes, 182–184
Microphase separation, block copolymers, 235
Microscopic images, photochromic liquid crystalline compounds, 326f
Military
applications, 75–76
camouflage data, 77f, 78f
NASA's ST5 mission, 74–75
sample camouflage data, 77f, 78f
See also Conducting polymer-based electrochromics
Modeling, phase shifts, 93
Models, space-filling, of substituted poly(thiophene) polymers, 227, 228f
Molecular gels, electrochromic, 179–180
Molecular glasses, amorphous, 290–291

Monosaccharide-containing polymers, behaviors, 352
Monosaccharide moieties, polyacetylenes, 343
Morphological stability, glassy liquid crystals, 295, 296f
Morphology
  bolamphiphilic polydiacetylene, 103–104
  labile metal crosslinks and temperature, 243–244
Multicolor electroluminescence. *See* Electroluminescence

N

Naphthodithiophene moieties, stability, 370
Near field scanning optical microscope (NSOM)
  mechanochromism, 84, 86
  poly(diacetylene)s (PDAs), 87–88
Near infrared (NIR). *See* Ruthenium complex polymers; Trialkoxysilane-capped poly(methyl methacrylate)-silica
Nematic order
  high-temperature glassy nematics, 297f
  molecular self-organization, 292f
  See also Glassy liquid crystals (GLCs)
Networks
  liquid crystalline gels, 282
  See also Chiral gels; Gel networks
Night vision safety (NVS), visible light attenuation, 52
Nonionic polymers, chemical structure, 21f
Nucleoside moieties, polyacetylenes, 343

O

Oligo(ethylene oxide) moiety, methacrylate copolymers with, 182–184
Oligomers. *See* Conjugated chromophores
Oligonucleotide hybridization
  colorimetric detection, 360–361, 364
  detection, 359
  electrochemical detection, 365–366
  fluorometric detection, 364–365
  water-soluble cationic polythiophene, 360
  See also Cationic polythiophene biosensors
Oligothiophenes
  charge transfer, 162
  electrochromic polymers containing pendant, 181–184
  methacrylate polymers with pendant, 182–184
  optical properties of chromophores, 166t
  vinyl polymers with pendant, 181, 182
  See also Conjugated chromophores
Optical absorption
  poly(diacetylene)s (PDAs), 83
  See also Photosensitive polymers
Optical data storage
  techniques, 123
  See also Photosensitive polymers
Optical planar waveguides
  optical losses of hybrid, 317, 318f
  preparation from hybrid materials, 309f, 310
  properties, 315t
  See also Trialkoxysilane-capped poly(methyl methacrylate)-silica
Optical properties
  cholesteric gels, 279
  cholesteric liquid-crystal films, 294–295

poly(thiophene) polymers, 225–229
reversible change, 2
tunability, 2–3
See also Chromogenic effects in polymers; Colloidal crystals; Poly(thiophene)s; Trialkoxysilane-capped poly(methyl methacrylate)-silica
Optical quality, polymer film preparation, 379–380
Optical switching experiment, ruthenium complex polymers, 61–62
Optical transformation, pH induced, in bolamphiphilic polydiacetylene, 104–105
Organic light-emitting diodes (OLEds) active materials, 212
See also Polarized electroluminescence
Organic photochemistry innovations, 123
See also Photosensitive polymers
Outgassing, conducting polymer-based electrochromics, 70, 73
Out-of-focus absorption, 123
Out-of-focus excitation, 123
Oxidative peak currents, poly(aniline) layer-by-layer films, 24, 26*f*

P

Packing arrangement, bolamphiphilic polydiacetylene, 103–104
Palladium-catalyzed coupling, poly(3-bromo-4-hexylthiophene) transformation, 222, 224
PCEA (*N*-(2-ethanol)-10,12-Pentacosadiynamide). *See* Poly(diacetylene)s (PDAs)
Pendant oligothiophenes, electrochromic polymers with, 181–184

10,12-Pentacosadiynoic acid (PCDA). *See* Poly(diacetylene)s (PDAs)
Perfluorinated ionomer Nafion®
chemical structure, 21*f*
polyanion, 20
See also Poly(aniline) (PANI) films
Peroxidase
oxidation of phenols and anilines, 377
reaction mechanism of peroxidase-catalyzed synthesis, 378
See also Electronic and photo active materials
pH
colorimetric response for polydiacetylene microstructures, 101*f*
optical and structural transformation in bolamphiphilic PDA, 104–105
tuning chain helicity of polyacetylenes, 354, 355*f*
Phase separation
block copolymers, 235
chiral gels upon polymerization, 285–286
Phase shifts
intermittent contact–atomic force microscope (IC–AFM), 91–92
modeling, 93
Phase transition temperatures, photochromic liquid crystalline compounds, 324*t*
Phenol Red
temperature dependence of transparency, 119*f*
UV-vis absorption spectra, 118*f*
See also Chromogenic gel networks
Phenols, peroxidase-catalyzed oxidation of, 377
Photoacid generator (PAG) protonation, 125
See also Photosensitive polymers
Photo active materials. *See* Electronic and photo active materials

Photochromic amorphous molecular materials
  azobenzene-based, 174–176
  dithienylethene-based, 177–179
  dual image formation, 178–179
  electronic absorption spectral change with irradiation, 176f, 178f
  formation of surface relief grating (SRG), 176
  glass-forming properties, 174, 177
  photochromic behavior, 175, 177–178
  synthesis, 174, 177
Photochromic behavior
  azobenzene-based photochromic materials, 175
  dithienylethene-based photochromic materials, 177–178
Photochromic liquid crystalline compounds
  change of reflection spectra of thin film of dimesogenic compounds, 324f
  chemical structures, 321
  cholesteryl group and azobenzene moiety, 321–322
  differential scanning calorimetry (DSC) of, 322f, 323f
  experimental, 322
  irradiation and cholesteric reflections, 325
  latent image recording, 326–327
  microscopic images of solid films after irradiation, 326f
  phase transition temperatures and λ_{max} of cholesteric reflection, 324t
  reflection spectra of films before and after irradiation, 325f
  temperature, 326–327
Photochromic polymers
  applications, 173
  description, 3–4
  See also Photosensitive polymers
Photoluminescence (PL)
  DO-PF (poly(9,9-dioctylfluorene)), 204f, 205f
  polarized PL of double layer polymer film, 216f
  triarylamine poly(paraphenylene), 205f
  See also Polarized electroluminescence; Poly(quinoline)s
Photonic band gap (PBG) materials
  colloidal crystals, 330
  polymer-based PBG composites, 12
Photonic papers
  application of ink to surface, 333, 334
  embedding colloidal crystals of polymer beads in elastomeric matrix, 335–336
  fabrication for color switch between visible and invisible state, 333, 335
  mechanism for color writing with colorless ink, 331–332
  See also Colloidal crystals
Photoresponsive, glassy liquid crystals (GLCs), 301–302
Photosensitive polymers
  fluorene, 125–129
  image formation upon photoacid generation, 130
  linear absorption and fluorescence emission spectra, 132f
  optical data storage, 123
  out-of-focus absorption, 123
  out-of-focus excitation, 123
  poly(styrene-*co*-maleic anhydride) modification, 129, 132f
  reaction of fluorene with acid, 124f
  synthesis and characterization of organic fluorescent dyes, 124–125
  two-photon absorbing fluorophore-labeled polymer, 132f

two-photon absorption, 123–124
two-photon fluorescence images of, films, 130
two-photon fluorescent images of multi-layered films, 131
two-photon induced photochromism of spiropyran derivatives, 124
two-photon upconverted fluorescence spectra of modified poly(styrene-*co*-maleic anhydride), 132*f*
writing and reading by two-photon excitation, 133
See also Fluorene
Phototunability, reversible, reflective coloration, 301, 302*f*
Piezochromic polymers
 description, 9–10
 examples, 10–11
 mechanism of color change, 10
 reversible, 11–12
Pigmentary mode, coloration, 299
Polarized electroluminescence
 differential scanning calorimetry (DSC) of poly[3-(6-methoxyhexyl)thiophene] (P6OMe) powders, 213
 double-layer device, 217*f*
 experimental, 218–219
 final thickness of P6OMe layer, 214
 heterostructure of active light-emitting diode (LED), 215
 P6OMe, 213
 polarized absorption spectra of double-layer film of oriented poly(p-phenylenevinylene) (PPV)/P6OMe, 215*f*
 polarized photoluminescence of double-layer film of oriented PPV/P6OMe, 216*f*
 procedure to prepare heterostructures, 213–214
 red emission of P6OMe at low voltages, 217–218
 red emitting polymer, 213
 segmented PPV, 212
 sketch of PPV/P6OMe interface, 214*f*
 yellow-green emitting polymer, 212
 See also Electroluminescence
Polyacetylenes. *See* Amphiphilic helical polyacetylenes
Poly(acrylamide) (PAAm)
 chemical structure, 21*f*
 polyanion, 20
 See also Poly(aniline) (PANI) films
Poly(2-acrylamido-2-methyl-1-propanesulfonic acid) (PAMPS)
 chemical structure, 21*f*
 polyanion, 20
 See also Poly(aniline) (PANI) films
Poly(acrylic acid) (PAA)
 chemical structure, 21*f*
 polyanion, 20
 See also Poly(quinoline)s
Poly(acrylic acid)–pyrene methanol (PAA–PM)
 fluorescence emission spectra of PAA–PM electrospun membrane vs. 2,4-dinitrotoluene (DNT) concentration, 396–397
 fluorescence emission spectra with varying DNT concentration, 397*f*
 scanning electron microscopy (SEM) images, 395*f*
 sensitivity, 397
 sensor fabrication, 394
 synthesis, 393, 394*f*
Poly(aniline) (PANI)
 applications, 19–20
 chemical structure, 21*f*
 enzymatically synthesized, 378
 enzymatically synthesized conducting, 380–383

layer-by-layer (LBL) films
containing, 20
leucoemeraldine PANI, 20
reversible reduction/oxidation
behavior of PANI/sulfonated
polystyrene (SPS) complex, 383
self-doping PANI (SPANI), 20,
21*f*
UV-vis spectra, 381*f*
UV-vis spectra change of
PANI/SPS complex, 382*f*
Poly(aniline) (PANI) films
absorbance switching profiles, 30*f*
assembly, 23–24
assembly method, 22
current and charge switching
profiles for electrostatic and H-
bond, 27*f*
cyclic voltammetry, 24, 25*f*, 26
double potential step
chronoamperometry, 26–27
electrochromic performance, 30–31
experimental, 22
measurement methods, 22
oxidative and reductive peak
currents for electrostatic and H-
bond, 26*f*
potential step absorptometry, 30
redox center concentration and
extinction, 29*t*
spectroelectrochemistry, 27–30
thickness of composites, 23*t*
Polyanions, chemical structures, 21*f*
Poly(aryleneethynylene)s (PAEs)
absorption and emission of
poly(dialkylfluorenyleneethynyl
ene)s (PFEs), 154*f*
absorption spectrum of PFEs, 153*f*
chromicities in
poly(heteroaryleneethynylene)s,
156–157
dialkyl-poly(p-
phenyleneethynylene)s (dialkyl-
PPEs), 147

interplay between chromicity and
rotational conformation of
dialkyl-PPEs, 151*f*
solid state structure of didodecyl-
PPEs, 151, 152
solvatochromicity in donor and
acceptor substituted PPEs, 154–
156
solvatochromicity of amide-
substituted PPE, 155*f*
solvatochromicity of dialkoxy-PPE
(6), 155*f*
solvatochromism of
poly(dialkylfluorenyleneethynyl
ene)s (PFEs), 153–154
synthesis, 148–149
thermochromicity of dialkyl-PPEs,
151, 152
thermochromism and
solvatochromism of dialkyl-
PPEs, 150–153
thick film of
poly(heteroaryleneethynylene)s,
157, 158
Poly(azophenol)s
atomic force microscope (AFM) of
surface relief grating (SRG),
384*f*
enzymatically synthesized, 383–
385
UV-vis spectra for polymer and
monomer, 384*f*
Poly(benzobisimidazobenzophenanthr
oline) (BBB)
electrochemical characterization,
37–38, 40*f*, 42*f*
ladder BBB (BBL), 36
morphological and structural
characterization, 38, 39, 41
spectroelectrochemical
characteristics, 41, 43
structure, 36*f*
V_2O_5 film as counter electrode, 44–
47

See also Electrochromic devices (ECD)
Poly(diacetylene)s (PDAs)
 atomic force microscope (AFM) topography and friction images of red monolayer, 90*f*
 amino acid-terminated amphiphilic, 98–102
 amphiphilic (monofunctional) and bolaamphiphilic (bis-functional), 97–98
 atomic force microscope (AFM), 84, 85–86
 blue-to-red transition, 87–89, 97
 bolaamphilphilic PDAs, 102–105
 characteristics, 83
 experimental, 84–86
 film preparation, 84–85
 film structure, 86–87
 friction anisotropy, 90–91
 imaging in-plane anisotropy with intermittent contact AFM (IC–AFM), 91–93
 instrumentation, 85–86
 lipid doping-induced structural transformation in bolaamphiphilic PDA, 105–107
 mechanochromism, 9, 87–89
 modeling phase shifts, 93
 modifications, 97
 near field scanning optical microscopy (NSOM), 84, 86
 NSOM shear force topography and simultaneous fluorescence images, 87*f*
 optical absorption, 83
 optical properties, 97
 PCDA (10,12-pentacosadiynoic acid), 84, 85*f*
 PCEA (*N*-(2-ethanol)-10,12-pentacosadiynamide), 84, 85*f*
 phase shifts in IC–AFM, 91–92
 promising sensor platforms, 96–97
 topographic AFM images of blue PCDA, 89*f*
 topographic AFM images showing tip-induced patterning of red PCDA domains, 88*f*
 topographic and phase images of PDA monolayer thin film, 92*f*
 See also Amphiphilic polydiacetylenes; Bolaamphiphilic polydiacetylenes
Poly(dialkylfluoryleneethynylene)s (PFEs)
 absorption spectrum, 153*f*
 solvatochromism, 153–154
 synthesis, 148–149
 See also Poly(aryleneethynylene)s (PAEs)
Poly(9,9-dioctylfluorene) (DO-PF). *See* Poly(fluorenes) (PFs); Poly(paraphenylene) (PPP)
Polyelectrolytes, chemical structure, 21*f*
Poly(ethylene oxide) (PEO)
 chemical structure, 21*f*
 conductivity of PEO gel electrolytes, 44, 45*f*
 polyanion, 20
 See also Poly(aniline) (PANI) films
Poly(fluorene)s (PFs)
 DO-PF (poly(9,9-dioctylfluorene), 203
 photoluminescence spectra of DO-PF, 204*f*, 205*f*
 properties, 202
 thermogravimetric analysis (TGA) of DO-PF, 206*f*
 See also Poly(paraphenylene) (PPP)
Poly(heteroaryleneethynylene)s
 chromicities, 156–157
 synthesis, 148–149
 think film of sample, 158
 UV-vis absorption spectra, 156*f*
 See also Poly(aryleneethynylene)s (PAEs)
Polyhydrazines

polymeric ligands, 52–53
See also Ruthenium complex polymers
Poly(1-hydroxypyrene)
  enzymatically synthesized, 385–386
  fluorescence spectra, 386*f*
Polymer hydrogels, embedding colloidal crystals, 330
Polymeric materials
  electrochromic devices, 35–36
  See also Chromogenic effects in polymers; Ruthenium complex polymers
Polymeric metal complexes
  block copolymers and other adducts, 240–243
  homopolymers, 239–240
  properties, 234–235
  See also Lanthanide complexes
Polymerization, substituted polyacetylenes, 344*t*
Polymer light-emitting diodes
  substituted poly(thiophene)s, 221
  See also Poly(thiophene)s
Poly[3-(6-methoxyhexyl)thiophene] (P6OMe)
  differential scanning calorimetry, 213
  structure, 213
  thickness of P6OMe layer, 214
  See also Polarized electroluminescence
Poly(methyl methacrylate) (PMMA)
  Fourier transform infrared spectra of PMMA and hybrids, 312*f*
  near infrared absorption spectra of PMMA and hybrids, 317*f*
  organic-inorganic hybrid materials, 308–309
  properties, 315*t*
  See also Trialkoxysilane-capped poly(methyl methacrylate)-silica
Poly(methyl methacrylate)–pyrene methanol (PMMA–PM)
  fluorescence emission spectra with varying 2,4-dinitrotoluene (DNT) concentration, 395*f*
  scanning electron microscopy (SEM) images, 395*f*
  sensitivity, 397
  sensor fabrication, 394
  synthesis, 392, 393*f*
Poly(2,2'-[10-methyl-3,7-phenothiazylene]-6,6'-bis[4-phenylquinoline]) (PPTZPQ)
  absorption spectra as function of potential, 41, 43*f*
  electrochemical characterization, 37–38, 40*f*, 42*f*
  electrochromic switching, 45, 46*f*
  morphological and structural characterization, 38, 39, 41
  spectroelectrochemical characteristics, 41, 43
  structure, 36*f*
  V_2O_5 film as counter electrode, 44–47
  visible spectra in transmittance mode, 48*f*
  See also Electrochromic devices (ECD)
Poly(paraphenylene) (PPP)
  atomic force microscope (AFM) image of thin film of triarylamine (TA)-PPP/poly(9,9-dioctylfluorene) (DO-PF) blend, 207*f*
  band diagram of materials in efficient blue LED, 208*f*
  blue electroluminescence, 202
  blue light emitting polymers, 207
  charge confinement at TA-PPP/2-(4-*t*-butylphenyl)-5-biphenyloxadiazole (*t*-PBD)interfaces, 208
  charge injection and carrier mobility, 202–203
  current-light intensity-voltage response, 208*f*

diarylamino side group or TA-PPP, 203
differential scanning calorimetry (DSC) of TA-PPP and DO-PF, 206*f*
fabrication of light-emitting diodes (LED), 207
PL spectra of DO-PF, 204*f*
PL spectrum of DO-PF thin film after thermal treatment, 205*f*
PL spectrum of TA-PPP thin film after thermal treatment, 205*f*
poly(fluorenes) (PFs), 202
repeat unit of TA-PPP, 203
solubility of TA-PPP, 203
solvatochromism, 203
thermogravimetric analysis (TGA) of TA-PPP and DO-PF, 206*f*
thin films of TA-PPP with *t*-PBD, 207–208
UV-vis absorption and photoluminescence (PL) spectra for TA-PPP, 204*f*
Poly(paraphenylene vinylene) (PPV) segmented, 212
solubility and photoluminescent efficiency, 202–203
See also Polarized electroluminescence; Poly(paraphenylene) (PPP)
Poly(p-phenyleneethynylene)s (PPEs)
interplay between chromicity and rotational conformation in dialkyl-PPEs, 151*f*
solid state structure of didodecyl-PPE, 152
solvatochromicity in donor and acceptor substituted PPEs, 154–156
solvatochromicity of amide-substituted PPE, 155*f*
solvatochromicity of dialkoxy-PPE, 155*f*
solvatochromicity of dialkyl-PPEs, 150*f*
synthesis, 148–149
thermochromicity of dialkyl-PPE, 152
thermochromism and solvatochromism, 150–153
See also Poly(aryleneethynylene)s (PAEs)
Poly(pyridine), blue light emitting diodes, 265
Poly(quinoline)s
blend with poly(acrylic acid), 268
blend with poly(vinyl alcohol) (PVA), 268–269
chemical structures of poly(2,6-[4-phenylquinoline]) (PPQ) and poly(2,6-[p-phenylene]-4-phenylquinoline) (PPPQ), 265
composite with tetraethyl orthosilicate (TEOS), 270–271
conjugated polymers, 265
emitting colors under UV radiation of PPQ-silica gels, 272
excimer formation, 271, 273
experimental, 266
fluorescence emission spectra of PPPQ/PVA blend films, 269*f*
fluorescence spectra of PPPQ/TEOS glass under fast mixing, 274*f*
fluorescence spectra of PPQ in formic acid, 267*f*
fluorescence spectra of PPQ/poly(acrylic acid) blend films, 270*f*
fluorescence spectra of PPQ/poly(acrylic acid) mixtures, 268*f*
fluorescence spectra of PPQ-silica gels, 271*f*
fluorescent properties in blended polymer films, 268–269
fluorescent properties in silica, 270–273
fluorescent properties in solutions, 266–268

future work, 275
interchain interaction, 273
poly(acrylic acid) addition, 267–268
reactions, 265
sketch of interaction between PPQ and silica, 273
See also Electroluminescence
Poly(styrene-co-maleic anhydride)
fluorescence emission spectra, 132*f*
fluorophore labeled modification, 129, 132*f*
linear absorption, 132*f*
two-photon absorbing fluorophore-labeled polymer, 132*f*
two-photon upconverted fluorescence spectra, 132*f*
See also Photosensitive polymers
Poly(thiophene)s
chromicity, 147–148
3,4-disubstituted, 225*t*
electroluminescence (EL), 229–230
electrophilic substitution of poly(3-hexylthiophene) (P3HT), 221–222
EL spectra of o-tolyl-substituted P3HT, 230*f*
experimental, 231
^{1}H NMR spectra of substituted products, 224*f*
measurements, 231
o-methoxyphenyl and 1-naphthyl substituents, 227, 228*f*
optical properties, 225–229
o-tolyl-PHT, 227, 229
partially substituted P3HT, 226*t*
Ph-PHT, 226–227
polymer light-emitting diodes (PLEDs), 221
schematic of post-functionalization of P3HT, 223*f*
structural analogs of o-tolyl-PHT, 228
synthesis methods, 231

transformation of poly(3-bromo-4-hexylthiophene) (Br-PHT), 222, 224
See also Cationic polythiophene biosensors; Conjugated polymers
Poly(vinyl alcohol) (PVA). *See* Poly(quinoline)s
Post-functionalization. *See* Poly(thiophene)s
Potential step absorptometry, poly(aniline) layer-by-layer films, 30

Q

Quantum amplified isomerization (QAI)
chain reaction mechanism of hexamethyl Dewar benzene (HBDB) conversion to hexamethylbenzene (HMB), 136
characterization data for Dewar benzene-containing copolymers, 140*t*
conversion of Dewar benzene to benzene derivatives in polymer films, 137, 139
DCA (9,10-dicyanoanthracene) sensitizer, 137
Dewar benzene conversion in solid poly(methyl methacrylate) (PMMA) film, 144*f*
Dewar benzene conversion on side chain copolymer film by FRS, 145*f*
forced Rayleigh scattering (FRS), 142–145
highest quantum yields, 137
irradiation of copolymer films, 140–142
irradiation of sensitizer, 136–137
photoinitiated conversion in PMMA film, 138*f*

photoinitiated conversion of Dewar benzene to benzene moieties in side chain polymer films, 142f, 143f
possible mechanism, 139
process, 136
side chain polymers, 139–142
synthesis of copolymers, 140, 141

R

Radiation, conducting polymer-based electrochromics, 70, 73
Reading, two-photon excitation of photosensitive polymer film, 133
Real time tunability, optical properties, 2
Reductive peak currents, poly(aniline) layer-by-layer films, 24, 26f
Reflectance, conducting polymer-based electrochromics, 69, 71–73
Reflection, selective, cholesteric film, 294f
Reflection band
 effect of excited state quencher in chiral gels, 287
 shifting to lower wavelengths, 282–283
 temperature dependence for chiral gels, 284, 287, 288f, 289
 time dependence, 286
 width of, as function of time, 286f
Reflection spectra, photochromic liquid crystalline compounds, 324f
Reflective coloration
 glassy cholesteric films, 300f
 photoresponsive glassy liquid crystals, 301–302
 reversible tunability, 301, 302f
 See also Glassy liquid crystals (GLCs)
Refractive index, equation for average, 330

Response time, electrochemical doping and dedoping of methacrylate homo- and copolymers, 184t
Reversible color control
 chromogenic gel networks, 117–120
 thermochromic gel networks, 115–117
Reversible transparency
 chromogenic gel networks, 117–120
 thermotropic gel networks, 111–114
Reversible tunability
 photoresponsive glassy liquid crystals, 301–302
 reflective coloration, 301, 302f
Ruthenium complex polymers
 coloration efficiency, 62–63
 electrochemical switching, 59t
 electrochemistry, 58–59
 electrochromism, 52
 experimental, 54–57
 extinction coefficients, 61
 isophthalic dihydrazide synthesis, 55
 measurements, 54
 metallized films, 52–53
 optical attenuation of coated ITO/glass, 62f
 optical switching experiment, 61–62
 polyhydrazines as ligands, 53f
 reagents and synthesis, 54–57
 spectral data for, in acetonitrile, 61t
 spectral data for, on ITO glass, 60t
 spectroscopic study, 59–63
 structures, 53f
 synthesis, 57
 synthesis of dicarbonylhydrazine-Ru polymers, 56–57
 synthesis of ligand polymers, 55–56
 thermal analysis, 57–58
 See also Electrochromic polymers

S

Salting in, 114
Salting out, 114
Scanning electron microscopy (SEM)
　PAA–PM (poly(acrylic acid)–pyrene methanol) and poly(methyl methacrylate)–pyrene methanol (PMMA–PM), 395f
　photonic paper before and after swelling matrix, 332f
　PMMA hybrid material, 313f
Selective reflection, cholesteric film, 294f
Self-doping poly(aniline) (SPANI)
　chemical structure, 21f
　polyanion, 20
　See also Poly(aniline) (PANI) films
Self-erasing processes, electrochromic devices, 44
Sensing devices
　conjugated polymers, 96
　poly(diacetylene)s (PDAs), 96–97
Sensitizer 9,10-dicyanoanthracene. See Quantum amplified isomerization (QAI)
Sensors. See Fluorescent optical chemical sensors
Silica
　composite of poly(quinoline) and, 270–273
　sol-gel technique, 270–271
　See also Poly(quinoline)s; Trialkoxysilane-capped poly(methyl methacrylate)-silica
Silicon-based alternating copolymers
　carbazole unit introduction, 251
　electroluminescence (EL) spectra, 258f, 259f
　electroluminescent device (ELD) applications, 255–262
　EL spectra as function of applied voltages, 256f

photoluminescent (PL) spectral features, 260
photophysical properties, 251–255
polymerization results, 249t
proposed scheme for formation of charge transfer complex, 256f, 261f
red electroluminescent materials, 253, 255
synthesis, 248–249, 250
thermal and photophysical properties, 249t
UV-vis absorption and emission spectra, 252f, 253f
UV-vis absorption spectra of Ru(II)-chelated, 254f
Single-layer polymer blends. See Electroluminescence
Smectic order
molecular self-organization, 292f
See also Glassy liquid crystals (GLCs)
Solar wind exposure, conducting polymer-based electrochromics, 70, 73
Solvatochromism
amphiphilic helical polyacetylenes, 346–348
dialkyl-poly(p-phenyleneethynylene)s (dialkyl-PPEs), 150–153
donor and acceptor substituted PPEs, 154–156
phenomenon, 2–3
poly(dialkylfluorenyleneethynylene)s (PFEs), 153–154
Solvent dependence
circular dichroism changes for polyacetylenes, 353f
polyacetylenes bearing D-monosaccharide acetonides, 352f
polyacetylenes bearing L-amino acid methyl esters pendants, 350f

Spacecraft
  electrochromic requirement, 67
  four-device panel, 72f
  NASA's ST5 mission, 74–75
  space durability tests, 70, 71, 73
  space-use controller, 74
  ST5 microsatellites, 76f
  ST5 mission, 74–75
  typical device, 72f
  See also Conducting polymer-based electrochromics
Space durability, conducting polymer-based electrochromics, 70, 71, 73
Space-filling models, substituted poly(thiophene) polymers, 227, 228f
Spectroelectrochemistry
  poly(aniline) layer-by-layer films, 27–30
  poly(benzobisimidazobenzophenanthroline) (BBB), ladder BBB (BBL), and poly(2,2'-[10-methyl-3,7-phenothiazylene]-6,6'-bis[4-phenylquinoline]) (PPTZPQ) films, 41, 43
  ruthenium complex polymers, 59–60
Spectroscopy, ruthenium complex polymers, 59–63
ST5 mission, spacecraft, 74–75
Static tunability, optical properties, 2
Stern–Volmer plots, electrospun membranes, 396
Stimuli, external, tuning chain helicity of polyacetylenes, 353–356
Structural mode, coloration, 299
Structural transformation
  pH induced, in bolamphiphilic polydiacetylene, 104–105
  ribbon-to-vesicle, 106–107
Sulfonated polystyrene, role of, template during polymerization, 381
Supercooled liquid crystals. *See* Glassy liquid crystals (GLCs)
Surface packing, bolamphiphilic polydiacetylene, 103–104
Surface relief grating (SRG)
  atomic force microscope (AFM) image of SRG on polyazophenol thin film, 384f
  azobenzene-based photochromic materials, 176
  formation, 380
Switching
  cholesterics, 279–280
  electric field inducing, 281
  See also Chiral gels

T

Temperature
  high-temperature glassy nematics, 297f
  labile metal crosslinks, 243–244
  photochromic liquid crystalline compounds, 326–327
  reflection band of chiral gels vs., 284, 287, 288f, 289
  tuning chain helicity of polyacetylenes, 354, 355f
Temperature, color control
  chromogenic gel networks, 117–120
  thermochromic gel networks, 115–117
Temperature, transparency control
  Bromothymol Blue, 118f
  chromogenic gel networks, 117–120
  Phenol Red, 119f
  thermotropic gel networks, 111–114
Tetraethyl orthosilicate (TEOS), composite with poly(quinoline), 270–271, 274f
Thermal analysis, ruthenium complex polymers, 57–58

Thermal vacuum, conducting polymer-based electrochromics, 69, 73
Thermochromic gel networks
  reversible color control with temperature, 115–117
  UV-vis absorption spectra of Phenol Red in, 116f
Thermochromism, dialkyl-poly(p-phenyleneethynylene)s (dialkyl-PPEs), 150–153
Thermogravimetric analysis
  poly(methyl methacrylate) hybrid material, 314f
  triaryl-poly(p-phenylene) and poly(9,9-dioctylfluorene), 206f
Thermotropic gel networks
  calorimetric measurements on aqueous, 112–113
  differential scanning calorimetry, 113f
  hybrid solar and electrically controlled light filters, 114
  incorporation of salts into aqueous polymer systems, 114
  preparation, 111–112
  reversible transparency control with temperature, 111–114
Thienyl-based monomers and polymers
  cyclic voltammetry, 371f
  cyclization-polymerization, 370
Thin films
  embedding colloidal crystals, 330–331
  See also Trialkoxysilane-capped poly(methyl methacrylate)-silica
Thiophenes
  charge transfer, 162
  See also Cationic polythiophene biosensors; Conjugated polymers; Poly(thiophene)s
Three-dimensional two-photon fluorescence imaging, multilayer structures, 128–129

Threshold voltage
  chiral gels, 282–283, 284
  See also Voltage
Transition metal oxides, electrochromic devices, 35–36
Transitions, liquid crystals, 292–293
Transparency
  Bromothymol Blue, 118f
  chromogenic gel networks, 117–120
  Phenol Red, 119f
  thermotropic gel networks, 111–114
Trialkoxysilane-capped poly(methyl methacrylate)-silica
  atomic force microscope image of prepared hybrid material, 313f
  characterization, 310–311
  differential scanning calorimetry of poly(methyl methacrylate) (PMMA) and prepared hybrid material, 314f
  experimental, 310–311
  Fourier transform infrared spectra of PMMA and prepared hybrid materials, 312f
  hybrid materials as waveguide materials, 308
  materials, 310
  near infrared (NIR) absorption spectra of PMMA and prepared hybrid materials, 317f
  NIR absorption spectrum of prepared hybrid material, 316f
  optical losses of hybrid planar optical waveguides, 317, 318f
  preparation, 308–309
  preparation of optical planar waveguides, 310
  properties of hybrid thin films, 312, 315–316
  properties of PMMA, hybrid materials, and planar optical waveguides, 315t

scanning electron microscopy of prepared hybrid material, 313f
scheme for preparing optical planar waveguides, 309f
structure of prepared hybrid thin films, 311–312
synthesis of precursor solutions, 310
thermogravimetric analysis of PMMA and prepared hybrid material, 314f
variation of optical loss of prepared optical planar waveguide, 318f
Tunable chiroptical properties. *See* Amphiphilic helical polyacetylenes; Chromogenic effects in polymers
Tunable photoluminescence. *See* Poly(quinoline)s
Tunable reflective coloration
glassy cholesteric films, 299–300
reversible, 301
See also Glassy liquid crystals (GLCs)
Two-photon absorption
advantage, 161
applications, 162
fluorene, 125, 126f
poly(styrene-*co*-maleic anhydride) modified polymer, 132f
properties of conjugated chromophores, 167–170
rate of light absorption, 161–162
See also Conjugated chromophores; Photosensitive polymers
Two-photon excitation, writing and reading in photosensitive polymer film, 133
Two-photon fluorescence imaging
multi-layered films between grid images, 129, 131
multilayer structures, 128–129
photosensitive films, 129, 130
photosensitive polymers, 123–124
three-dimensional imaging of layered structure, 130

U

Ultrathin films
preparation, 83
See also Poly(diacetylene)s (PDAs)
Ultraviolet (UV) spectra, polyacetylenes containing D-glucose pendants, 345f
Upconverted fluorescence
fluorene, 125, 127, 128f
poly(styrene-*co*-maleic anhydride) modified polymer, 132f
UV-vis absorption spectra
arylene-ethynylene (AE) polymer, 374f
Bromothymol Blue, 117f
cationic polythiophene derivative solutions, 363f
double-layer film of oriented poly(p-phenylenevinylene) (PPV)/poly[3-(6-methoxyhexyl)thiophene] (P6OMe), 215
Phenol Red, 118f
4-phenylazophenol monomer and polymer, 384f
poly(aniline), 381f
poly(aniline)/sulfonated polystyrene, 382f
poly(heteroaryleneethynylene)s, 156f
poly(thiophene) derivatives, 375f
Ru(II)-chelated polymers, 254f
silicon-based alternating copolymers, 252f, 253f
triarylamine or diarylamino side poly(paraphenylene), 204f
UV-vis transmission spectra
photonic paper before and after swelling matrix, 332, 334

V

V_2O_5 film, counter electrode in electrochromic devices, 44–47
Valine-containing polyacetylene, circular dichroism changes with solvent, 353f, 354
Variable emittance materials, conducting polymers, 67
Vinyl polymers, pendant oligothiophenes, 181, 182
Visible light attenuation, electrochromic materials, 52
Voltage
  critical, 283
  photographs of cholesteric gel at various, 285
  photonic paper before and after swelling with isopropanol, 335f
  reflection band of chiral gels vs., 288f
  threshold, 282–283, 284
  transmission of chiral gels vs., 282f, 283f
  See also Chiral gels
Voltage-tunable electroluminescence. See Electroluminescence

W

Waveguides, optical. See Trialkoxysilane-capped poly(methyl methacrylate)-silica
Wavelength of diffracted light, Bragg equation, 330
Writing, two-photon excitation of photosensitive polymer film, 133